Symmetry and Quantum Mechanics

MONOGRAPHS AND RESEARCH NOTES IN MATHEMATICS

Series Editors

John A. Burns
Thomas J. Tucker
Miklos Bona
Michael Ruzhansky

Published Titles

Actions and Invariants of Algebraic Groups, Second Edition, Walter Ferrer Santos
and Alvaro Rittatore

Analytical Methods for Kolmogorov Equations, Second Edition, Luca Lorenzi

Application of Fuzzy Logic to Social Choice Theory, John N. Mordeson, Davender S. Mal
and Terry D. Clark

*Blow-up Patterns for Higher-Order: Nonlinear Parabolic, Hyperbolic Dispersion and
Schrödinger Equations*, Victor A. Galaktionov, Enzo L. Mitidieri, and Stanislav Pohozaev

Complex Analysis: Conformal Inequalities and the Bieberbach Conjecture, Prem K. Kythe

*Computational Aspects of Polynomial Identities: Volume l, Kemer's Theorems, 2nd Editio
Alexei Kanel-Belov, Yakov Karasik, and Louis Halle Rowen

A Concise Introduction to Geometric Numerical Integration, Fernando Casas
and Sergio Blanes

Cremona Groups and Icosahedron, Ivan Cheltsov and Constantin Shramov

Delay Differential Evolutions Subjected to Nonlocal Initial Conditions
Monica-Dana Burlică, Mihai Necula, Daniela Roşu, and Ioan I. Vrabie

Diagram Genus, Generators, and Applications, Alexander Stoimenow

Difference Equations: Theory, Applications and Advanced Topics, Third Edition
Ronald E. Mickens

Dictionary of Inequalities, Second Edition, Peter Bullen

Finite Element Methods for Eigenvalue Problems, Jiguang Sun and Aihui Zhou

Introduction to Abelian Model Structures and Gorenstein Homological Dimensions
Marco A. Pérez

Iterative Methods without Inversion, Anatoly Galperin

Iterative Optimization in Inverse Problems, Charles L. Byrne

Line Integral Methods for Conservative Problems, Luigi Brugnano and Felice Iavernaro

Lineability: The Search for Linearity in Mathematics, Richard M. Aron,
Luis Bernal González, Daniel M. Pellegrino, and Juan B. Seoane Sepúlveda

Modeling and Inverse Problems in the Presence of Uncertainty, H. T. Banks, Shuhua Hu,
and W. Clayton Thompson

Monomial Algebras, Second Edition, Rafael H. Villarreal

*Nonlinear Functional Analysis in Banach Spaces and Banach Algebras: Fixed Point
Theory Under Weak Topology for Nonlinear Operators and Block Operator Matrices with
Applications,* Aref Jeribi and Bilel Krichen

Published Titles Continued

Partial Differential Equations with Variable Exponents: Variational Methods and Qualitative Analysis, Vicenţiu D. Rădulescu and Dušan D. Repovš

A Practical Guide to Geometric Regulation for Distributed Parameter Systems Eugenio Aulisa and David Gilliam

Reconstruction from Integral Data, Victor Palamodov

Signal Processing: A Mathematical Approach, Second Edition, Charles L. Byrne

Sinusoids: Theory and Technological Applications, Prem K. Kythe

Special Integrals of Gradshteyn and Ryzhik: the Proofs – Volume I, Victor H. Moll

Special Integrals of Gradshteyn and Ryzhik: the Proofs – Volume II, Victor H. Moll

Stochastic Cauchy Problems in Infinite Dimensions: Generalized and Regularized Solutions, Irina V. Melnikova

Submanifolds and Holonomy, Second Edition, Jürgen Berndt, Sergio Console, and Carlos Enrique Olmos

Symmetry and Quantum Mechanics, Scott Corry

The Truth Value Algebra of Type-2 Fuzzy Sets: Order Convolutions of Functions on the Unit Interval, John Harding, Carol Walker, and Elbert Walker

Forthcoming Titles

Geometric Modeling and Mesh Generation from Scanned Images, Yongjie Zhang

Groups, Designs, and Linear Algebra, Donald L. Kreher

Handbook of the Tutte Polynomial, Joanna Anthony Ellis-Monaghan and Iain Moffat

Microlocal Analysis on R^n and on NonCompact Manifolds, Sandro Coriasco

Practical Guide to Geometric Regulation for Distributed Parameter Systems, Eugenio Aulisa and David S. Gilliam

MONOGRAPHS AND RESEARCH NOTES IN MATHEMATICS

Symmetry and Quantum Mechanics

Scott Corry

Lawrence University
Appleton, Wisconsin, USA

CRC Press is an imprint of the
Taylor & Francis Group, an **informa** business

A CHAPMAN & HALL BOOK

CRC Press
Taylor & Francis Group
6000 Broken Sound Parkway NW, Suite 300
Boca Raton, FL 33487-2742

© 2017 by Taylor & Francis Group, LLC
CRC Press is an imprint of Taylor & Francis Group, an Informa business

No claim to original U.S. Government works

Printed on acid-free paper
Version Date: 20161021

International Standard Book Number-13: 978-1-4987-0116-7 (Hardback)

This book contains information obtained from authentic and highly regarded sources. Reasonable efforts have been made to publish reliable data and information, but the author and publisher cannot assume responsibility for the validity of all materials or the consequences of their use. The authors and publishers have attempted to trace the copyright holders of all material reproduced in this publication and apologize to copyright holders if permission to publish in this form has not been obtained. If any copyright material has not been acknowledged please write and let us know so we may rectify in any future reprint.

Except as permitted under U.S. Copyright Law, no part of this book may be reprinted, reproduced, transmitted, or utilized in any form by any electronic, mechanical, or other means, now known or hereafter invented, including photocopying, microfilming, and recording, or in any information storage or retrieval system, without written permission from the publishers.

For permission to photocopy or use material electronically from this work, please access www.copyright.com (http://www.copyright.com/) or contact the Copyright Clearance Center, Inc. (CCC), 222 Rosewood Drive, Danvers, MA 01923, 978-750-8400. CCC is a not-for-profit organization that provides licenses and registration for a variety of users. For organizations that have been granted a photocopy license by the CCC, a separate system of payment has been arranged.

Trademark Notice: Product or corporate names may be trademarks or registered trademarks, and are used only for identification and explanation without intent to infringe.

Visit the Taylor & Francis Web site at
http://www.taylorandfrancis.com

and the CRC Press Web site at
http://www.crcpress.com

Printed and bound in the United States of America by Edwards Brothers Malloy on sustainably sourced paper

To Sebastian

Contents

Author Biography	xiii
Preface	xv
Plan of the Book	xix
List of Figures	xxiii

I Spin   1

1 Physical Space   3

1.1	Modeling space	3
1.2	Real linear operators and matrix groups	7
1.3	$SO(3)$ is the group of rotations	11

2 Spinor Space   17

2.1	Angular momentum in classical mechanics	17
2.2	Modeling spin	22
2.3	Complex linear operators and matrix groups	27
2.4	The geometry of $SU(2)$	31
	2.4.1 The tangent space to the circle $U(1) = S^1$	31
	2.4.2 The tangent space to the sphere $SU(2) = S^3$	33
	2.4.3 The exponential of a matrix	34
	2.4.4 $SU(2)$ is the universal cover of $SO(3)$	40
2.5	Back to spinor space	43

3 Observables and Uncertainty   47

3.1	Spin observables	47
3.2	The Lie algebra $\mathfrak{su}(2)$	50
3.3	Commutation relations and uncertainty	55
3.4	Some related Lie algebras	60
	3.4.1 Warmup: The Lie algebra $\mathfrak{u}(1)$	60
	3.4.2 The Lie algebra $\mathfrak{sl}_2(\mathbb{C})$	61
	3.4.3 The Lie algebra $\mathfrak{u}(2)$	65

Contents

	3.4.4 The Lie algebra $\mathfrak{gl}_2(\mathbb{C})$	66

4 Dynamics

69

4.1 Time-independent external fields	70
4.2 Time-dependent external fields	74
4.3 The energy-time uncertainty principle	75
4.3.1 Conserved quantities	77

5 Higher Spin

79

5.1 Group representations	80
5.2 Representations of $SU(2)$	83
5.3 Lie algebra representations	87
5.4 Representations of $\mathfrak{su}(2)_{\mathbb{C}} = \mathfrak{sl}_2(\mathbb{C})$	92
5.5 Spin-s particles	98
5.6 Representations of $SO(3)$	106
5.6.1 The $\mathfrak{so}(3)$-action	109
5.6.2 Comments about analysis	114

6 Multiple Particles

121

6.1 Tensor products of representations	121
6.2 The Clebsch-Gordan problem	128
6.3 Identical particles—spin only	130

II Position & Momentum

133

7 A One-Dimensional World

135

7.1 Position	136
7.2 Momentum	140
7.3 The Heisenberg Lie algebra and Lie group	143
7.3.1 The meaning of the Heisenberg group action	145
7.4 Time-evolution	147
7.4.1 The free particle	149
7.4.2 The infinite square well	150
7.4.3 The simple harmonic oscillator	152

8 A Three-Dimensional World

161

8.1 Position	161
8.2 Linear momentum	164
8.2.1 The Heisenberg group H_3 and its algebra $\mathfrak{h}_3$	166
8.3 Angular momentum	168
8.4 The Lie group $G = H_3 \rtimes SO(3)$ and its Lie algebra $\mathfrak{g}$	171

Contents

8.5	Time-evolution	173
	8.5.1 The free particle	174
	8.5.2 The three-dimensional harmonic oscillator	174
	8.5.3 Central potentials	176
	8.5.4 The infinite spherical well	178
8.6	Two-particle systems	180
	8.6.1 The Coulomb potential	182
8.7	Particles with spin	185
	8.7.1 The hydrogen atom	189
8.8	Identical particles	191

9 Toward a Relativistic Theory **197**

9.1	Galilean relativity	197
9.2	Special relativity	206
9.3	$SL_2(\mathbb{C})$ is the universal cover of $SO^+(1,3)$	216
9.4	The Dirac equation	220

A Appendices **231**

A.1	Linear algebra	231
	A.1.1 Vector spaces and linear transformations	231
	A.1.2 Inner product spaces and adjoints	236
A.2	Multivariable calculus	239
A.3	Analysis	242
	A.3.1 Hilbert spaces and adjoints	242
	A.3.2 Some big theorems	243
A.4	Solutions to selected exercises	245

Bibliography **251**

Index **253**

Author Biography

Scott Corry obtained his Ph.D. in mathematics from the University of Pennsylvania in 2007 after earning a B.A. in mathematics from Reed College in 2001. He joined the faculty of Lawrence University in 2007 and was promoted to Associated Professor of mathematics in 2013. Originally working in the areas of Galois theory and p-adic algebraic geometry, his interests have expanded to encompass a broad range of topics, including combinatorics and mathematical physics. The unifying theme in all of his work is a focus on symmetry.

Preface

This book began in summer 2011 as an attempt to understand the mathematical framework underlying quantum mechanics. Although I have long been interested in physics, the immediate impetus for writing came from a reading project with my friend Doug Martin in the Physics Department at Lawrence University. In effect, this text is a first course in quantum mechanics from the mathematical point of view, emphasizing the role of symmetry, and inspired by J.S. Townsend's *A Modern Approach to Quantum Mechanics* [22].

The history of mathematics is deeply entwined with the history of physics, and the two subjects continue to influence each other in dramatic and inspiring ways. Nevertheless, it is an unfortunate fact that physicists and mathematicians often speak past each other and sometimes fail to appreciate the value of each others' concerns.[1] A physicist colleague once sent me a small poster for my office inscribed with the pithy phrase: "Mathematics: Physics without Purpose." I can imagine some of my mathematical colleagues retorting with "Physics: Mathematics without Rigor." While these taunts can be great fun, they do not help bridge the gap between two powerful worldviews. Mathematicians decry physicists' desire to choose coordinates, and we tell them about the beauty of abstract objects, generally defined as sets with further structure. Physicists often don't see the point of these abstractions, and in any case have little intuition for working with them. Instead, they are delighted with a "debauch of indices" (E. Cartan) and are quite skilled at computing, which is their real aim. After all, a physical theory is only justified qua physical theory by its agreement with experiment.

This text takes a middle road, and is loosely structured as a conversation between M(athematician) and P(hysicist). Starting with some basic physical intuitions and experimental results, M and P set out to make a model of the physical world. M introduces abstract mathematical objects, but she always motivates them with reference to experiment and appeals to simplicity. In this way, I hope that physicists already comfortable with the computations of quantum mechanics will gain an appreciation for the natural way in which these abstract objects arise. In response to these abstractions, P tends to

---

[1] Of course, there are mathematicians who work in the area of mathematical physics, and physicists who identify primarily as mathematical physicists. These two communities presumably understand each other reasonably well, and I am not thinking of them. Instead, I refer to the majority of working mathematicians (who may not know much about modern physics) and the bulk of experimental physicists (who may not know much about modern mathematics).

xv

choose coordinates, but M is careful to make him account for all the *other* choices he could have made. P's instinct is to say: "of course I could have chosen differently, but then I would just need to do some bookkeeping to translate between the resulting computations." But M insists that they study the particular *structure* of the collection of possible choices at each stage. In general, this is a group structure, and M and P are naturally led to build a model of the physical world based on group representations. Remarkably, much of the mathematical structure of quantum mechanics falls out from this procedure, giving it an aura of inevitability and extreme beauty. In this way, I hope that mathematicians already comfortable with Lie groups and their representations will gain an appreciation for quantum mechanics and its myriad connections to pure mathematics. Of course, the main audience for this book is the advanced undergraduate or beginning graduate student whose understanding of both physics and mathematics is just beginning to grow.

Indeed, the student I have in mind will have taken courses in multivariable calculus, linear algebra, abstract algebra, real analysis, and perhaps topology. But she may not have seen any truly rich connections between these various subjects, and in any case would benefit from an opportunity to review them in a new context, where she will gain exposure to some graduate-level topics: smooth manifolds, group representations, Lie algebras, and Hilbert spaces. My aim is to introduce these new topics in a natural way, as an outgrowth of a compelling physical and mathematical exploration. Moreover, an introduction is all that I intend, leaving a deeper and fuller account to other texts and future courses. Especially with regard to the analytic subtleties that arise in the context of self-adjoint operators on Hilbert spaces, I am content to raise awareness of the difficulties while avoiding getting bogged down in the details. Students will be better able to comprehend a graduate course in real analysis if they have some prior understanding of why one should bother with those technical details in the first place. A great place to learn about the details in the context of quantum mechanics is B.C. Hall's excellent text *Quantum Theory for Mathematicians* [10].

This book is *not* intended as a replacement for introductory physics texts such as [9, 22]: the reader will not learn perturbation theory nor gain proficiency at computing the energy levels or eigenstates of any but the simplest quantum mechanical systems. Nevertheless, M and P *do* try out their model on systems such as the infinite spherical well, the harmonic oscillator, and the hydrogen atom. But the point is always to illustrate the underlying mathematical structure, not the explicit form of the solutions or their physical consequences. A highly recommended undergraduate text for students of mathematics that does present perturbation theory with applications to scattering problems is [3], which has a more analytic focus than our text, and also provides a fuller discussion of the relationship between quantum and classical mechanics.

Likewise, this text is not meant as a replacement for more advanced mathematical treatments of quantum mechanics such as [10, 20]. In particular, M introduces a piece of mathematics only if she feels it is demanded by physi-

Preface

xvii

cal considerations. And even then, she resists the temptation to develop the ideas in even modest generality, preferring to stay close to the physical model under development. Overall, one might read this book as a motivating introduction to Lie groups and their representations, with focus on the quantum mechanically relevant Heisenberg group H_3 and special unitary group $SU(2)$.

While the results described herein are well known, the presentation is somewhat novel. In any case, my aim is to whet the appetite for further study, and I hope this text serves to reveal the simplicity and beauty of a subject that is often perceived as complicated and intimidating. Exercises occur throughout, and I have provided solutions to those tagged by the symbol ♣ in Appendix A.4. In an effort to bridge the gap between the physics and mathematics literature, I have adopted some notation that may be more familiar to physicists than mathematicians. In particular, time-derivatives are denoted by $\dot{c}(t)$ rather than $c'(t)$, and primes instead decorate objects viewed from M's point of view as compared with P's. In addition, I denote complex conjugation by α^* rather than $\overline{\alpha}$, and use $L^\dagger$ to denote the adjoint/Hermitian conjugate of an operator rather than L^*. A brief review of key material from linear algebra, multivariable calculus, and analysis is provided in Appendices A.1–A.3.

This book is dedicated to my son, Sebastian, and I certainly could not have written it without the love, support, and patience of my wife Madera. I would also like to thank my students Karl Mayer, Sanfer D'souza, and Daniel Martinez Zambrano for working through early drafts of this material as part of their Senior Experiences at Lawrence University—their questions and comments have been extremely helpful. My colleague Allison Fleshman from the Chemistry Department provided enthusiastic conversations and insightful comments about the periodic table, and Doug Martin in the Physics Department has given me continual encouragement and inspiration. Finally, many thanks go to an anonymous reviewer for excellent suggestions that have substantially improved the exposition.

Scott Corry
Appleton, WI
May 2016

Plan of the Book

Part I: Spin

Chapter 1: Physical Space *... in which M and P discover the group of rotations, $SO(3)$.*

The book begins with a description of physical space as isomorphic to $\mathbb{R}^3$, but care is taken to note that the *choice* of isomorphism is arbitrary. The argument by which this freedom of choice leads to an action of $SO(3)$ is carefully rehearsed so that it may serve as a template for later discussions in less familiar contexts.

Chapter 2: Spinor Space *... in which M and P discover the special unitary group $SU(2)$ and its relation to the group of rotations $SO(3)$.*

On the basis of the Stern-Gerlach experiment, spinor space is described as isomorphic to $\mathbb{C}^2$, but once again the choice of isomorphism is arbitrary. Following closely the pattern of Chapter 1, this freedom of choice leads to an action of $SU(2)$. The relationship between physical space and spinor space is established by showing that $SU(2)$ is the universal cover of $SO(3)$.

Chapter 3: Observables and Uncertainty *... in which M and P discover the Lie algebra $\mathfrak{su}(2)$ and its complexification $\mathfrak{sl}_2(\mathbb{C})$.*

A discussion of quantum observables leads to the definition of several Lie algebras and an exploration of their relationship to the corresponding Lie groups. The Lie bracket is shown to have a physical interpretation in terms of uncertainty.

Chapter 4: Dynamics *... in which M and P discover the Schrödinger equation.*

The time-evolution of spin-states is modeled as a curve in the unitary group $U(2)$, and this curve is shown to be determined by the Hamiltonian of the physical system, obtained by quantizing the classical expression for the energy.

xix

xx   *Plan of the Book*

Chapter 5: Higher Spin *...in which M and P classify the representations of $SU(2)$.*

The complex irreducible representations of $SU(2)$ are classified by working with the corresponding representations of the Lie algebra $\mathfrak{sl}_2(\mathbb{C})$. These representations are described explicitly and a physical interpretation is provided in terms of higher spin particles measured by a Stern-Gerlach apparatus. The final section studies the irreducible representations of $SO(3)$ as preparation for the theory of orbital angular momentum in the second part of the book.

Chapter 6: Multiple Particles *...in which M and P learn about the tensor product.*

The tensor product of spinor spaces provides a model for the spin-states of a system of two particles. This leads to the Clebsch-Gordan problem for $SU(2)$, whose solution describes how the tensor product of irreducible representations decomposes as a direct sum of irreducibles.

Part II: Position & Momentum

Chapter 7: A One-Dimensional World *...in which M and P discover the Heisenberg group, H_1.*

Position space $L^2(\mathbb{R})$ is introduced in order to model the position of a particle in one dimension. The freedom of choice of an origin in physical space leads to an action of the group $(\mathbb{R}, +)$. This translation action extends to an action of the Heisenberg group H_1, and the corresponding Lie algebra action provides the position and momentum operators. The resulting framework is applied to several physical systems: the free particle, the infinite square well, and the harmonic oscillator.

Chapter 8: A Three-Dimensional World *...in which M and P combine their studies of the Heisenberg group H_3 and the rotation group $SO(3)$.*

Following the pattern developed for the one-dimensional world, position space $L^2(\mathbb{R}^3)$ is introduced for three dimensions. This space carries a translation action of the group $(\mathbb{R}^3, +)$ which extends to an action of the Heisenberg group H_3. Hearkening back to Chapter 1, the choice of a basis for physical space leads to an action of $SO(3)$ on position space, which combines with the Heisenberg action to yield an action of the group $G = H_3 \rtimes SO(3)$. The corresponding Lie algebra action provides the position, linear momentum, and orbital angular momentum operators. Several physical systems are studied: the free particle, the infinite spherical well, the harmonic oscillator, and the Coulomb potential. In

Plan of the Book

xxi

order to incorporate spin, the G-action is extended to an action of the group $\widetilde{G} = H_3 \rtimes SU(2)$ on spinor-valued wavefunctions. This framework is applied to the hydrogen atom, and a discussion of identical particles leads to the Pauli exclusion principle and some insight into the structure of the periodic table of the elements.

Chapter 9: Toward a Relativistic Theory ... *in which M and P discover the central extension of the Galilean group, the restricted Lorentz group $SO^+(1,3)$, and the Dirac equation.*

The final chapter considers the relationship between the time-evolutions of wavefunctions for observers in uniform relative motion. This leads to an action of the central extension of the Galilean group in the non-relativistic context, and to actions of the restricted Lorentz and Poincaré groups in special relativity. The text ends with a discussion of the Dirac equation describing a relativistic spin-$\frac{1}{2}$ particle.

List of Figures

1.1 The basis on the left is left-handed; the basis on the right is right-handed. 5

1.2 The 2×2 special orthogonal matrix L_θ defines a counter-clockwise rotation through the angle θ. 11

1.3 Observer M's rotated coordinate system (dashed) drawn on top of P's coordinate system (solid). The vector $\mathbf{L}$ represents a spinning top's angular momentum. 15

1.4 A rotation of $\pi/2$ radians about the $\mathbf{u}_2$-axis. The dashed lines in both pictures show P's coordinate axes. 16

2.1 The position, linear momentum, and angular momentum of a particle moving in three-dimensional space. 17

2.2 A Stern-Gerlach device oriented in the positive z-direction. The vertical arrows indicate the z-component of the resulting inhomogeneous magnetic field. 20

2.3 The classical expectation for the behavior of an electron-beam in a Stern-Gerlach device. 21

2.4 The actual behavior of an electron-beam in a Stern-Gerlach device. 21

2.5 Orthogonal projection onto the complex line spanned by $|\psi\rangle$, with one real dimension suppressed. 26

2.6 The tangent line to the circle S^1 at the identity. 32

2.7 A rotation by α in the xz-plane. The dashed lines show P's coordinate axes. 45

3.1 The Implicit Function Theorem for the circle $U(1)$. 61

5.1 The behavior of a beam of spin-s particles in a Stern-Gerlach device. 100

5.2 A rotation by α in the xz-plane. The dashed lines show P's coordinate axes. 101

5.3 Spherical coordinates on $\mathbb{R}^3$. 112

7.1 Observer M's location is translated from P's through a distance $w \in \mathbb{R}$. 135

xxiii

xxiv   *List of Figures*

7.2   Observer M's coordinate system is translated from P's through a distance $w \in \mathbb{R}$, so $x' = x - w$. 136

7.3   The same potential viewed by P (bottom) and by the translated observer M (top). The potential has a local minimum at the origin for P, but for M the local minimum is at the position-value $x' = -w$. 148

7.4   The first four energy eigenfunctions for the infinite spherical well. The plots on the left are the wavefunctions ψ_n; the plots on the right are the corresponding probability densities $|\psi_n|^2$. 151

7.5   A simple harmonic oscillator with mass m and spring-constant k. The picture on the left shows the equilibrium position of the mass at $x = 0$; the picture on the right shows the stretched spring exerting a restoring force $F = -kx$ on the mass m at position x. 152

7.6   The first four energy eigenfunctions for the simple harmonic oscillator, expressed in terms of the dimensionless position variable $\widetilde{x} = \sqrt{\frac{m\omega}{\hbar}}x$. The plots on the left are the wavefunctions ψ_n; the plots on the right are the corresponding probability densities $|\psi_n|^2$. The shaded portions indicate the classical region for an oscillator with energy $E_n = \hbar\omega(n + \frac{1}{2})$. Since the wavefunction extends outside the classical region, there is a non-zero probability of finding the oscillator at a classically forbidden position. 159

8.1   Projection of the position vector $\boldsymbol{\lambda}$ onto the line spanned by the unit vector $\mathbf{u}$. 162

8.2   The Coulomb force on an electron due to a nucleus of positive charge Ze. The vector $\mathbf{r}$ (not shown) points *from* the nucleus *to* the electron. 183

8.3   The Periodic Table of the Elements. The groups of columns marked with letters indicate elements with a common value of azimuthal quantum number l for their highest-energy electron in the ground state: s corresponds to $l = 0$, p to $l = 1$, d to $l = 2$, f to $l = 3$. 192

9.1   Observer M's rotated coordinate system moving away from observer P at a constant velocity $\mathbf{v}$. 198

9.2   The motion $\mathbf{b}(t)$ of a ball traveling with velocity $\mathbf{v}_b$ with respect to observer P. The vector $\mathbf{b}'(t)$ describes the motion in observer M's rotated coordinate system, which is moving away from P with velocity $\mathbf{v}$. 199

9.3   Observer M's rotated coordinate system moving away from observer P at constant velocity $\mathbf{v}$, starting at the location $\mathbf{x} = \mathbf{w}$ at time $t = s$. 200

List of Figures

xxv

9.4   The sphere of light at time $t > 0$ resulting from a flash at P's space-time origin. The sphere has radius ct. 207

9.5   The sphere of light at time $t' > 0$ resulting from a flash at M's space-time origin. The sphere has radius ct'. 208

9.6   M and P's farewell diagram, illustrating the relationship between some of the various Lie groups (top) and Lie algebras (bottom) that have featured in the book. 229

Part I

Spin

Chapter 1

Physical Space

In which M and P discover the group of rotations, SO(3).

1.1 Modeling space

Let us imagine two observers, M(athematician) and P(hysicist). They are located together, looking at empty space, armed with meter sticks and protractors. Their sense impressions will probably lead them to agree about the following statements:

1. Space is three-dimensional and flat (i.e., not curved);

2. Their meter sticks are identical;

3. Their protractors are identical.

Of course, statement number 1 is imprecise. What does three-dimensional really mean? What about flat? Nonetheless, these shared intuitions lead M to propose the following model of the empty physical space surrounding them:

Definition 1.1. Physical space *is a three-dimensional real inner product space* $(V, \langle, \rangle)$.

P politely asks M to motivate her choice of model. M responds by unpacking her concise definition[1]: V is a vector space over the field of real numbers $\mathbb{R}$. To say that V is three-dimensional means that there exists an ordered set of three vectors $\{\mathbf{v}_1, \mathbf{v}_2, \mathbf{v}_3\} \subset V$ such that every $\mathbf{v}$ in V can be written uniquely as

$$\mathbf{v} = c_1\mathbf{v}_1 + c_2\mathbf{v}_2 + c_3\mathbf{v}_3$$

for some particular real numbers c_1, c_2, c_3. The uniqueness means that two distinct 3-tuples of real numbers will yield distinct linear combinations of the basis vectors $\mathbf{v}_i$. Now P understands that M is using a vector space to capture the intuition that space is "flat," and in this context "three-dimensional" acquires a precise meaning that meshes with physical intuition: there are exactly three independent directions in space, no more and no less.

An *inner product* on V is a function $\langle, \rangle \colon V \times V \to \mathbb{R}$ such that if $c \in \mathbb{R}$ and $\mathbf{v}, \mathbf{v}', \mathbf{w} \in V$, then:

---

[1]See Appendix A.1 for a summary of basic concepts in linear algebra.

3

4   *Symmetry and Quantum Mechanics*

    i) $\langle c\mathbf{v} + \mathbf{v}', \mathbf{w}\rangle = c\,\langle \mathbf{v}, \mathbf{w}\rangle + \langle \mathbf{v}', \mathbf{w}\rangle$ (*linearity in first component*);

    ii) $\langle \mathbf{v}, \mathbf{w}\rangle = \langle \mathbf{w}, \mathbf{v}\rangle$ (*symmetry*);

    iii) $\langle \mathbf{v}, \mathbf{v}\rangle \geq 0$ with equality if and only if $\mathbf{v} = \mathbf{0}$ (*positive definite*).

Exercise 1.2. *Show that conditions i) and ii) for an inner product imply linearity in the second component:* $\langle \mathbf{w}, c\mathbf{v} + \mathbf{v}'\rangle = c\,\langle \mathbf{w}, \mathbf{v}\rangle + \langle \mathbf{w}, \mathbf{v}'\rangle$.

M summarizes by saying that an inner product is a bilinear, symmetric, positive definite function on $V \times V$. So far, P isn't very impressed with this as motivation for M's definition of space. But M continues: define the *length* of a vector $\mathbf{v}$ in V to be $|\mathbf{v}| := \sqrt{\langle \mathbf{v}, \mathbf{v}\rangle}$, which is a non-negative real number by condition iii). Furthermore, define the *angle*[2] between two nonzero vectors $\mathbf{v}$ and $\mathbf{w}$ to be $\theta(\mathbf{v}, \mathbf{w}) := \arccos\left(\frac{\langle \mathbf{v}, \mathbf{w}\rangle}{|\mathbf{v}||\mathbf{w}|}\right)$. Observer P now understands: an inner product on V is an algebraic gadget that captures the notion of length and angle. Since M and P have identical meter sticks and protractors, they have the same notion of length and angle, hence they agree about the inner product on V.

Exercise 1.3. *Suppose that $(X, \langle,\rangle)$ is a real inner product space. Show that the inner product $\langle,\rangle$ is uniquely determined by the corresponding length function. That is, show that if two inner products define the same length function, then they are the same. (Hint: Compute $|\mathbf{x}_1 + \mathbf{x}_2|^2$.)*

Example 1.4. *Fix an integer $n \geq 1$, and consider the set $\mathbb{R}^n$ of all n-tuples of real numbers:*
$$\mathbb{R}^n := \{(x_1, x_2, \ldots, x_n) \mid x_i \in \mathbb{R}\}.$$

Then $\mathbb{R}^n$ is an n-dimensional real vector space under the operations of component-wise addition and scalar multiplication. The standard basis *for $\mathbb{R}^n$ is given by $\{\mathbf{e}_1, \ldots, \mathbf{e}_n\}$, where the vector $\mathbf{e}_i$ has a 1 in the ith slot and zeros elsewhere. Define the* dot product *of two vectors in $\mathbb{R}^n$ by the formula*

$$(x_1, x_2, \ldots, x_n) \cdot (y_1, y_2, \ldots, y_n) := \sum_{i=1}^{n} x_i y_i.$$

The reader should check that $\cdot : \mathbb{R}^n \times \mathbb{R}^n \to \mathbb{R}$ defines an inner product on $\mathbb{R}^n$. The resulting real inner product space $(\mathbb{R}^n, \cdot)$ is called real Euclidean n-space.

    P is tired of all this formalism, and wants to start doing experiments. So he begins to set up his lab. He is going to want to make measurements, so his first order of business is to set up a coordinate system. In terms of the model, this requires a specific choice of ordered basis $\{\mathbf{v}_1, \mathbf{v}_2, \mathbf{v}_3\}$ for the vector space V.

---

[2]Here $\arccos : [-1, 1] \to [0, \pi]$ is the inverse of the cosine function. This definition makes sense because of the *Cauchy-Schwarz inequality* which states that $|\langle \mathbf{v}, \mathbf{w}\rangle| \leq |\mathbf{v}||\mathbf{w}|$ for all vectors $\mathbf{v}, \mathbf{w} \in V$.

Physical Space

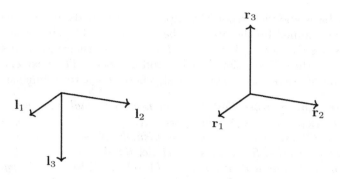

FIGURE 1.1: The basis on the left is left-handed; the basis on the right is right-handed.

The fact that V is three-dimensional guarantees that at least one such basis exists, but P quickly realizes that there are in fact infinitely many distinct bases to choose from, each of which corresponds to a different coordinate system for his lab. However, P doesn't like all coordinate systems equally. He prefers one in which the coordinate axes are orthogonal. Moreover, it strikes P as convenient to normalize his basis vectors to have length 1 (i.e., the same length as his meter stick). Thus he decides to choose an *orthonormal* basis $\{\mathbf{u}_1, \mathbf{u}_2, \mathbf{u}_3\}$ for the inner product space $(V, \langle, \rangle)$, which means that the basis vectors are pairwise-orthogonal and have unit length[3]. Observer M agrees that this is a reasonable condition to impose on a choice of basis: after all, since M and P agree about the inner product on V, they will also agree about whether a given basis is orthonormal.

P has one more preference about coordinate systems: because he is right-handed, he wants to use a right-handed coordinate system. By a right-handed system he means the following (see Figure 1.1): if he points the fingers of his right hand along the direction of his first coordinate axis, and curls them toward his second axis, then his thumb will point in the direction of his third axis (rather than in the opposite direction). Using physical intuition, P observes that any orthogonal coordinate system is either right-handed or left-handed, and that any two right-handed systems are related by a rotation, and similarly for any two left-handed systems. However, to move a right-handed system onto a left-handed system requires a reflection. At first, M is reluctant to build these considerations into their model of space, but she finally relents after specifying the following implication: the distinction between right- and left-handed coordinate systems divides the collection of all bases for V into two disjoint subsets. P's preference for right-handed systems corresponds to the selection of one of these two subsets as the *positively oriented* bases. If you flip the sign of the third basis vector in a positively oriented basis, you get a negatively

---

[3]The *Gram-Schmidt orthogonalization algorithm* guarantees that such a basis exists.

6   *Symmetry and Quantum Mechanics*

oriented basis and vice-versa (this flipping corresponds to the reflection across the plane spanned by the first two basis vectors). The specification of which subset is positive is called an *orientation* on V, so their model of space is now an *oriented* three-dimensional inner product space. The next exercise makes the notion of orientation rigorous and generalizes it to n dimensions.

Exercise 1.5. *Suppose that X is an n-dimensional real vector space, and that $\gamma := \{\mathbf{x}_1, \mathbf{x}_2, \ldots, \mathbf{x}_n\}$ and $\gamma' := \{\mathbf{x}'_1, \mathbf{x}'_2, \ldots, \mathbf{x}'_n\}$ are two ordered bases for X. Then there is a unique isomorphism $\Phi : X \to X$ with the property that $\Phi(\mathbf{x}_i) = \mathbf{x}'_i$ for all i. Since Φ is invertible, the determinant of Φ is a nonzero real number. Define a relation $\sim$ on the set of all bases of X by saying that $\gamma \sim \gamma'$ if and only if $\det(\Phi) > 0$. Show that $\sim$ is an equivalence relation that partitions the set of all bases for X into two subsets. The choice of one of these subsets as positive is called an* orientation *on X.*

After all of this discussion, P finally chooses a positively oriented orthonormal basis $\beta := \{\mathbf{u}_1, \mathbf{u}_2, \mathbf{u}_3\}$ for the oriented inner product space $(V, \langle, \rangle)$. Having made this choice, he can think about space in more concrete terms as follows. Given an arbitrary vector $\mathbf{v} \in V$, there exist unique real numbers a, b, c such that $\mathbf{v} = a\mathbf{u}_1 + b\mathbf{u}_2 + c\mathbf{u}_3$. This correspondence defines an isomorphism of vector spaces $\varphi \colon V \to \mathbb{R}^3$ defined by $\varphi(\mathbf{v}) = (a, b, c)$. Moreover, since P's basis is orthonormal, φ actually yields an isomorphism of inner product spaces $\varphi \colon (V, \langle, \rangle) \to (\mathbb{R}^3, \cdot)$. That is, φ not only preserves the vector space structure, it also preserves the inner products.

Exercise 1.6. *Suppose that $(X, \langle, \rangle)$ is an n-dimensional real inner product space, and that $\gamma := \{\mathbf{u}_1, \mathbf{u}_2, \ldots, \mathbf{u}_n\}$ is an orthonormal basis for X. For any vector $\mathbf{x} \in X$, the fact that γ is a basis means that there exist unique real numbers $a_1, a_2, \ldots, a_n$ such that $\mathbf{x} = \sum_{i=1}^{n} a_i \mathbf{u}_i$. Show that the function $\varphi \colon X \to \mathbb{R}^n$ defined by $\varphi(\mathbf{x}) := (a_1, a_2, \ldots, a_n)$ is an isomorphism of vector spaces. Further, show that φ preserves the inner products, hence is an isomorphism of inner product spaces: for all $\mathbf{x}_1, \mathbf{x}_2 \in X$, we have $\langle \mathbf{x}_1, \mathbf{x}_2 \rangle = \phi(\mathbf{x}_1) \cdot \phi(\mathbf{x}_2)$.*

Thus, *once P chooses a positively oriented orthonormal basis*, he has identified space with $(\mathbb{R}^3, \cdot)$ endowed with the familiar right-handed orientation in which the standard basis is positive. With this, P feels like he is back on firm ground and starts thinking about some experiments he wants to perform. But M interrupts with an annoying question: how does this more concrete description of space depend on P's choice of orthonormal basis?

To make the question more precise, M chooses a different positively oriented orthonormal basis for $(V, \langle, \rangle)$, which she denotes by $\beta' := \{\mathbf{u}'_1, \mathbf{u}'_2, \mathbf{u}'_3\}$. As above, this choice of basis defines an isomorphism $\varphi' \colon (V, \langle, \rangle) \to (\mathbb{R}^3, \cdot)$. But note that φ' is *not* the same isomorphism as φ, so when P and M each decide to describe space as $(\mathbb{R}^3, \cdot)$, their descriptions do not agree. Nevertheless, they are both working with $(V, \langle, \rangle)$, so there must be a way of translating

Physical Space

between the two descriptions. To discover the translation, consider the composed function $\varphi' \circ \varphi^{-1} \colon (\mathbb{R}^3, \cdot) \to (\mathbb{R}^3, \cdot)$. This function is an isomorphism, since it is the composition of two isomorphisms. Since it maps $\mathbb{R}^3$ to itself, we say that it is an *automorphism* of the inner product space $(\mathbb{R}^3, \cdot)$. We can picture the situation via the commutative diagram below:

$$
\begin{array}{ccc}
V & = & V \\
\varphi \downarrow & & \downarrow \varphi' \\
\mathbb{R}^3 & \xrightarrow{\varphi' \circ \varphi^{-1}} & \mathbb{R}^3.
\end{array}
$$

In order to better understand the automorphism $\varphi' \circ \varphi^{-1}$, we pause to provide a brief review of certain classes of linear operators on real Euclidean n-space.

1.2   Real linear operators and matrix groups

Suppose that $L \colon \mathbb{R}^n \to \mathbb{R}^n$ is a linear operator, and let $\varepsilon := \{\mathbf{e}_1, \mathbf{e}_2, \ldots, \mathbf{e}_n\}$ denote the standard basis of $\mathbb{R}^n$ (see example 1.4). Then L is represented (with respect to ε) by an $n \times n$ matrix of real numbers, which we also denote by L:

$$
L = [L_{ij}] \qquad \text{where} \qquad L(\mathbf{e}_j) = \sum_{i=1}^{n} L_{ij} \mathbf{e}_i.
$$

This means that the jth column of the matrix L is the vector $L(\mathbf{e}_j)$. We may now express the effect of the linear operator L as left-multiplication on column vectors: if $\mathbf{v} = (v_1, v_2, \ldots, v_n)$, then viewing $\mathbf{v}$ as a column yields $L(\mathbf{v}) = L\mathbf{v} \in \mathbb{R}^n$, where the ith component of $L\mathbf{v}$ is

$$
(L\mathbf{v})_i = \sum_{j=1}^{n} L_{ij} v_j,
$$

the dot product of the ith row of L with $\mathbf{v}$.

The next few propositions express properties of the linear operator L in terms of the corresponding matrix.

Proposition 1.7. *The linear operator L is an automorphism of the vector space $\mathbb{R}^n$ if and only if the matrix L is invertible.*

Proof. The operator L is an automorphism if and only if it possesses an inverse: a linear operator $M \colon \mathbb{R}^n \to \mathbb{R}^n$ such that $L \circ M = M \circ L = \mathrm{id}$, the identity transformation. But composition of linear operators corresponds to multiplication of the corresponding matrices, so the existence of an inverse operator is equivalent to the existence of a matrix M such that $LM = ML = I_n$, the $n \times n$ identity matrix. $\qquad\square$

8 *Symmetry and Quantum Mechanics*

Denote by $GL(n, \mathbb{R})$ the set of all invertible $n \times n$ matrices with real entries. Note that this set has the following properties with respect to the operation of matrix multiplication:

i) (*closure*) If A and B are in $GL(n, \mathbb{R})$, then so is their matrix product AB, since the inverse of AB is equal to $B^{-1}A^{-1}$.

ii) (*identity*) The identity matrix $I_n \in GL(n, \mathbb{R})$, and it satisfies $I_n A = AI_n = A$ for all $A \in GL(n, \mathbb{R})$.

iii) (*inverses*) If A is in $GL(n, \mathbb{R})$, then so is A^{-1}, since $(A^{-1})^{-1} = A$.

iv) (*associativity*) Matrix multiplication is associative: $(AB)C = A(BC)$ for all matrices A, B, C of compatible sizes. This follows from the fact that matrix multiplication corresponds to the composition of linear operators, and composition of functions is associative.

These statements mean that $GL(n, \mathbb{R})$ forms a *group* under matrix multiplication; it is non-abelian since $AB \neq BA$ for matrices in general.

Definition 1.8. *The group of all invertible $n \times n$ real matrices, denoted $GL(n, \mathbb{R})$, is called the* real general linear group. *It is the symmetry group of the vector space $\mathbb{R}^n$.*

Proposition 1.9. *Suppose that $L \in GL(n, \mathbb{R})$. Then L preserves the dot product on $\mathbb{R}^n$ if and only if $L^{-1} = L^T$, the transpose of the matrix L. (By "preserves the dot product" we mean that $L\mathbf{v} \cdot L\mathbf{w} = \mathbf{v} \cdot \mathbf{w}$ for all $\mathbf{v}, \mathbf{w} \in \mathbb{R}^n$.)*

Proof. Recall that the transpose of a matrix is obtained by reflecting across the main diagonal: $(L^T)_{ij} := L_{ji}$. We first show that for all $\mathbf{v}, \mathbf{w} \in \mathbb{R}^n$, we have $L\mathbf{v} \cdot \mathbf{w} = \mathbf{v} \cdot L^T\mathbf{w}$. Indeed, note that if we think of $\mathbf{v}$ and $\mathbf{w}$ as column vectors, then $\mathbf{w}^T$ is a row vector, and we may express the dot product as matrix multiplication: $\mathbf{v} \cdot \mathbf{w} = \mathbf{w}^T\mathbf{v}$. Replacing $\mathbf{v}$ by $L\mathbf{v}$ yields

$$L\mathbf{v} \cdot \mathbf{w} = \mathbf{w}^T(L\mathbf{v}) = (\mathbf{w}^T L)\mathbf{v} = (L^T\mathbf{w})^T\mathbf{v} = \mathbf{v} \cdot L^T\mathbf{w}.$$

(Here we have used the fact that $(AB)^T = B^T A^T$ for any two matrices of compatible sizes.) We now apply this identity to the dot product of $L\mathbf{v}$ and $L\mathbf{w}$:

$$L\mathbf{v} \cdot L\mathbf{w} = \mathbf{v} \cdot L^T L\mathbf{w}.$$

Clearly, if $L^{-1} = L^T$, then $L^T L = I_n$ and L preserves the dot product as claimed. Going the other direction, if L preserves the dot product, then we find that $\mathbf{v} \cdot \mathbf{w} = \mathbf{v} \cdot L^T L\mathbf{w}$ for all $\mathbf{v}, \mathbf{w}$. Subtraction yields

$$0 = \mathbf{v} \cdot \mathbf{w} - \mathbf{v} \cdot L^T L\mathbf{w} = \mathbf{v} \cdot (I_n - L^T L)\mathbf{w}.$$

Since this equation holds for all $\mathbf{v}$ and $\mathbf{w}$, we may take $\mathbf{v} = (I_n - L^T L)\mathbf{w}$ to find that

$$(I_n - L^T L)\mathbf{w} \cdot (I_n - L^T L)\mathbf{w} = 0.$$

Since the dot product is positive definite, it follows that $(I_n - L^T L)\mathbf{w} = 0$ for all $\mathbf{w}$. Thus, $I_n - L^T L$ is the zero operator, so that $L^T L = I_n$ and $L^{-1} = L^T$. $\square$

Physical Space

Denote by $O(n) \subset GL(n, \mathbb{R})$ the subset of matrices satisfying the condition $L^{-1} = L^T$. Such matrices are called *orthogonal*. The reader should check that $O(n)$ is a *subgroup* of $GL(n, \mathbb{R})$: a subset containing the identity matrix and closed under matrix multiplication and inverses.

Definition 1.10. *The group of all invertible $n \times n$ real matrices satisfying $L^{-1} = L^T$, denoted $O(n)$, is called the* orthogonal group. *It is the symmetry group of real Euclidean n-space $(\mathbb{R}^n, \cdot)$.*

Exercise 1.11. *Suppose that $L: \mathbb{R}^n \to \mathbb{R}^n$ is a linear operator. Show that $L \in O(n)$ if and only if L preserves the lengths of vectors: $|L\mathbf{v}| = |\mathbf{v}|$ for all $\mathbf{v} \in \mathbb{R}^n$. (Hint: use proposition 1.9 and compute $|L(\mathbf{v} + \mathbf{w})|^2$.)*

Proposition 1.12. *The determinant of an orthogonal matrix is ± 1.*

Proof. If L is orthogonal, then $I_n = L^T L$. Taking the determinant of both sides yields $1 = \det(I_n) = \det(L^T L) = \det(L^T)\det(L) = \det(L)^2$. It follows that $\det(L) = \pm 1$. $\qquad\square$

Now endow $(\mathbb{R}^n, \cdot)$ with the orientation for which the standard basis ε is positive. If $L \in GL(n, \mathbb{R})$ is an invertible matrix, we say that L is *orientation preserving* if L sends positively oriented bases to positively oriented bases. By exercise 1.5, we see that L is orientation preserving if and only if $\det L > 0$. It follows from the previous proposition that an orthogonal matrix L is orientation preserving if and only if $\det L = 1$; such matrices are called *special orthogonal*, and they form a subgroup of the orthogonal group.

Definition 1.13. *The* special orthogonal group *is the subgroup $SO(n) \subset O(n)$ of orthogonal matrices with determinant 1. It is the symmetry group of oriented real Euclidean n-space.*

Before returning to our observers M and P in three dimensions, we take a close look at all of these groups in the cases $n = 1$ and $n = 2$.

Example 1.14. *For $n = 1$ we have the following groups:*

- *$GL(1, \mathbb{R}) = \mathbb{R}^\times$, the group of nonzero real numbers;*

- *$O(1) = \{\pm 1\}$, the sign group;*

- *$SO(1) = \{1\}$, the trivial group.*

Example 1.15. *The real general linear group $GL(2, \mathbb{R})$ consists of 2×2 real matrices with nonzero determinant. Hence, we have*

$$GL(2, \mathbb{R}) = \left\{ \begin{bmatrix} a & b \\ c & d \end{bmatrix} \;\middle|\; ad - bc \neq 0 \right\}.$$

Equivalently, a 2×2 real matrix is an element of $GL(2, \mathbb{R})$ if and only if its

10   *Symmetry and Quantum Mechanics*

columns form a basis for $\mathbb{R}^2$. Concretely, this means that neither column is a scalar multiple of the other.

Writing out the orthogonality condition $I_2 = L^T L$ we find

$$\begin{bmatrix} 1 & 0 \\ 0 & 1 \end{bmatrix} = \begin{bmatrix} a & c \\ b & d \end{bmatrix} \begin{bmatrix} a & b \\ c & d \end{bmatrix} = \begin{bmatrix} a^2 + c^2 & ab + cd \\ ab + cd & b^2 + d^2 \end{bmatrix}.$$

Equating entries, we obtain the conditions $a^2 + c^2 = b^2 + d^2 = 1$ and $ab + cd = 0$. Interpreting these relations as dot products, we see that L is orthogonal exactly when the columns of L have unit length and are orthogonal to each other. Thus, a 2×2 real matrix is an element of $O(2)$ if and only if its columns form an orthonormal basis for $(\mathbb{R}^2, \cdot)$.

Note that for any choice of a, c such that $a^2 + c^2 = 1$, the point (a, c) lies on the unit circle in $\mathbb{R}^2$. Hence, there exists a unique angle $\theta \in [0, 2\pi)$ such that $a = \cos(\theta)$ and $c = \sin(\theta)$ (see Figure 1.2). The remaining two conditions on b and d then imply that $(b, d) = \pm(-\sin(\theta), \cos(\theta))$. Thus, we have the following description of the orthogonal group $O(2)$:

$$O(2) = \left\{ \begin{bmatrix} \cos(\theta) & -\sin(\theta) \\ \sin(\theta) & \cos(\theta) \end{bmatrix}, \begin{bmatrix} \cos(\theta) & \sin(\theta) \\ \sin(\theta) & -\cos(\theta) \end{bmatrix} \mid 0 \leq \theta < 2\pi \right\}.$$

Finally, if we demand that the determinant is $+1$, we find that only half of the matrices in $O(2)$ remain as special orthogonal matrices:

$$SO(2) = \left\{ L_\theta = \begin{bmatrix} \cos(\theta) & -\sin(\theta) \\ \sin(\theta) & \cos(\theta) \end{bmatrix} \mid 0 \leq \theta < 2\pi \right\}.$$

Hence, each element of $SO(2)$ is uniquely determined by an angle $\theta \in [0, 2\pi)$. In fact, the special orthogonal matrix L_θ describes a counter-clockwise rotation through the angle θ (see Figure 1.2). The first standard basis vector $\mathbf{e}_1 = (1, 0)$ is sent to $(\cos(\theta), \sin(\theta))$, while the second standard basis vector $\mathbf{e}_2 = (0, 1)$ is sent to $(-\sin(\theta), \cos(\theta))$.

Exercise 1.16. *Generalize the analysis in the preceding example to show that for all $n \geq 1$:*

a) *an $n \times n$ real matrix L is in $GL(n, \mathbb{R})$ if and only if the columns of L form a basis for $\mathbb{R}^n$;*

b) *an $n \times n$ real matrix L is in $O(n)$ if and only if the columns of L form an orthonormal basis for $(\mathbb{R}^n, \cdot)$.*

Part b) of the previous exercise has the following important consequence. We have defined a linear operator L on real Euclidean n-space to be orthogonal if its matrix with respect to the standard basis satisfies $L^T = L^{-1}$. But in fact, the next proposition shows that this relation between the transpose and inverse of an orthogonal operator holds for the matrix of L with respect to *any* orthonormal basis. We will make use of this fact in proposition 1.19 in the next section.

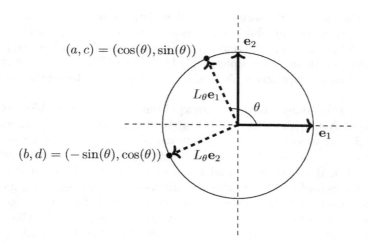

FIGURE 1.2: The 2 × 2 special orthogonal matrix L_θ defines a counter-clockwise rotation through the angle θ.

Proposition 1.17. *Let $L\colon \mathbb{R}^n \to \mathbb{R}^n$ be a linear operator, and $\gamma := \{\mathbf{u}_1, \ldots, \mathbf{u}_n\}$ be any orthonormal basis for $(\mathbb{R}^n, \cdot)$. Denote by $[L]_\gamma$ the matrix representing L in the basis γ. Then L is an orthogonal transformation if and only if $[L]_\gamma^T = [L]_\gamma^{-1}$.*

Proof. Let Q be the change of basis matrix from γ to ε, the standard basis. Recall that the jth column of Q is simply the basis vector $\mathbf{u}_j$. Hence, by part b) of the previous exercise, Q is orthogonal and $Q^{-1} = Q^T$. Then writing L for $[L]_\varepsilon$ as usual, we have $[L]_\gamma = QLQ^{-1} = QLQ^T$. We may then compute

$$[L]_\gamma^T [L]_\gamma = (QLQ^T)^T QLQ^T = QL^T Q^T QLQ^T = Q(L^T L)Q^T.$$

From this equation we see that $[L]_\gamma^T [L]_\gamma = I_n$ if and only if $L^T L = I_n$ as claimed. □

1.3 SO(3) is the group of rotations

Recall the situation of M and P, illustrated by the following diagram:

$$\begin{array}{ccc} V & = & V \\ {\scriptstyle \varphi}\downarrow & & \downarrow{\scriptstyle \varphi'} \\ \mathbb{R}^3 & \xrightarrow{\varphi' \circ \varphi^{-1}} & \mathbb{R}^3. \end{array}$$

12   *Symmetry and Quantum Mechanics*

P has chosen an orthonormal basis $\beta = \{\mathbf{u}_1, \mathbf{u}_2, \mathbf{u}_3\}$ for the inner product space $(V, \langle, \rangle)$, thereby obtaining an isomorphism $\varphi \colon (V, \langle, \rangle) \to (\mathbb{R}^3, \cdot)$. Meanwhile, M has chosen a different orthonormal basis $\beta' = \{\mathbf{u}'_1, \mathbf{u}'_2, \mathbf{u}'_3\}$, thereby obtaining a different isomorphism $\varphi' \colon (V, \langle, \rangle) \to (\mathbb{R}^3, \cdot)$. To study the difference in their descriptions of physical space, we are considering the composition $\varphi' \circ \varphi^{-1}$.

Since this composition of isomorphisms is an automorphism of $\mathbb{R}^3$, we may identify it with an invertible matrix of real numbers, i.e., an element of $GL(3, \mathbb{R})$. Moreover, because the automorphism $\varphi' \circ \varphi^{-1}$ preserves the dot product, the corresponding matrix is actually an element of the orthogonal group $O(3)$. But because M and P both chose positively oriented bases, the automorphism $\varphi' \circ \varphi^{-1}$ preserves the orientation on $(\mathbb{R}^3, \cdot)$, which means that its determinant is 1. As a concise summary, M says that the difference between the two descriptions of space is the automorphism $\varphi' \circ \varphi^{-1}$, which may be identified with an element of the special orthogonal group $SO(3)$.

Exercise 1.18. *Show that the matrix representing $\varphi' \circ \varphi^{-1}$ with respect to the standard basis on $\mathbb{R}^3$ is simply the change of basis matrix from the basis β to the basis β'. Recall that this is the matrix representing the identity transformation* id$\colon V \to V$ *with respect to the bases β and β':*

$$[\mathrm{id}]_\beta^{\beta'} := [b_{ij}] \qquad where \qquad \mathbf{u}_j = \sum_{i=1}^3 b_{ij} \mathbf{u}'_i.$$

In example 1.15, we saw that the group $SO(2)$ consists entirely of rotations. In the next proposition, we establish the corresponding result for three dimensions.

Proposition 1.19. *The special orthogonal group $SO(3)$ consists of rotations in real Euclidean 3-space. More precisely, for each non-identity element $A \in SO(3)$, there is a unique line of $\mathbb{R}^3$ that is fixed pointwise by A. Moreover, A acts as a rotation through some angle θ around this fixed axis.*

Proof. Let $A \in SO(3)$ be an arbitrary special orthogonal matrix. We begin by showing that 1 is an eigenvalue for A, so that A fixes the line spanned by a corresponding eigenvector in $\mathbb{R}^3$; this line will turn out to be the axis of rotation.

Note that the characteristic polynomial $p_A(\lambda) := \det(A - \lambda I_3)$ is a degree 3 polynomial with real coefficients and hence has at least one real root, say $\lambda_0 \in \mathbb{R}$. Thus, λ_0 is an eigenvalue for A, and we may choose a unit length eigenvector $\mathbf{u} \in \mathbb{R}^3$ such that $A\mathbf{u} = \lambda_0 \mathbf{u}$. Now use the fact that A preserves the dot product:

$$1 = \mathbf{u} \cdot \mathbf{u} = A\mathbf{u} \cdot A\mathbf{u} = \lambda_0 \mathbf{u} \cdot \lambda_0 \mathbf{u} = \lambda_0^2 \mathbf{u} \cdot \mathbf{u} = \lambda_0^2.$$

Thus, we see that any real eigenvalue of A satisfies $\lambda_0 = \pm 1$.

There are two cases depending on the factorization of the characteristic polynomial over $\mathbb{R}$:

Physical Space

i) the polynomial $p_A(\lambda)$ factors into linear factors over $\mathbb{R}$, so that A has 3 real eigenvalues $\lambda_0, \lambda_1, \lambda_2$, each of which is ± 1 by the previous argument. Then $1 = \det(A) = \lambda_0 \lambda_1 \lambda_2$, which implies that at least one of the eigenvalues is $+1$. If all three of them are $+1$ then A is the identity. Otherwise, only one of the eigenvalues is $+1$, and the corresponding 1-dimensional eigenspace is the unique line that is fixed pointwise by A.

ii) the polynomial $p_A(\lambda)$ factors into a linear factor $(x - \lambda_0)$ and an irreducible quadratic. By the quadratic formula, the complex roots of the irreducible quadratic factor are complex conjugates, say $\lambda_1 = \lambda_2^* \in \mathbb{C}$. It follows that $1 = \det(A) = \lambda_0 \lambda_1 \lambda_2 = \lambda_0 |\lambda_1|^2$. Since $|\lambda_1|^2 > 0$, we see that $\lambda_0 = -1$ is impossible in this case, so that $\lambda_0 = 1$. Again, the corresponding 1-dimensional eigenspace is the unique line that is fixed pointwise by A.

Thus, we have established that 1 is an eigenvalue for A, and we have chosen a unit length eigenvector $\mathbf{u}$ such that $A\mathbf{u} = \mathbf{u}$. Expand $\{\mathbf{u}\}$ to an orthonormal basis $\gamma := \{\mathbf{u}, \mathbf{v}, \mathbf{w}\}$ for $(\mathbb{R}^3, \cdot)$. In the basis γ, the linear operator A is represented by a matrix of the following form:

$$[A]_\gamma = \begin{bmatrix} 1 & v & w \\ 0 & a & b \\ 0 & c & d \end{bmatrix}.$$

In fact, $v = w = 0$. Indeed, we simply need to use the orthonormality of the basis together with the orthogonality of the matrix A to compute:

$$v = A\mathbf{v} \cdot \mathbf{u} = \mathbf{v} \cdot A^T \mathbf{u} = \mathbf{v} \cdot A^{-1}\mathbf{u} = \mathbf{v} \cdot \mathbf{u} = 0.$$

The proof that $w = 0$ is the same, with $\mathbf{w}$ in place of $\mathbf{v}$.

Therefore, in the basis γ, our special orthogonal transformation A is represented by a matrix

$$[A]_\gamma = \begin{bmatrix} 1 & 0 & 0 \\ 0 & a & b \\ 0 & c & d \end{bmatrix} = \begin{bmatrix} 1 & \mathbf{0}^T \\ \mathbf{0} & B \end{bmatrix},$$

where we have written B for the 2×2 matrix in the lower right. Now by proposition 1.17, the matrix $[A]_\gamma$ satisfies $[A]_\gamma^T [A]_\gamma = I_3$. But explicit computation then shows that

$$I_3 = [A]_\gamma^T [A]_\gamma = \begin{bmatrix} 1 & \mathbf{0}^T \\ \mathbf{0} & B^T B \end{bmatrix},$$

so that $B^T B = I_2$ and $B \in O(2)$. But we also have that $1 = \det(A) = \det(B)$, so that in fact $B \in SO(2)$. As revealed in example 1.15, the elements of $SO(2)$ describe rotations, and we may write $[A]_\gamma$ explicitly as

$$[A]_\gamma = \begin{bmatrix} 1 & 0 & 0 \\ 0 & \cos(\theta) & -\sin(\theta) \\ 0 & \sin(\theta) & \cos(\theta) \end{bmatrix}$$

14 *Symmetry and Quantum Mechanics*

for some angle $0 \leq \theta < 2\pi$. The matrix clearly reveals the effect of the linear operator A: it fixes the axis spanned by the eigenvector $\mathbf{u}$ while rotating the plane perpendicular to that axis through the angle θ. $\qquad\square$

Once P recognizes $SO(3)$ as the group of rotations, he adds to the previous discussion as follows: his choice of the basis β leads to the identification $\varphi \colon (V, \langle, \rangle) \to (\mathbb{R}^3, \cdot)$, sending β to the standard basis ε of $\mathbb{R}^3$. Then M's choice of basis β' determines an automorphism $\mathcal{A}$ of $(V, \langle, \rangle)$ defined by $\mathcal{A}(\mathbf{u}_i) = \mathbf{u}_i'$. In physical terms, $\mathcal{A}$ rotates P's coordinate axes onto M's. In terms of P's description, the automorphism $\mathcal{A}$ is represented with respect to the basis β by the matrix $A := [\mathcal{A}]_\beta \in SO(3)$. Explicitly, we have

$$A := [a_{ij}] \qquad \text{where} \qquad \mathcal{A}(\mathbf{u}_j) = \mathbf{u}_j' = \sum_{i=1}^{3} a_{ij} \mathbf{u}_i.$$

It follows from exercise 1.18 that A is the inverse of the matrix representing $\varphi' \circ \varphi^{-1}$. Thus, the matrix $A \in SO(3)$ that describes (with respect to β) the rotation that moves P's coordinate axes onto M's is the inverse of the change of basis matrix describing the translation from P's description to M's. Conversely, an arbitrary element of $SO(3)$ will describe for P a way of rotating his coordinate axes to obtain a new right-handed coordinate system, hence a new positively oriented orthonormal basis for $(V, \langle, \rangle)$. Thus, the group $SO(3)$ acts as the group of rotations on P's copy of space $(\mathbb{R}^3, \cdot)$, serving to connect P's basis with all other possible choices of positive orthonormal basis. The next definition specifies exactly what it means for a group to *act* on a set.

Definition 1.20. *Let G be a group, and X a set. Then a G-action on X is a function $G \times X \to X$ (denoted by $(g, x) \mapsto g \star x$) with the following properties:*

- *$(g_1 g_2) \star x = g_1 \star (g_2 \star x)$ for all $g_1, g_2 \in G$ and $x \in X$;*

- *$e \star x = x$ for all $x \in X$, where $e \in G$ is the identity element.*

In the case where X is a vector space, we can make the additional requirement that each $g \in G$ acts linearly on X:

$$g \star (cx + y) = c(g \star x) + (g \star y) \qquad \text{for all } x, y \in X \text{ and scalars } c.$$

Such a linear G-action is called a representation *of the group G. Almost all of the group actions considered in this text will be linear representations, and we will have much more to say about them in Chapter 5.*

Exercise 1.21. *Show that left-multiplication $(A, \mathbf{x}) \mapsto A \star \mathbf{x} := A\mathbf{x}$ defines an action of $SO(3)$ on $\mathbb{R}^3$. This is clearly a linear action, and we refer to it as the* defining representation *of $SO(3)$. We will study the other representations of $SO(3)$ in Section 5.6.*

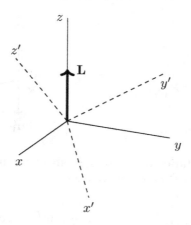

FIGURE 1.3: Observer M's rotated coordinate system (dashed) drawn on top of P's coordinate system (solid). The vector **L** represents a spinning top's angular momentum.

Observer P wants to try all this out to make sense of it. So having laid out his coordinate system, he starts a top spinning at the origin, with its axis of rotation along the third axis. When he looks down on it from the positive third axis, it is spinning counter-clockwise. The angular momentum[4] of the top is represented by a vector $\mathbf{L} = (0, 0, c)$, where $c > 0$ is the magnitude (see Figure 1.3). This entire description derives from P's initial choice of the basis β, which yielded the identification with $\mathbb{R}^3$. As above, suppose that M's basis β' is obtained from β via the rotation $\mathcal{A}$. What is the column vector[5] representing the top's angular momentum under M's identification of space with $\mathbb{R}^3$? As explained above, since the rotation sending P's basis to M's is represented by the orthogonal matrix $A := [\mathcal{A}]_\beta$, the observed coordinates transform according to $A^{-1} = A^T$. Hence, M will measure $A^T[0, 0, c]^T$ for the angular momentum of the top. Take a simple example: suppose that M's coordinate system is obtained by rotating P's through an angle of $\frac{\pi}{2}$ counter-clockwise around P's second axis (see Figure 1.4). Then $\mathbf{u}'_1 = -\mathbf{u}_3, \mathbf{u}'_2 = \mathbf{u}_2$, and $\mathbf{u}'_3 = \mathbf{u}_1$. Thus, the rotation $\mathcal{A}$ is defined by

$$\mathcal{A}(\mathbf{u}_1) = -\mathbf{u}_3, \qquad \mathcal{A}(\mathbf{u}_2) = \mathbf{u}_2, \qquad \mathcal{A}(\mathbf{u}_3) = \mathbf{u}_1.$$

The matrix of $\mathcal{A}$ with respect to P's basis β is

$$A = \begin{bmatrix} 0 & 0 & 1 \\ 0 & 1 & 0 \\ -1 & 0 & 0 \end{bmatrix}.$$

---

[4] See Section 2.1 for a brief discussion of angular momentum.

[5] For reasons of typographical economy, we have been writing elements of $\mathbb{R}^3$ as row vectors (a, b, c), although they are really columns $[a, b, c]^T$ for the purposes of matrix multiplication.

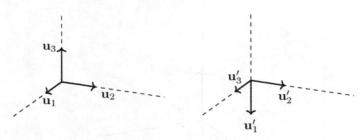

FIGURE 1.4: A rotation of $\pi/2$ radians about the $\mathbf{u}_2$-axis. The dashed lines in both pictures show P's coordinate axes.

Hence, M will measure $A^T[0,0,c]^T = [-c,0,0]^T$ for the angular momentum of the top. This is what we should expect, because P's positive third axis points along M's negative first axis.

P thinks of all this as an elaborate bookkeeping device. M understands this point of view, but advocates for a richer viewpoint. Namely, because P's choice of positively oriented orthonormal basis is arbitrary, the only way to make his identification of space with $(\mathbb{R}^3, \cdot)$ independent of this choice is to remember that $SO(3)$ acts on this inner product space. Moreover, any physically meaningful mathematical object connected with this model of space should also be independent of the choice of basis, hence must support an action of $SO(3)$. That is, we expect to find that the mathematical gadgets that serve as models for physical systems will support a natural action of the group $SO(3)$. Observer P nods politely and changes the subject ... he wants to tell M about something called the Stern-Gerlach experiment.

Chapter 2

Spinor Space

In which M and P discover the special unitary group $SU(2)$ and its relation to the group of rotations $SO(3)$.

In 1922, Otto Stern and Walter Gerlach sent a beam of silver atoms through an inhomogeneous magnetic field and measured the resulting deflection of the atoms. Before we can understand the surprising results of their experiment, we need just a bit of information about the classical theory of angular momentum.

2.1 Angular momentum in classical mechanics

Suppose that the function $\mathbf{r} \colon \mathbb{R} \to \mathbb{R}^3$ describes the position of a particle with mass m, so that at time t, the particle is at the location $\mathbf{r}(t) = (x(t), y(t), z(t))$. Then the velocity of the particle is given by the time-derivative $\dot{\mathbf{r}}$, and its *linear momentum* is defined to be $\mathbf{p} := m\dot{\mathbf{r}}$. Thus, the linear momentum is a measure of the particle's linear motion, taking into account both its velocity and mass. For a similar measure of rotational motion, we define the *angular momentum* of the particle with respect to the origin as $\mathbf{L} := \mathbf{r} \times \mathbf{p}$, the cross product of the particle's position and linear momentum (see Figure 2.1).

Now let's revisit the top, T, that observer P started spinning at the end of the previous chapter. The top has its axis of symmetry aligned with the z-axis, and is spinning counter-clockwise at the rate of ω radians per second—so T

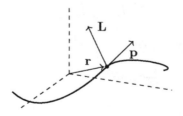

FIGURE 2.1: The position, linear momentum, and angular momentum of a particle moving in three-dimensional space.

17

18 *Symmetry and Quantum Mechanics*

makes a full revolution every $2\pi/\omega$ seconds. Consider a point of T at height z and distance $r \geq 0$ from the z-axis. As T spins, the point follows a circular trajectory:

$$\mathbf{r}(t) = (r\cos(\omega t), r\sin(\omega t), z). \tag{2.1}$$

If we assume that T has a uniform mass density, ρ, then a small volume ΔV centered at our point will have mass $\rho \Delta V$. Treating this small volume as a single particle, it has angular momentum $\mathbf{r} \times \dot{\mathbf{r}}\rho \Delta V$. Integrating over the spatial extent of $T \subset \mathbb{R}^3$ at a given instant yields the angular momentum of the spinning object T:

$$\mathbf{L} := \int_T \mathbf{r} \times \dot{\mathbf{r}}\rho dV.$$

Let's work this out explicitly in the case where $T = B_R$ is the ball of radius $R > 0$ centered at the origin. Note that the derivative of the circular trajectory (2.1) is

$$\dot{\mathbf{r}}(t) = (-r\omega\sin(\omega t), r\omega\cos(\omega t), 0),$$

so the cross product is $\mathbf{r} \times \dot{\mathbf{r}} = r\omega(-\cos(\omega t)z, -\sin(\omega t)z, r)$. Using cylindrical coordinates, we compute that at any instant of time we have:

$$\begin{aligned}
\mathbf{L} &:= \int_{B_R} \mathbf{r} \times \dot{\mathbf{r}}\rho dV \\
&= \omega\rho \int_{z=-R}^{R} \int_{r=0}^{\sqrt{R^2-z^2}} \int_{\theta=0}^{2\pi} (-r\cos(\theta)z, -r\sin(\theta)z, r^2)rd\theta drdz.
\end{aligned}$$

The first two components of this integral are zero due to the inner integration over θ, while for the z-component we have

$$\begin{aligned}
L_z &= \omega\rho \int_{z=-R}^{R} \int_{r=0}^{\sqrt{R^2-z^2}} \int_{\theta=0}^{2\pi} r^3 d\theta drdz \\
&= 2\pi\omega\rho \int_{z=-R}^{R} \int_{r=0}^{\sqrt{R^2-z^2}} r^3 drdz \\
&= \frac{\pi}{2}\omega\rho \int_{z=-R}^{R} (R^2 - z^2)^2 dz \\
&= \frac{8}{15}\pi\omega\rho R^5.
\end{aligned}$$

Note that the total mass of the ball is $M = \frac{4}{3}\pi R^3 \rho$, so that the z-component of the angular momentum may be rewritten as $L_z = I\omega$, where $I := \frac{2}{5}MR^2$ is the *moment of inertia* of the spinning ball. Finally, defining the *angular velocity vector* as $\boldsymbol{\omega} := (0, 0, \omega)$, we may write the angular momentum as $\mathbf{L} = I\boldsymbol{\omega}$. We

Spinor Space

see that the angular momentum is a vector quantity in $\mathbb{R}^3$ that incorporates both the rate of rotation and the distribution of mass around the axis of rotation. Observe that spinning the ball clockwise rather than counter-clockwise corresponds to replacing $\omega > 0$ by $-\omega < 0$, which has the effect of reversing the direction of the angular momentum $\mathbf{L}$. In particular, for any angular speed ω, the vector $\mathbf{L}$ points along the z-axis, with magnitude determined by the absolute value of ω, and direction ("up" or "down") determined by the sign of ω.

Of course, there is nothing special about the z-axis here. If ω is an arbitrary vector in $\mathbb{R}^3$, then it spans a line of rotational symmetry for the ball B_R, and $\mathbf{L} = I\omega$ is the angular momentum of B_R when it spins counter-clockwise around ω at the angular speed of $|\omega|$ radians per second. Of course, a counter-clockwise rotation around ω is a clockwise rotation around $-\omega$, so changing the rotational sense (without changing the angular speed) simply replaces $\mathbf{L}$ by $-\mathbf{L}$.

We would like to find a way of measuring the angular momentum of the ball in a laboratory. It turns out that if the ball is small and electrically charged, then there is an ingenious way of measuring its angular momentum that relies on the classical theory of magnetic fields. We will describe the details below, but the upshot is that with the right experimental setup, the observed deflection of the ball in the z-direction will be directly proportional to the z-component, L_z, of its angular momentum, so that we can measure the ball's angular momentum by instead measuring the magnitude of its spatial deflection.

So suppose that our spinning ball is small, and that it carries a distribution of electric charge. The rotating charge turns the ball into a little magnet, characterized by its *magnetic dipole moment*, μ, a vector quantity proportional to the angular momentum, $\mathbf{L}$:

$$\mu = \gamma \mathbf{L}.$$

Here, the constant of proportionality, γ, depends on the charge distribution on the ball. The important point for us is that this magnetic dipole moment determines the force that the ball will experience if placed within an external magnetic field.

In particular, if $\mathbf{B}$ is an inhomogeneous magnetic field with dominant direction z, and strength increasing linearly in the positive z-direction, then the ball will experience a force in the z-direction, proportional to the z-component of μ (i.e., proportional to the z-component of $\mathbf{L}$). Hence, if we were to send the ball down the x-axis through the field $\mathbf{B}$, then the ball would be deflected in the z-direction, traveling up if $L_z > 0$, down if $L_z < 0$ (if $L_z = 0$, then the ball would experience no deflection). Moreover, if we confine the field $\mathbf{B}$ to a region of fixed length along the x-axis, and if we know the constant velocity of the ball as it travels down the x-axis, then the magnitude of the z-deflection will be directly proportional to the magnitude of L_z. This is the fact that

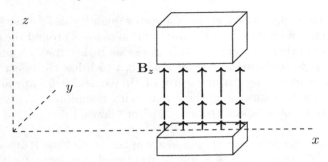

FIGURE 2.2: A Stern-Gerlach device oriented in the positive z-direction. The vertical arrows indicate the z-component of the resulting inhomogeneous magnetic field.

Stern and Gerlach used as the basis of their experiment with silver atoms[1] in 1922. By a *Stern-Gerlach device* (see Figure 2.2), we will mean a device of fixed length that produces such an inhomogeneous field **B**.

Now imagine sending a beam of these spinning balls down the x-axis, through a Stern-Gerlach device oriented in the positive z-direction as described in the previous paragraph. We assume that all of the balls have the same linear velocity, but that their angular momenta are distributed among all directions and a large range of angular speeds. That is, we have carefully prepared the translational motion, but have made no special preparation of the rotational motions. In particular, the z-components of the angular momenta, L_z, will form a continuous range of values, positive and negative. Since the z-deflection of an individual ball is proportional to the z-component of its angular momentum, we should find a continuous spread of the beam in the positive and negative z-directions.

Being small and charged, we expect our spinning ball to provide a crude classical model of the electron, considered as a charged point particle. So, applying the previous thought experiment to a beam of electrons, we record our conclusion as a

> **Classical Expectation:** *We should observe a continuous spread of the electron beam in the positive and negative z-directions, reflecting a continuous range of values for L_z among the individual electrons (see Figure 2.3).*

But the experimental results are strikingly at odds with this expectation:

---

[1] The experiment described actually requires a neutral particle in order to avoid the Lorentz force that a moving charge would experience. However, the magnetic moment of a silver atom is due almost entirely to the magnetic moment of its outermost electron, so in effect, Stern and Gerlach were detecting the angular momentum of the electron (see [22, pp. 1–4]). Hence, for the idealized Stern-Gerlach thought experiments discussed here, we speak in terms of the negatively charged electron, even though the actual historical experiment requires a neutral particle.

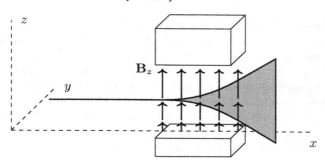

FIGURE 2.3: The classical expectation for the behavior of an electron-beam in a Stern-Gerlach device.

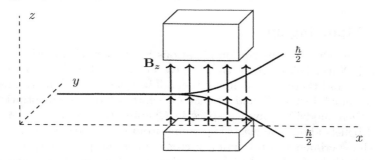

FIGURE 2.4: The actual behavior of an electron-beam in a Stern-Gerlach device.

Experimental Result: *The electron beam splits into two discrete pieces, with half the electrons deflecting upward as if they have $L_z = \frac{\hbar}{2}$, while the other half deflects downward by the same amount, as if they have $L_z = -\frac{\hbar}{2}$ (see Figure 2.4). Here, $\hbar = 1.054573 \times 10^{-34}$ kg $\cdot$ m^2/s is the reduced Planck constant.*

Faced with this experimental fact, the only conclusion we can draw is that the crude classical model of the electron as a spinning charged ball is wrong. Instead of displaying a continuous range of angular momenta, the beam behaves as if it contains a 50-50 mix of two types of electrons: those that are "spin up" and those that are "spin down" along the z-direction. While the *sign* for each electron appears to be random, the *magnitude* of the z-component of angular momentum is fixed at $\frac{\hbar}{2}$. But the phenomenon is actually stranger still, because there is nothing special about the z-direction!

Indeed, suppose that we turn on the electron beam before establishing the magnetic field **B**. Then we could choose any unit vector, **u**, orthogonal to the beam's direction, and set up a Stern-Gerlach device with orientation **u**, inducing an inhomogeneous magnetic field as described above, but now with dominant direction **u**. From the rotational symmetry of physical space, the electron beam will behave just as before: half of the electrons will be deflected in the $+\mathbf{u}$-direction, half in the $-\mathbf{u}$-direction, but the *amount* of deflection will always be the same, corresponding to a $\mathbf{u}$-component of angular momentum of magnitude $\frac{\hbar}{2}$. This is quite bizarre, because (thinking classically) it seems to suggest that each electron is spinning at the same rate around every axis, with half of them spinning clockwise, half counter-clockwise! As described below (belief 2), one way out of this difficulty is to give up on the idea that individual electrons possess a definite angular momentum, and instead think in terms of definite probabilities for measurement outcomes.

2.2 Modeling spin

As P finishes his description of various Stern-Gerlach experiments[2], M is stunned. Nevertheless, these things have been revealed through careful experiment, and there is no denying them. The question is: how to model this phenomenon? Just as P and M made a short list of shared intuitions that led to their model of physical space in Chapter 1, they now make a list of shared beliefs about the electron, coming from their knowledge of the Stern-Gerlach experiments. If $\mathbf{u}$ is a unit vector in physical space $(V, \langle, \rangle)$, then $SG\mathbf{u}$ denotes a Stern-Gerlach device producing an inhomogeneous magnetic field with dominant direction $\mathbf{u}$.

1. An electron passing through an $SG\mathbf{u}$ will return an angular momentum measurement of $\pm\frac{\hbar}{2}$, which we think of as "spin up" and "spin down" along the direction $\mathbf{u}$;

2. Until we make a measurement with an $SG\mathbf{u}$, a particular electron may have no definite spin along $\mathbf{u}$, but it *does* have a definite probability of returning each of the values $\pm\frac{\hbar}{2}$ when measured by an $SG\mathbf{u}$;

3. If an electron exits an $SG\mathbf{u}$ spin up, then it will measure spin up if measured immediately by a successive $SG\mathbf{u}$. Likewise, if an electron exits an $SG\mathbf{u}$ spin down, then it will measure spin down if measured immediately by a successive $SG\mathbf{u}$;

4. More generally, if the angle between $\mathbf{u}$ and $\mathbf{u}'$ is α, then an electron that exits an $SG\mathbf{u}$ spin up will measure spin up with probability $\cos^2\left(\frac{\alpha}{2}\right)$ if measured immediately by an $SG\mathbf{u}'$. Similarly, an electron that exits an $SG\mathbf{u}$ spin down will measure spin down with probability $\cos^2\left(\frac{\alpha}{2}\right)$ if measured immediately by an $SG\mathbf{u}'$.

---

[2]One can imagine a number of different experiments involving multiple Stern-Gerlach devices arranged in sequence, with different orientations (see [22, pp. 5–9]). Beliefs 3 and 4 come from the results of such experiments

Spinor Space

P thinks about this list for a while, and observes that together these beliefs imply that measurement with a Stern-Gerlach device *does something* to the electron. Indeed, by beliefs 1 and 3 he can produce electrons that will measure spin up along $\mathbf{u}$ with probability 1, by blocking the electrons that emerge spin down from the $SG\mathbf{u}$. Given such a beam of electrons, belief 4 says that if $\mathbf{u}'$ is orthogonal to $\mathbf{u}$, then half of the electrons will measure spin up when measured by an $SG\mathbf{u}'$. But again by 4, the electrons in this spin up along $\mathbf{u}'$ stream will each measure spin up along $\mathbf{u}$ with probability $\frac{1}{2}$. Thus, the measurement with $SG\mathbf{u}'$ has changed the definite probabilities announced in belief 2. M agrees with P and concludes that any successful model of spin will have to include a concept of measurements as "operating" on spin-states.

Since belief 1 says that there are two distinct measurement outcomes (spin up and spin down), while belief 2 suggests that general states are some kind of combination of these possibilities, it seems reasonable to look for a model based on a 2-dimensional vector space. Roughly speaking, a basis should correspond to states of definite spin up and spin down (announced by belief 3), while a general state should be a linear combination of the basis. Moreover, belief 4 (which arises from experimental data) reminds P of Malus' law about the intensity of polarized light transmitted through a linear polarizer. M and P discuss this for a while, and after trying and failing with real vector spaces, they instead propose the following model involving the complex numbers.[3]

Definition 2.1. Spinor space *is a two-dimensional complex inner product space* $(W, \langle | \rangle)$. *The* spin-states *of an electron are represented by unit vectors in* W, *and two unit vectors represent the same spin-state if one is a scalar multiple of the other. That is, the unit vectors* $\mathbf{w}$ *and* $\mathbf{w}'$ *represent the same spin-state if and only if* $\mathbf{w} = e^{i\theta}\mathbf{w}'$ *for some* $\theta \in \mathbb{R}$.

Recall that a *complex inner product space* is a complex vector space W together with a function $\langle | \rangle \colon W \times W \to \mathbb{C}$ such that if $\alpha \in \mathbb{C}$ and $\mathbf{a}, \mathbf{a}', \mathbf{b} \in W$, then[4]

i) $\langle \alpha\mathbf{a} + \mathbf{a}' | \mathbf{b} \rangle = \alpha^* \langle \mathbf{a}|\mathbf{b} \rangle + \langle \mathbf{a}'|\mathbf{b} \rangle$ *(conjugate-linear in first component)*[5];

---

[3]The reader may well wonder why it is necessary to employ complex numbers. While there are a variety of reasons for the use of complex numbers in quantum mechanics, the immediate reason in terms of our current story is that if we were to use real numbers, then we would need to find a natural way of translating rotations of physical space $\mathbb{R}^3$ into rotations of $\mathbb{R}^2$. But there is no relationship between the rotation groups $SO(3)$ and $SO(2)$ that is suitable for this purpose. As we will see in the course of this chapter, there is an extremely elegant relationship between the groups $SO(3)$ and $SU(2)$, the analogue of the rotation group for $\mathbb{C}^2$, and this relationship plays a central role in the theory of quantum mechanical spin.

[4]Here, α^* denotes the complex conjugate of a complex number α.

[5]We follow the convention common in the physics literature of defining a complex inner product to be conjugate-linear in the first component and linear in the second. Mathematicians usually do the opposite, and take inner products to be linear in the first component and conjugate-linear in the second. This difference in conventions can lead to confusion, so beware.

24 *Symmetry and Quantum Mechanics*

ii) $\langle \mathbf{a}|\mathbf{b}\rangle = \langle \mathbf{b}|\mathbf{a}\rangle^*$ (*conjugate symmetry*);

iii) $\langle \mathbf{a}|\mathbf{a}\rangle \geq 0$ with equality if and only if $\mathbf{a} = 0$ (*positive definite*).

Exercise 2.2. *Show that conditions i) and ii) for an inner product imply linearity in the second component:* $\langle \mathbf{a}|\alpha\mathbf{b} + \mathbf{b}'\rangle = \alpha \langle \mathbf{a}|\mathbf{b}\rangle + \langle \mathbf{a}|\mathbf{b}'\rangle$.

As in the real case, an inner product on W determines a *norm* $\|\mathbf{a}\| := \sqrt{\langle \mathbf{a}|\mathbf{a}\rangle}$. Moreover, the Cauchy-Schwarz inequality holds: $|\langle \mathbf{a}|\mathbf{b}\rangle| \leq \|\mathbf{a}\|\|\mathbf{b}\|$ for all $\mathbf{a}, \mathbf{b} \in W$. Two vectors $\mathbf{a}$ and $\mathbf{b}$ are defined to be *orthogonal* in $(W, \langle|\rangle)$ if and only if $\langle \mathbf{a}|\mathbf{b}\rangle = 0$.

Exercise 2.3. *Suppose that* $(X, \langle|\rangle)$ *is a complex inner product space. Show that the inner product* $\langle|\rangle$ *is uniquely determined by the corresponding norm. (Hint: compute* $\|\mathbf{x}_1 + \mathbf{x}_2\|^2$ *and* $\|\mathbf{x}_1 + i\mathbf{x}_2\|^2$. *Compare exercise 1.3).*

Example 2.4. *Fix an integer* $n \geq 1$, *and consider the set* $\mathbb{C}^n$ *of all n-tuples of complex numbers:*

$$\mathbb{C}^n := \{(\alpha_1, \alpha_2, \ldots, \alpha_n) \mid \alpha_i \in \mathbb{C}\}.$$

The set $\mathbb{C}^n$ *is an n-dimensional complex vector space under the operations of component-wise addition and scalar multiplication. Moreover, define the* dot product *of two vectors in* $\mathbb{C}^n$ *by the formula*

$$(\alpha_1, \alpha_2, \ldots, \alpha_n) \cdot (\beta_1, \beta_2, \ldots, \beta_n) := \sum_{i=1}^{n} \alpha_i^* \beta_i.$$

Then $\cdot : \mathbb{C}^n \times \mathbb{C}^n \to \mathbb{C}$ *defines an inner product on* $\mathbb{C}^n$. *The complex inner product space* $(\mathbb{C}^n, \cdot)$ *is called* complex Euclidean *n-space.*

After unpacking all of this terminology, M reiterates the meaning of definition 2.1: the possible spin-states of an electron are given by the unit vectors in $(W, \langle|\rangle)$, where two unit vectors correspond to the same spin-state if and only if they differ by a complex number of modulus 1, called a *phase*. This phase ambiguity is somewhat mysterious at this point, and the first order of business is to find a way of producing some quantities associated to spin-states in a phase-independent way.

To this end, we now present some convenient and powerful notation, introduced by P.A.M. Dirac in 1939 and popularized in his classic textbook [2]. If we use the symbol ψ to denote a spin-state, then ψ is actually an equivalence class of unit vectors in $(W, \langle|\rangle)$. Nevertheless, we will generally think of ψ as an actual unit vector, always remembering that the vector is only well defined up to a phase $e^{i\theta}$. We will often write $|\psi\rangle$ instead of ψ to emphasize that we have chosen a unit vector in W to represent the spin-state. The unit vector $|\psi\rangle$ is called a *ket*, being the latter half of a bracket $\langle \mathbf{a}|\mathbf{b}\rangle$. Hence, every ket determines a unique spin-state, but each spin-state is represented by infinitely many kets, any two of which differ by a phase $e^{i\theta}$. For each ket $|\psi\rangle$, there is a

Spinor Space   25

corresponding *bra* $\langle\psi|$, which is the linear mapping from W to $\mathbb{C}$ given by the inner product[6]:

$$\langle\psi|(\mathbf{a}) := \langle\psi|\mathbf{a}\rangle = \langle\mathbf{a}|\psi\rangle^*.$$

We must remember that the inner product $\langle\psi|\mathbf{a}\rangle$ depends on a choice of a representing ket $|\psi\rangle$, and not only on the spin-state.

Exercise 2.5 ($\clubsuit$[7]). *Show that the bra corresponding to the ket $e^{i\theta}|\psi\rangle$ is $e^{-i\theta}\langle\psi|$. More generally, if $|\psi\rangle = c_1|\phi_1\rangle + c_2|\phi_2\rangle$ is a complex linear combination, then $\langle\psi| = c_1^*\langle\phi_1| + c_2^*\langle\phi_2|$.*

The next proposition eliminates the phase ambiguity in our description by showing that spin-states are in one-to-one correspondence with orthogonal projections onto lines in W.

Proposition 2.6. *There is a one-to-one correspondence between spin-states and rank one orthogonal projections on W:*

$$\psi \longleftrightarrow P_\psi = |\psi\rangle\langle\psi|.$$

Here, $P_\psi\colon W \to W$ is given by the formula $P_\psi(\mathbf{a}) = \langle\psi|\mathbf{a}\rangle|\psi\rangle$.

Proof. Suppose that ψ is a spin-state, and choose a representing ket $|\psi\rangle$. Any other ket representing ψ is of the form $e^{i\theta}|\psi\rangle$, and hence spans the same complex line $\mathbb{C}|\psi\rangle$ contained in W. Any $\mathbf{a} \in W$ can be written uniquely as $\mathbf{a} = \mathbf{b} + \mathbf{b}^\perp$ for $\mathbf{b} \in \mathbb{C}|\psi\rangle$ and $\mathbf{b}^\perp \in (\mathbb{C}|\psi\rangle)^\perp$, the orthogonal complement to $\mathbb{C}|\psi\rangle$. In terms of this decomposition, the orthogonal projection onto the line $\mathbb{C}|\psi\rangle$ is defined as $P_\psi(\mathbf{a}) = P_\psi(\mathbf{b} + \mathbf{b}^\perp) := \mathbf{b}$ (see Figure 2.5). But note that $\mathbf{b} = \langle\psi|\mathbf{a}\rangle|\psi\rangle =: |\psi\rangle\langle\psi|(\mathbf{a})$, so that $P_\psi = |\psi\rangle\langle\psi|$ as claimed.

Conversely, suppose that $P\colon W \to W$ is any rank one orthogonal projection on W. Choose a basis ket $|\psi\rangle$ for the 1-dimensional range of P, and note that any other choice differs from $|\psi\rangle$ by a phase $e^{i\theta}$. It follows that P determines a unique spin-state ψ such that $P = P_\psi$. $\qquad\square$

Note that if two kets differ by a phase, then so will their projections onto the line spanned by ψ:

$$P_\psi(e^{i\theta}|\phi\rangle) = e^{i\theta}P_\psi(|\phi\rangle).$$

We can eliminate this phase dependence by taking the squared norm of the projections:

$$\|P_\psi(e^{i\theta}|\phi\rangle)\|^2 = \|P_\psi(|\phi\rangle)\|^2 = |\langle\psi|\phi\rangle|^2|\langle\psi|\psi\rangle|^2 = |\langle\psi|\phi\rangle|^2,$$

where we have used the fact that $e^{i\theta}$ has modulus 1 and $|\psi\rangle$ has unit norm. Thus, we have succeeded in producing a quantity that depends only on the

---

[6]That is, $\langle\psi| := \langle\psi|-\rangle\colon W \to \mathbb{C}$ is an element of the *dual space* of W.

[7]$\clubsuit$ indicates an exercise with a solution in Appendix A.4.

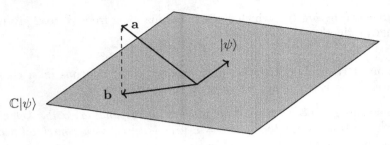

FIGURE 2.5: Orthogonal projection onto the complex line spanned by $|\psi\rangle$, with one real dimension suppressed.

spin-states ψ and ϕ, and not on the choice of representing kets. Note that the real number $|\langle\psi|\phi\rangle|^2$ is between 0 and 1, since it is the squared length of the orthogonal projection of a unit vector (alternatively, this follows from the Cauchy-Schwarz inequality). Hence, we may interpret this number as a probability, as recorded in the following interpretation, fundamental to all that follows.

Probability Interpretation: *Given two spin-states ψ and ϕ, the probability that ϕ, when measured with a "ψ-device," will be found in the state ψ is given by the squared modulus of the inner product $|\langle\psi|\phi\rangle|^2$.*

But what kind of thing is a "ψ-device"? To answer this and to make a connection with their list of beliefs coming from the Stern-Gerlach experiments, M and P need to establish a connection between spinor space $(W, \langle|\rangle)$ and physical space $(V, \langle,\rangle)$. Observer P is eager to help, so as in the previous chapter, he chooses a right-handed orthonormal basis $\{\mathbf{u}_1, \mathbf{u}_2, \mathbf{u}_3\}$ for V which yields the identification of physical space with $(\mathbb{R}^3, \cdot)$ together with the rotation action of $SO(3)$. Henceforth, we will denote a Stern-Gerlach device $SG\mathbf{u}_3$ by SGz since such a device is oriented along P's positive z-axis if P labels his three coordinate axes by x, y, z as usual.

Having thus installed his SGz, every electron that P measures comes out either spin up or spin down along the z-direction (beliefs 1 and 3). These two outcomes correspond to two distinct spin-states, represented by kets $|+z\rangle$ and $|-z\rangle$ in W, uniquely determined up to individual multiplication by phases. Moreover, these kets form an orthonormal basis of W. Indeed, belief 3 says that the probability of $|+z\rangle$ being found in the state $|+z\rangle$ upon measurement is 1, while the probability of it being found in the state $|-z\rangle$ is 0. Since these probabilities are given by the squared absolute values of the inner products, the kets are orthonormal as claimed.

Thus, an arbitrary ket $|\phi\rangle$ in W can be written uniquely as $|\phi\rangle = c_+|+z\rangle + c_-|-z\rangle$, where c_+, c_- are complex numbers satisfying $|c_+|^2 + |c_-|^2 = 1$. Moreover, the probability that an electron with spin-state

Spinor Space

27

ϕ will be spin up when measured with P's SGz is given by

$$|\langle +z|\phi\rangle|^2 = |\langle +z|c_+|+z\rangle + \langle +z|c_-|-z\rangle|^2 = |c_+|^2.$$

Similarly, $|c_-|^2$ is the probability that ϕ will be spin down when measured by an SGz. Thus, the model nicely captures belief 2: general spin-states are *superpositions* of the spin up and spin down states, with coefficients that determine the definite probabilities for measurements. The complex coefficients c_+, c_- are called *probability amplitudes*.

The preceding discussion shows that every unit vector in physical space yields an ordered pair of orthogonal spin-states. In particular, P's choice of basis for physical space (together with his installation of an SGz) has determined an orthonormal basis for spinor-space W, up to phases. Continuing with her investigation from Chapter 1, observer M wonders how a different choice of basis for V would change the basis for W? Before we take up her question, we pause to remind the reader of some different types of linear operators on complex Euclidean n-space.

2.3 Complex linear operators and matrix groups

This section follows the pattern of Section 1.2, extending the results obtained there for real linear operators to the complex case. Most of the proofs extend easily to the complex situation, so we only briefly mention the necessary changes.

Suppose that $L\colon \mathbb{C}^n \to \mathbb{C}^n$ is a linear operator, and let $\varepsilon := \{e_1, e_2, \ldots, e_n\}$ denote the standard basis of $\mathbb{C}^n$ (see example 2.4) . Then L is represented (with respect to ε) by an $n \times n$ matrix of complex numbers, which we also denote by L:

$$L = [L_{ij}] \qquad \text{where} \qquad L(e_j) = \sum_{i=1}^{n} L_{ij} e_i.$$

The proof of proposition 1.7 works just as well in the complex case to show that L is an automorphism of $\mathbb{C}^n$ if and only if the matrix L is invertible.

Definition 2.7. *The group of all invertible $n \times n$ complex matrices, denoted $GL(n, \mathbb{C})$, is called the* complex general linear group. *It is the symmetry group of the vector space $\mathbb{C}^n$.*

Definition 2.8. *If B is a complex $m \times n$ matrix, then its* conjugate transpose $B^\dagger$ *is the $n \times m$ matrix obtained by taking the ordinary transpose of B and then replacing each entry with its complex conjugate:*

$$(B^\dagger)_{ij} := B_{ji}^*.$$

Proposition 2.9. *Suppose that* $L \in GL(n, \mathbb{C})$. *Then* L *preserves the dot product on* $\mathbb{C}^n$ *if and only if* $L^{-1} = L^\dagger$, *the conjugate transpose of the matrix* L.

Proof. Adapt the proof of proposition 1.9 by replacing the ordinary transpose with the conjugate transpose. $\square$

Definition 2.10. *The group of all invertible* $n \times n$ *complex matrices satisfying* $L^{-1} = L^\dagger$, *denoted* $U(n)$, *is called the* unitary group. *It is the symmetry group of complex Euclidean* n-*space* $(\mathbb{C}^n, \cdot)$.

Exercise 2.11. *Suppose that* $L \colon \mathbb{C}^n \to \mathbb{C}^n$ *is a linear operator. Show that* $L \in U(n)$ *if and only if* L *preserves the norm of vectors:* $\|L\mathbf{w}\| = \|\mathbf{w}\|$ *for all* $\mathbf{w} \in \mathbb{C}^n$. *(Hint: use proposition 2.9 and compute* $\|L(\mathbf{v} + \mathbf{w})\|^2$ *and* $\|L(\mathbf{v} + i\mathbf{w})\|^2$. *Compare exercise 1.11.)*

Proposition 2.12. *The determinant of any unitary matrix is a complex number of modulus 1.*

Proof. If L is unitary, then $I_n = L^\dagger L$. Taking the determinant of both sides yields $1 = \det(I_n) = \det(L^\dagger L) = \det(L^\dagger)\det(L) = \det(L)^* \det(L) = |\det(L)|^2$. It follows that $|\det(L)| = 1$ as claimed. $\square$

Definition 2.13. *The* special unitary group *is the subgroup* $SU(n) \subset U(n)$ *of unitary matrices with determinant 1.*

Example 2.14. *The case* $n = 1$ *is somewhat more interesting in the complex case than in the real case:*

- $GL(1, \mathbb{C}) = \mathbb{C}^\times$, *the group of nonzero complex numbers;*

- $U(1) = \{e^{i\theta} \mid \theta \in \mathbb{R}\}$, *the phase group;*

- $SU(1) = \{1\}$, *the trivial group.*

Exercise 2.15. *Show that the unitary group* $U(1)$ *is isomorphic to the special orthogonal group* $SO(2)$.

Example 2.16. *The complex general linear group* $GL(2, \mathbb{C})$ *consists of* 2×2 *complex matrices with nonzero determinant:*

$$GL(2, \mathbb{C}) = \left\{ \begin{bmatrix} \alpha & \beta \\ \gamma & \delta \end{bmatrix} \mid \alpha\delta - \beta\gamma \neq 0 \right\}.$$

As in the real case, a 2×2 *complex matrix is an element of* $GL(2, \mathbb{C})$ *if and only if its columns form a basis for* $\mathbb{C}^2$.

Writing out the unitarity condition $I_2 = L^\dagger L$ *we find*

$$\begin{bmatrix} 1 & 0 \\ 0 & 1 \end{bmatrix} = \begin{bmatrix} \alpha^* & \gamma^* \\ \beta^* & \delta^* \end{bmatrix} \begin{bmatrix} \alpha & \beta \\ \gamma & \delta \end{bmatrix} = \begin{bmatrix} |\alpha|^2 + |\gamma|^2 & \alpha^*\beta + \gamma^*\delta \\ \beta^*\alpha + \delta^*\gamma & |\beta|^2 + |\delta|^2 \end{bmatrix}.$$

Thus, the conditions for L to be unitary are $|\alpha|^2 + |\gamma|^2 = |\beta|^2 + |\delta|^2 = 1$ and $\alpha^*\beta + \gamma^*\delta = 0$. Interpreting these relations as dot products, we see that L is unitary exactly when the columns of L have unit norm and are orthogonal to each other. Thus, a 2×2 complex matrix is an element of $U(2)$ if and only if its columns form an orthonormal basis for $(\mathbb{C}^2, \cdot)$.

If in addition we require that $\det(L) = 1$, then using the formula for the inverse of a 2×2 matrix, we find that the unitarity condition $L^\dagger = L^{-1}$ becomes

$$\begin{bmatrix} \alpha^* & \gamma^* \\ \beta^* & \delta^* \end{bmatrix} = \begin{bmatrix} \delta & -\beta \\ -\gamma & \alpha \end{bmatrix} \iff \alpha = \delta^* \quad and \quad \beta = -\gamma^*.$$

Thus, we find that the special unitary group is

$$SU(2) = \left\{ \begin{bmatrix} \alpha & \beta \\ -\beta^* & \alpha^* \end{bmatrix} \mid |\alpha|^2 + |\beta|^2 = 1 \right\}.$$

This group will play a major role in the remainder of our story.

Exercise 2.17. *Generalize the analysis in the preceding example to show that for all $n \geq 1$:*

a) an $n \times n$ complex matrix L is in $GL(n, \mathbb{C})$ if and only if the columns of L form a basis for $\mathbb{C}^n$;

b) an $n \times n$ complex matrix L is in $U(n)$ if and only if the columns of L form an orthonormal basis for $(\mathbb{C}^n, \cdot)$.

c) Let $L: \mathbb{C}^n \to \mathbb{C}^n$ be a linear operator, and $\gamma := \{\mathbf{u}_1, \ldots, \mathbf{u}_n\}$ be any orthonormal basis for $(\mathbb{C}^n, \cdot)$. Denote by $[L]_\gamma$ the matrix representing L in the basis γ. Then L is a unitary transformation if and only if $[L]_\gamma^\dagger = [L]_\gamma^{-1}$. (Hint: adapt the proof of proposition 1.17.)

Now we rejoin observers M and P, who are still puzzling over spinor space. Recall M's question: how would a different choice of basis for physical space V affect the basis for spinor space W obtained by the installation of a Stern-Gerlach device along the third axis? To study this question, P continues to consider the basis $\gamma := \{|+z\rangle, |-z\rangle\}$ for W coming from his third basis vector $\mathbf{u}_3$ in physical space. But M considers the different orthonormal basis $\gamma' := \{|+z'\rangle, |-z'\rangle\}$ for W that arises from her basis β' for V, together with her installation of an SGz' oriented along her third basis vector $\mathbf{u}_3'$. Just as in Chapter 1, these two orthonormal bases for W determine distinct isomorphisms $\Phi, \Phi': (W, \langle | \rangle) \to (\mathbb{C}^2, \cdot)$, defined by sending γ, γ' respectively to the standard basis of $\mathbb{C}^2$. As before, in order to determine the translation between their two descriptions of spinor space, we consider the automorphism of $(\mathbb{C}^2, \cdot)$ defined by the composition $\Phi' \circ \Phi^{-1}$. The situation is pictured in the following diagram:

$$\begin{array}{ccc} W & = & W \\ \Phi \downarrow & & \downarrow \Phi' \\ \mathbb{C}^2 & \xrightarrow{\Phi' \circ \Phi^{-1}} & \mathbb{C}^2. \end{array}$$

30   *Symmetry and Quantum Mechanics*

By the discussion above, the automorphism $\Phi' \circ \Phi^{-1}$ may be identified with an invertible matrix of complex numbers, i.e., an element of $GL(2, \mathbb{C})$. Moreover, because the automorphism $\Phi' \circ \Phi^{-1}$ preserves the dot product, the corresponding matrix is actually an element of the unitary group $U(2)$.

Exercise 2.18. *Show that the matrix of $\Phi' \circ \Phi^{-1}$ with respect to the standard basis on $\mathbb{C}^2$ is the change of basis matrix from γ to γ'. (Compare exercise 1.18.)*

Going further, M observes that by exploiting phases, she can change her basis kets (without changing the corresponding spin states) to arrange for the matrix to be an element of the special unitary group $SU(2)$. Indeed, setting $\delta := \det(\Phi' \circ \Phi^{-1})$, proposition 2.12 shows that $|\delta| = 1$.

Exercise 2.19. *Choose a square root of δ, and denote it by $\sqrt{\delta}$. Show that $\sqrt{\delta} \in U(1)$, and hence may be used as a phase. Then consider the orthonormal basis of W given by $\gamma'' := \left\{ \sqrt{\delta} \, |+z'\rangle, \sqrt{\delta} \, |-z'\rangle \right\}$. This basis, while distinct from γ', corresponds to the same pair of orthogonal spin-states. As usual, sending the basis γ'' to the standard basis of $\mathbb{C}^2$ determines an isomorphism $\Phi'' \colon (W, \langle | \rangle) \to (\mathbb{C}^2, \cdot)$. Use exercise 2.18 to show that*

$$\Phi'' \circ \Phi^{-1} = \frac{1}{\sqrt{\delta}} \Phi' \circ \Phi,$$

and conclude that $\Phi'' \circ \Phi^{-1}$ is an element of $SU(2)$.

We now assume (after adjustment by a phase as above) that M's basis kets $|\pm z'\rangle$ yield an automorphism $\Phi' \circ \Phi^{-1}$ which is an element of the special unitary group $SU(2)$. Following in the pattern of his comments about $SO(3)$ in Chapter 1, observer P summarizes the situation as follows: his choice of the z-basis γ leads to the identification $\Phi \colon (W, \langle | \rangle) \to (\mathbb{C}^2, \cdot)$, sending γ to the standard basis of $\mathbb{C}^2$. Then M's choice of basis γ' determines an automorphism $\mathcal{B}$ of $(W, \langle | \rangle)$ defined by $\mathcal{B}(|+z\rangle) = |+z'\rangle$ and $\mathcal{B}(|-z\rangle) = |-z'\rangle$. In terms of P's description, the automorphism $\mathcal{B}$ is represented with respect to the basis γ by a matrix $B := [\mathcal{B}]_\gamma$. Explicitly, we have $B := [\beta_{ij}]$, where

$$|+z'\rangle = \beta_{11}|+z\rangle + \beta_{21}|-z\rangle \quad \text{and} \quad |-z'\rangle = \beta_{12}|+z\rangle + \beta_{22}|-z\rangle.$$

It follows from exercise 2.18 that B is the inverse of the special unitary matrix $\Phi' \circ \Phi^{-1}$. Thus, the matrix $B \in SU(2)$ that describes (with respect to γ) the automorphism that moves the z-basis onto the z'-basis is the inverse of the change of basis matrix describing the translation from P's description to M's. Conversely, an arbitrary element of $SU(2)$ will describe for P a way of superposing the z-basis to obtain a new orthonormal basis for W, hence a new pair of orthogonal spin-states.

Thus, $SU(2)$ acts on spinor space $(\mathbb{C}^2, \cdot)$ similarly to the way $SO(3)$ acts on physical space $(\mathbb{R}^3, \cdot)$. But M wants an answer to the following question:

if the rotation $A \in SO(3)$ connects P's right-handed orthonormal basis for V to M's, how can we determine the corresponding matrix $B \in SU(2)$ that connects P's z-basis for W to M's z'-basis? This is really a question about the relationship between the action of $SO(3)$ on $\mathbb{R}^3$ and the action of $SU(2)$ on $\mathbb{C}^2$. Before considering the actions, we should ask a more basic question: what is the relationship between the groups themselves?

2.4 The geometry of $SU(2)$

The group $SU(2)$ consists of unitary 2×2 matrices with determinant 1. Thus, the 2×2 complex matrix B is in $SU(2)$ if and only if $B^\dagger = B^{-1}$ and $\det(B) = 1$. In example 2.16, we discovered the explicit form of these matrices:

$$ B = \begin{bmatrix} \alpha & \beta \\ -\beta^* & \alpha^* \end{bmatrix} \qquad \alpha, \beta \in \mathbb{C} \text{ and } |\alpha|^2 + |\beta|^2 = 1. $$

If we write $\alpha = a_1 + ia_2$ and $\beta = b_1 + ib_2$ for $a_j, b_j \in \mathbb{R}$, the condition on α and β becomes

$$ a_1^2 + a_2^2 + b_1^2 + b_2^2 = 1, $$

which defines the unit sphere $S^3 \subset \mathbb{R}^4$. Thus, as a topological space, $SU(2)$ is the three-dimensional unit sphere. In particular, it is path connected and simply connected[8].

In an effort to establish a connection between the groups $SU(2)$ and $SO(3)$, we would like to discover a natural way in which $SU(2)$ acts on $\mathbb{R}^3$ via rotations. We begin with the observation that $SU(2)$ acts on itself by conjugation: $(B, M) \mapsto B \star M := BMB^{-1}$ defines an $SU(2)$-action on $SU(2)$ in the sense of definition 1.20. Thinking of the copy of $SU(2)$ being acted upon as the 3-sphere S^3, we have an action of $SU(2)$ on S^3. This is close to what we want, because the *tangent space* to S^3 at any point is a copy of $\mathbb{R}^3$.

2.4.1 The tangent space to the circle $U(1) = S^1$

In order to motivate and clarify the notion of the tangent space to $SU(2) = S^3 \subset \mathbb{R}^4$, consider the simpler case of the group $U(1)$ consisting of complex numbers of modulus 1. Note that a complex number $z = x + iy \in U(1)$ if and only if $x^2 + y^2 = 1$, so $U(1)$ is the unit circle $S^1 \subset \mathbb{R}^2$.

From Figure 2.6, it is clear that the tangent space to the circle S^1 at the point $(1, 0)$ is the vertical line $x = 1$. Since we want our tangent spaces

---

[8]Path connected means that any two elements of S^3 may be joined by a continuous path in S^3. Simply connected means that all loops in S^3 may be continuously contracted to a point, which formalizes the idea that S^3 has "no holes."

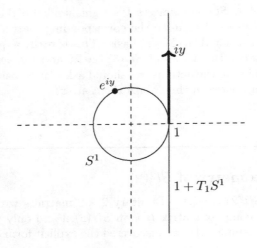

FIGURE 2.6: The tangent line to the circle S^1 at the identity.

to be closed under vector addition and scalar multiplication, we will refer to the vertical line $x = 1$ as the *translated tangent space*, and use the term *tangent space* for the vertical line $x = 0$, which is a vector subspace of $\mathbb{R}^2$. Since the point $(1,0)$ corresponds to the identity element $z = 1$ of the group $U(1)$, we denote this tangent space by $T_1 S^1$ and write

$$T_1 S^1 = \{iy \mid y \in \mathbb{R}\} = i\mathbb{R} \subset \mathbb{C}.$$

The translated tangent space is then given by

$$1 + T_1 S^1 = 1 + i\mathbb{R} \subset \mathbb{C}.$$

But how could we determine this tangent space without relying on the picture? Well, suppose that $c : (-\epsilon, \epsilon) \to \mathbb{C} = \mathbb{R}^2$ is a one-to-one differentiable curve with the property that $c(t) \in U(1) = S^1$ for all t and $c(0) = 1 \in U(1)$. That is, c is a parametrization of the curve S^1 near the identity. Then the derivative $\dot{c}(0) \in \mathbb{R}^2$ is a tangent vector to S^1 at the identity. The totality of all such tangent vectors forms the tangent line $T_1 S^1$. But note that we have $1 = c(t)c(t)^*$ for all t, since each $c(t) \in U(1)$ is a complex number of modulus 1. Taking the derivative with respect to t and evaluating at $t = 0$ yields:

$$\begin{aligned} 0 &= \dot{c}(0)c(0)^* + c(0)\dot{c}(0)^* \\ &= \dot{c}(0) + \dot{c}(0)^*. \end{aligned}$$

We see that the derivative $\dot{c}(0)$ must be a purely imaginary complex number: $\dot{c}(0) = iy$ for some $y \in \mathbb{R}$. Thus, this computation reproduces the description of the tangent space $T_1 S^1$ provided above.

Note that every element of $i\mathbb{R}$ does indeed arise from a curve c. Indeed, for any $y \in \mathbb{R}$, consider the curve $c(t) := e^{ity} \in U(1)$. Then $c(0) = 1$ and $\dot{c}(0) = iy$.

Spinor Space

2.4.2 The tangent space to the sphere $SU(2) = S^3$

We wish to determine the tangent space at the identity of $SU(2)$ by following the strategy described for the circle group $U(1)$ in the previous section. Recall that $SU(2) = S^3 \subset \mathbb{R}^4$. It will be convenient to make the explicit identification of the element $(x_0, x_1, x_2, x_3) \in \mathbb{R}^4$ with the matrix

$$\begin{bmatrix} x_0 + ix_3 & x_2 + ix_1 \\ -x_2 + ix_1 & x_0 - ix_3 \end{bmatrix}.$$

Such a matrix is in $SU(2)$ if and only if $x_0^2 + x_1^2 + x_2^2 + x_3^2 = 1$, which defines the three-dimensional sphere $S^3 \subset \mathbb{R}^4$. The identity matrix I corresponds to the point $(1, 0, 0, 0)$, and we denote the tangent space to S^3 at this point by $T_I S^3$.

So suppose that $c : (-\epsilon, \epsilon) \to \mathbb{R}^4$ is a one-to-one differentiable curve satisfying $c(t) \in SU(2) = S^3$ for all t and $c(0) = I \in SU(2)$. Then as in the case of the circle, the derivative $\dot{c}(0) \in \mathbb{R}^4$ is a tangent vector to S^3 at the identity, and the totality of all such tangent vectors forms the tangent space $T_I S^3$. But since the curve c lies entirely within $SU(2)$, we have $I = c(t)c(t)^\dagger$ for all t. Taking the derivative and evaluating at $t = 0$ yields

$$\begin{aligned} 0 &= \dot{c}(0)c(0)^\dagger + c(0)\dot{c}(0)^\dagger \\ &= \dot{c}(0) + \dot{c}(0)^\dagger. \end{aligned}$$

From this, we see that the derivative must be a skew-Hermitian matrix: $\dot{c}(0)^\dagger = -\dot{c}(0)$. But $\dot{c}(0)$ corresponds to a vector $(x_0, x_1, x_2, x_3) \in \mathbb{R}^4$, and this vector yields a skew-Hermitian matrix if and only if $x_0 = 0$. It follows that

$$\dot{c}(0) = \begin{bmatrix} ix_3 & x_2 + ix_1 \\ -x_2 + ix_1 & -ix_3 \end{bmatrix} \qquad \text{for some } x_1, x_2, x_3 \in \mathbb{R}.$$

Note that this matrix has trace zero in addition to being skew-Hermitian. We will show below that every such matrix is the tangent vector of some curve c, so that the tangent space is

$$T_I S^3 = \left\{ \begin{bmatrix} ix_3 & x_2 + ix_1 \\ -x_2 + ix_1 & -ix_3 \end{bmatrix} \mid x_1, x_2, x_3 \in \mathbb{R} \right\}.$$

Recall that in the case of $U(1) = S^1$, we found that the tangent space was i times the vector space $\mathbb{R}$. Following in this pattern, consider the real vector space $H_0(2)$ of 2×2 Hermitian matrices with trace zero. Thus, $X \in H_0(2)$ if and only if $X^\dagger = X$ and $\text{tr}(X) = 0$.

Exercise 2.20. *Show that a general element of $H_0(2)$ looks like*

$$X = \begin{bmatrix} x_3 & x_1 - ix_2 \\ x_1 + ix_2 & -x_3 \end{bmatrix} \qquad x_1, x_2, x_3 \in \mathbb{R}.$$

Check that the tangent space to S^3 at the identity may be described as $T_I S^3 = iH_0(2)$.

34   *Symmetry and Quantum Mechanics*

As promised, we now wish show that every element of $iH_0(2)$ does arise from a curve c in the manner described above. Recall how we did this for the circle at the end of the previous section: given an element $iy \in i\mathbb{R}$, we wrote down the curve $c(t) := e^{ity}$ with tangent vector iy at $t = 0$. We can make an analogous argument for the sphere S^3 provided we have a suitable exponential function for matrices. In the next section we show that such a function exists. For now we will simply assume the existence of the matrix exponential function together with the properties listed below in proposition 2.21. So suppose that $X \in H_0(2)$ is an arbitrary 2×2 Hermitian matrix with trace zero. Then define $c(t) := \exp(itX)$, which defines a differentiable curve in the space of 2×2 matrices with complex entries. In fact, in proposition 2.31 we will see that $c(t) \in SU(2)$ for all t, so that we have a curve in $SU(2) = S^3$ as desired. Moreover, $c(0) = I$ and $\dot{c}(0) = iX \in iH_0(2)$, thus showing that $iX \in T_I S^3$.

2.4.3   The exponential of a matrix

If A is an arbitrary $n \times n$ complex matrix, we want to define a matrix $\exp(A)$ with properties that generalize the usual exponential for complex numbers. We will do so by making use of the power series for the ordinary exponential function, replacing the scalar argument by a matrix. In order to justify this construction, we will need to extend some familiar analytic results to the setting of matrices, which is the purpose of this section. For the convenience of readers wishing to skip the analytic justification, we begin with a proposition that lists the basic properties of the matrix exponential; after reading proposition 2.21, the reader can safely jump to proposition 2.31.

Proposition 2.21. *There exists a function* $\exp \colon M(n, \mathbb{C}) \to GL(n, \mathbb{C})$ *that assigns to each* $n \times n$ *matrix* A *an invertible matrix* $\exp(A)$ *defined by the following absolutely convergent power series:*

$$\exp(A) := \sum_{k=0}^{\infty} \frac{1}{k!} A^k.$$

This matrix exponential function satisfies the following properties:

a) $\exp(0) = I_n$;

b) *If* $AB = BA$ *then* $\exp(A + B) = \exp(A)\exp(B) = \exp(B)\exp(A)$;

c) $\exp(A)^{-1} = \exp(-A)$;

d) *If* B *is invertible, then* $\exp(BAB^{-1}) = B\exp(A)B^{-1}$;

e) *For a fixed matrix* A, *the function* $c \colon \mathbb{R} \to GL(n, \mathbb{C})$ *defined by* $c(t) := \exp(tA)$ *is differentiable, and* $\dot{c}(t) = Ac(t)$ *for all* $t \in \mathbb{R}$. *In particular,* $\dot{c}(0) = A$.

Spinor Space 35

Before we can give the proof, we have some preliminary work to do. Suppose that $\{A_m\}$ is a sequence of $n \times n$ complex matrices. We define convergence of matrices entrywise: $A_m \to A \in M(n, \mathbb{C})$ if and only if $(A_m)_{ij} \to A_{ij}$ for all $1 \le i, j \le n$. In order to study convergence, it will be useful to introduce a norm on the space of matrices.

Definition 2.22. *The* norm *of a matrix $A \in M(n, \mathbb{C})$ is the supremum of the vector norms $\|Ax\|$, where x ranges over all unit vectors in $(\mathbb{C}^n, \cdot)$:*

$$\|A\| := \sup_{\|x\|=1} \|Ax\|.$$

Proposition 2.23. *The matrix norm has the following properties:*

   i) $\|A\| \ge 0$ *with equality if and only if $A = 0$*   *(positive definite);*

  ii) $\|\alpha A\| = |\alpha| \|A\|$ *for all $\alpha \in \mathbb{C}$*   *(scalars);*

 iii) $\|A + B\| \le \|A\| + \|B\|$ *(triangle inequality);*

 iv) $\|AB\| \le \|A\| \|B\|$ *(submultiplicative).*

Proof. We prove only iv), leaving the first three to the reader. The key observation is that for any non-zero vector x, we have the inequality

$$\|A\| \ge \left\| A \frac{x}{\|x\|} \right\| = \frac{\|Ax\|}{\|x\|}.$$

Multiplying through by $\|x\|$, we see that $\|A\| \|x\| \ge \|Ax\|$ for all x. Then for any unit vector x we have

$$\|A\| \|B\| = \|A\| \|B\| \|x\| \ge \|A\| \|Bx\| \ge \|ABx\|.$$

It follows that $\|A\| \|B\| \ge \|AB\|$ as claimed. $\qquad\qquad\square$

Exercise 2.24. *Prove parts i)–iii) of proposition 2.23.*

The next proposition provides a relationship between the norm of a matrix and the size of its entries.

Proposition 2.25. *Let $A \in M(n, \mathbb{C})$ be an $n \times n$ matrix. Then for all $1 \le i, j \le n$ we have the inequalities*

$$|A_{ij}| \le \|A\| \le \sum_{k,l=1}^{n} |A_{kl}|.$$

Proof. Denote by $a_j \in \mathbb{C}^n$ the jth column of the matrix A. Then applying A to the jth standard basis vector yields:

$$\|A\|^2 \ge \|Ae_j\|^2 = \|a_j\|^2 = \sum_{k=1}^{n} |A_{kj}|^2 \ge |A_{ij}|^2.$$

36   *Symmetry and Quantum Mechanics*

For the other inequality, let $\mathbf{x}$ be any unit vector in $\mathbb{C}^n$. Then using the fact that the norm of a vector is at most the sum of the moduli of its components, we find that

$$\|A\mathbf{x}\| \leq \sum_{k=1}^{n} |(A\mathbf{x})_k|.$$

But for each $1 \leq k \leq n$ we have

$$|(A\mathbf{x})_k| = \left| \sum_{l=1}^{n} A_{kl} x_l \right| \leq \sum_{l=1}^{n} |A_{kl}||x_l| \leq \sum_{l=1}^{n} |A_{kl}|.$$

Putting this together with the previous inequality yields:

$$\|A\mathbf{x}\| \leq \sum_{k,l=1}^{n} |A_{kl}|.$$

This inequality holds for all unit vectors $\mathbf{x}$, so $\|A\| \leq \sum_{k,l=1}^{n} |A_{kl}|$ as claimed. $\square$

Proposition 2.26. *Let $\{A_m\}$ be a sequence of matrices in $M(n, \mathbb{C})$. Then $A_m \to A$ entrywise if and only if $\|A_m - A\| \to 0$ in $\mathbb{R}$.*

Proof. First suppose that $A_m \to A$ entrywise, and define $B_m := A_m - A$. We wish to show that $\|B_m\| \to 0$. So let $\epsilon > 0$ be given, and choose $\delta = \epsilon/n^2$. Since $A_m \to A$ entrywise, it follows that $B_m \to 0$ entrywise, so there exists $M > 0$ such that for $m \geq M$ all entries of B_m have modulus less than δ. It then follows from proposition 2.25 that

$$\|B_m\| \leq \sum_{k,l=1}^{n} |(B_m)_{kl}| < n^2 \delta = \epsilon.$$

For the other direction, suppose that $\|B_m\| \to 0$, and consider the ij-entry of the sequence $\{B_m\}$. By proposition 2.25, we have $|(B_m)_{ij}| \leq \|B_m\| \to 0$, showing that the sequence $\{(B_m)_{ij}\}$ converges to zero. Since this argument holds for all i, j, it follows that $B_m \to 0$ entrywise, so $A_m \to A$. $\square$

Exercise 2.27. *Adapt the proof of the previous proposition to show that a sequence of matrices $\{A_m\}$ is Cauchy with respect to the matrix norm if and only if for all $1 \leq i, j \leq n$, the sequence of ij-entries $\{(A_m)_{ij}\}$ forms a Cauchy sequence in $\mathbb{C}$.*

Recall that $\mathbb{C}$ is complete, so that every Cauchy sequence of complex numbers converges. By proposition 2.26 and exercise 2.27, we see that the same is true for $M(n, \mathbb{C})$ with respect to the matrix norm: Cauchy sequences of matrices converge.

Now suppose that $\sum_{k=0}^{\infty} A_k$ is an infinite series of matrices, with partial sums $S_m := A_0 + A_1 + \cdots + A_m$. As in the case of scalars, we say that the infinite

Spinor Space 37

series converges if and only if the corresponding sequence of partial sums converges. Moreover, the next proposition shows that absolute convergence implies convergence for matrices, just as for scalars.

Proposition 2.28. *Suppose that $\sum_{k=0}^{\infty} A_k$ is an infinite series of matrices with the property that the corresponding series of norms converges: $\sum_{k=0}^{\infty} \|A_k\| < \infty$. Then $\sum_{k=0}^{\infty} A_k$ also converges in $M(n, \mathbb{C})$. Such a series is called* absolutely convergent.

Proof. By definition, the convergence of $\sum_{k=0}^{\infty} \|A_k\|$ means that the sequence of partial sums $T_m := \|A_0\| + \cdots + \|A_m\|$ converges in $\mathbb{R}$. Thus, $\{T_m\}$ is a Cauchy sequence. We wish to show that the sequence of matrix partial sums $S_m := A_0 + \cdots + A_m$ converges in $M(n, \mathbb{C})$. It suffices to show that $\{S_m\}$ is Cauchy with respect to the matrix norm. So let $\epsilon > 0$ be given. Then since $\{T_m\}$ is Cauchy, there exists $M > 0$ so that for all $m > l \geq M$ we have $|T_m - T_l| < \epsilon$. But then

$$\|S_m - S_l\| = \|A_{l+1} + \cdots + A_m\| \leq \|A_{l+1}\| + \cdots + \|A_m\| = |T_m - T_l| < \epsilon.$$

Hence, $\{S_m\}$ is Cauchy and therefore convergent. $\qquad\square$

Exercise 2.29. *Adapt the proof of proposition 2.26 to show that a matrix series $\sum_{k=0}^{\infty} A_k$ is absolutely convergent if and only if for all $1 \leq i, j \leq n$, the series of ij-entries $\sum_{k=0}^{\infty} (A_k)_{ij}$ is absolutely convergent in $\mathbb{C}$.*

The following proposition shows that absolutely convergent series of matrices can be multiplied term-by-term.

Proposition 2.30. *Suppose that $\sum_{k=0}^{\infty} A_k$ and $\sum_{k=0}^{\infty} B_k$ are absolutely convergent series of matrices in $M(n, \mathbb{C})$. For each $k \geq 0$ define a matrix*

$$C_k := \sum_{l=0}^{k} A_l B_{k-l} = A_0 B_k + A_1 B_{k-1} + \cdots + A_k B_0.$$

Then $\sum_{k=0}^{\infty} C_k$ is absolutely convergent and

$$\sum_{k=0}^{\infty} C_k = \left(\sum_{k=0}^{\infty} A_k \right) \left(\sum_{k=0}^{\infty} B_k \right).$$

Proof. We start by showing absolute convergence:

$$\begin{aligned}
\sum_{k=0}^{m} \|C_k\| &\leq \sum_{k=0}^{m} \sum_{l=0}^{k} \|A_l\| \|B_{k-l}\| \\
&\leq \left(\sum_{k=0}^{m} \|A_k\| \right) \left(\sum_{k=0}^{m} \|B_k\| \right) \\
&\leq \left(\sum_{k=0}^{\infty} \|A_k\| \right) \left(\sum_{k=0}^{\infty} \|B_k\| \right).
\end{aligned}$$

Symmetry and Quantum Mechanics

Thus, the partial sums of the series $\sum_{k=0}^{\infty}\|C_k\|$ are bounded above. By the monotone convergence theorem, it follows that $\sum_{k=0}^{\infty} C_k$ is absolutely convergent. By exercise 2.29, each matrix entry series $\sum_{k=0}^{\infty}(C_m)_{ij}$ is an absolutely convergent series of complex numbers.

Write A, B, C for the matrix sums of the three infinite series. To show that $C = AB$, we will show that $C_{ij} = (AB)_{ij}$ for all $1 \leq i, j \leq n$. Fixing i and j, we know that the ij-entry of C is

$$C_{ij} = \lim_{m \to \infty} \sum_{k=0}^{m}(C_m)_{ij} = \lim_{m \to \infty} \sum_{k=0}^{m}\sum_{l=0}^{k}(A_l B_{k-l})_{ij}.$$

In fact, slightly more is true. Note that each term $(A_l B_{k-l})_{ij}$ occurs exactly once in this limit for $k \geq l \geq 0$. Let $\{c_t\}$ denote the ordering of these terms obtained by first listing all those corresponding to $k = 0$, then $k = 1$, then $k = 2$, etc. Within each value of k we list the terms in order of increasing l. Then $\sum_{t=0}^{\infty} c_t$ is an absolutely convergent series converging to C_{ij}. Indeed, for each $m > 0$, let $k(m)$ be the integer such that $c_m = (A_l B_{k(m)-l})_{ij}$. Then

$$\sum_{t=0}^{m}|c_t| \leq \sum_{k=0}^{k(m)}\sum_{l=0}^{k}|(A_l B_{k-l})_{ij}|$$

$$\leq \sum_{k=0}^{k(m)}\sum_{l=0}^{k}\sum_{s=1}^{n}|(A_l)_{is}(B_{k-l})_{sj}|$$

$$\leq \sum_{s=1}^{n}\left(\sum_{k=0}^{k(m)}|(A_k)_{is}|\right)\left(\sum_{k=0}^{k(m)}|(B_k)_{sj}|\right)$$

$$\leq \sum_{s=1}^{n}\left(\sum_{k=0}^{\infty}|(A_k)_{is}|\right)\left(\sum_{k=0}^{\infty}|(B_k)_{sj}|\right).$$

Thus, the partial sums of $\sum_{t=0}^{\infty}|c_t|$ are bounded above, yielding absolute convergence. But the partial sums of $\sum_{k=0}^{\infty}(C_m)_{ij}$ form a subsequence of the partial sums of $\sum_{t=0}^{\infty} c_t$, so both these series converge to the same limit C_{ij}.

Now consider the ij-entry of the product of the mth partial sums of A and B:

$$\left(\left(\sum_{k=0}^{m} A_k\right)\left(\sum_{k=0}^{m} B_k\right)\right)_{ij} = \sum_{k=0}^{m}\sum_{l=0}^{m}(A_k B_l)_{ij}.$$

This sequence (indexed by $m \geq 0$) is a subsequence of the partial sums of a rearrangement of $\sum_{t=0}^{\infty} c_t$. Since rearrangements of absolutely convergent series converge to the same limit, it follows that this sequence also converges to C_{ij}:

$$C_{ij} = \lim_{m \to \infty}\left(\left(\sum_{k=0}^{m} A_k\right)\left(\sum_{k=0}^{m} B_k\right)\right)_{ij} = (AB)_{ij}.$$

Spinor Space — 39

$\square$

We are now ready to prove proposition 2.21:

Proof of Proposition 2.21. We begin by showing that the infinite series $\exp(A) := \sum_{k=0}^{\infty} \frac{1}{k!} A^k$ is absolutely convergent. Note that by the submultiplicativity of the matrix norm, we have $\|A^k\| \le \|A\|^k$ for all $k \ge 0$. Thus, the infinite series $\sum_{k=0}^{\infty} \frac{1}{k!} \|A^k\|$ is dominated by the series $\sum_{k=0}^{\infty} \frac{1}{k!} \|A\|^k$ which converges to $e^{\|A\|}$, the ordinary exponential of the real number $\|A\|$. By the comparison test, it follows that $\exp(A)$ is absolutely convergent.

The fact that $\exp(0) = I_n$ is immediate from the series definition, while $\exp(A)^{-1} = \exp(-A)$ follows from taking $B = -A$ in part b).

To prove part b), we use proposition 2.30:

$$\exp(A)\exp(B) = \left(\sum_{k=0}^{\infty} \frac{1}{k!} A^k\right)\left(\sum_{k=0}^{\infty} \frac{1}{k!} B^k\right)$$

$$= \sum_{k=0}^{\infty} \sum_{l=0}^{k} \frac{1}{l!(k-l)!} A^l B^{k-l}.$$

However, because we are assuming that $AB = BA$, we have the binomial formula:

$$(A+B)^k = \sum_{l=0}^{k} \binom{k}{l} A^l B^{k-l} = k! \sum_{l=0}^{k} \frac{1}{l!(k-l)!} A^l B^{k-l}.$$

Returning to the expression for the product of the exponentials, we find that

$$\exp(A)\exp(B) = \sum_{k=0}^{\infty} \frac{1}{k!} (A+B)^k = \exp(A+B).$$

Since $A + B = B + A$, it follows that $\exp(B)\exp(A) = \exp(A+B)$ as well.

The proof of part d) is a straightforward computation using the fact that $BA^k B^{-1} = (BAB^{-1})^k$:

$$B\exp(A)B^{-1} = \sum_{k=0}^{\infty} \frac{1}{k!} BA^k B^{-1} = \sum_{k=0}^{\infty} \frac{1}{k!} (BAB^{-1})^k = \exp(BAB^{-1}).$$

Finally, we prove that $c(t) := \exp(tA)$ is differentiable by checking each matrix entry. So fix i and j and consider the corresponding matrix entry function:

$$c_{ij}(t) = \exp(tA)_{ij} = \sum_{k=0}^{\infty} \frac{t^k}{k!} (A^k)_{ij}.$$

This is a power series with infinite radius of convergence, defining an analytic

40 *Symmetry and Quantum Mechanics*

function of the real variable t. Power series may be differentiated term-by-term within their radius of convergence, so we see that

$$\dot{c}_{ij}(t) = \sum_{k=1}^{\infty} \frac{t^{k-1}}{(k-1)!}(A^k)_{ij} = \sum_{k=0}^{\infty} \frac{t^k}{k!}(A^{k+1})_{ij}.$$

But $(A^{k+1})_{ij} = \sum_{l=0}^{n} A_{il}(A^k)_{lj}$, so

$$\dot{c}_{ij}(t) = \sum_{k=0}^{\infty} \frac{t^k}{k!} \sum_{l=0}^{n} A_{il}(A^k)_{lj} = \sum_{l=0}^{n} A_{il} \sum_{k=0}^{\infty} \frac{t^k}{k!}(A^k)_{lj} = (A \exp(tA))_{ij}.$$

This holds for all of the matrix entries, so $\dot{c}(t) = A \exp(tA)$ as claimed. $\square$

As a first application of the properties of the matrix exponential, we show in the next proposition that $\exp(iX)$ is unitary if X is Hermitian.

Proposition 2.31. *Let X be an $n \times n$ Hermitian matrix, so that iX is skew-Hermitian. Then $\exp(iX) \in U(n)$ is a unitary matrix. If in addition X has trace zero, then $\exp(iX) \in SU(n)$.*

Proof. Recall that by the Spectral Theorem A.32, the Hermitian matrix X has real eigenvalues $\lambda_i \in \mathbb{R}$ and is diagonalized by a unitary matrix U. That is, $UXU^{-1} = \mathrm{diag}(\lambda_1, \dots, \lambda_n)$. Now compute the exponential:

$$\begin{aligned}
\exp(iX) &= \exp(iU^{-1}\mathrm{diag}(\lambda_1, \dots, \lambda_n)U) \\
&= U^{-1}\exp(\mathrm{diag}(i\lambda_1, \dots, i\lambda_n))U \\
&= U^{-1}\left(\sum_{k=0}^{\infty} \frac{1}{k!}\mathrm{diag}((i\lambda_1)^k, \dots, (i\lambda_n)^k)\right)U \\
&= U^{-1}\mathrm{diag}(e^{i\lambda_1}, \dots, e^{i\lambda_n})U.
\end{aligned}$$

Hence, $\exp(iX)$ is unitary, being the product of three unitary matrices.

If the trace of X is zero, then $\lambda_1 + \cdots + \lambda_n = 0$. Moreover,

$$\begin{aligned}
\det(\exp(iX)) &= \det(U^{-1}\mathrm{diag}(e^{i\lambda_1}, \dots, e^{i\lambda_n})U) \\
&= \det(U)^{-1}\det(\mathrm{diag}(e^{i\lambda_1}, \dots, e^{i\lambda_n}))\det(U) \\
&= e^{i(\lambda_1 + \cdots + \lambda_n)} \\
&= e^0 \\
&= 1,
\end{aligned}$$

so $\exp(iX)$ is an element of $SU(n)$. $\square$

2.4.4 $SU(2)$ is the universal cover of $SO(3)$

The space of 2×2 traceless skew-Hermitian matrices $iH_0(2)$ came to our attention in Section 2.4.2 as the tangent space to $SU(2)$ at the identity. A basis

Spinor Space

for this real vector space is given by $\{\frac{1}{2i}\sigma_1, \frac{1}{2i}\sigma_2, \frac{1}{2i}\sigma_3\}$, where the $\sigma_j \in H_0(2)$ are the *Pauli matrices*

$$\sigma_1 = \begin{bmatrix} 0 & 1 \\ 1 & 0 \end{bmatrix}, \qquad \sigma_2 = \begin{bmatrix} 0 & -i \\ i & 0 \end{bmatrix}, \qquad \sigma_3 = \begin{bmatrix} 1 & 0 \\ 0 & -1 \end{bmatrix}.$$

In Chapter 3 we will provide an explanation for the scalar factor of $\frac{1}{2i}$ (see also example 2.35 below). For now, we define an inner product $\langle, \rangle$ on $iH_0(2)$ by taking $\{\frac{1}{2i}\sigma_1, \frac{1}{2i}\sigma_2, \frac{1}{2i}\sigma_3\}$ to be orthonormal and extending bilinearly.

Exercise 2.32. *Show that, taking $\frac{1}{2i}\sigma_j$ as an orthonormal basis, the squared length of an element $iY \in iH_0(2)$ is given by $\langle iY, iY \rangle = 4\det(iY)$.*

Recall that $SU(2)$ acts on itself by conjugation. The following exercise shows that $SU(2)$ also acts by conjugation on the tangent space $T_I S^3$.

Exercise 2.33. *Suppose that $B \in SU(2)$ and $iY \in iH_0(2)$. Then define $B \star iY := B(iY)B^{-1}$. Show that $\star$ defines an $SU(2)$-action on $iH_0(2)$ in the sense of definition 1.20 from Chapter 1.*

Our ultimate goal is to obtain a relationship between the group $SU(2)$ and the group of rotations $SO(3)$, and we are now close, because we have an action of $SU(2)$ on the three-dimensional real vector space $iH_0(2)$. Moreover, our choice of orthonormal basis $\{\frac{1}{2i}\sigma_j\}$ determines an isomorphism $F: (iH_0(2), \langle, \rangle) \to (\mathbb{R}^3, \cdot)$ defined by $F(\frac{1}{2i}\sigma_j) := \mathbf{e}_j$. We may use this identification to transfer the $SU(2)$-action to $\mathbb{R}^3$.

Exercise 2.34. *Show that the function $SU(2) \times \mathbb{R}^3 \to \mathbb{R}^3$ defined by*

$$(B, \mathbf{x}) \mapsto F\left(\frac{1}{2i}B \begin{bmatrix} x_3 & x_1 - ix_2 \\ x_1 + ix_2 & -x_3 \end{bmatrix} B^{-1}\right) \tag{2.2}$$

defines an action of $SU(2)$ on $\mathbb{R}^3$.

The situation is summarized in the following diagram, where the top row is the conjugation action of $SU(2)$ on its tangent space, and the bottom row is the action (2.2):

$$
\begin{array}{ccc}
SU(2) \times iH_0(2) & \xrightarrow{\ \star\ } & iH_0(2) \\
{\scriptstyle \mathrm{id} \times F} \downarrow & & \downarrow {\scriptstyle F} \\
SU(2) \times \mathbb{R}^3 & \xrightarrow{F \star (\mathrm{id} \times F^{-1})} & \mathbb{R}^3.
\end{array}
$$

Note that the $SU(2)$-action on $iH_0(2)$ preserves the determinant:

$$\det(B \star iY) = \det(B(iY)B^{-1}) = \det(B)\det(iY)\det(B)^{-1} = \det(iY).$$

It follows from exercise 2.32 that the action of $SU(2)$ on $\mathbb{R}^3$ obtained via the isomorphism F preserves lengths in $(\mathbb{R}^3, \cdot)$. By exercise 1.11, we find that

42   *Symmetry and Quantum Mechanics*

$SU(2)$ acts on $(\mathbb{R}^3, \cdot)$ via orthogonal operators. Thus, we get a continuous group homomorphism $f\colon SU(2) \to O(3)$ defined explicitly as follows: given $B \in SU(2)$, write

$$B \begin{bmatrix} x_3 & x_1 - ix_2 \\ x_1 + ix_2 & -x_3 \end{bmatrix} B^{-1} = \begin{bmatrix} y_3 & y_1 - iy_2 \\ y_1 + iy_2 & -y_3 \end{bmatrix}.$$

Then $f(B)$ is the orthogonal matrix that maps (x_1, x_2, x_3) to (y_1, y_2, y_3).

Example 2.35. *Consider the matrix*

$$B = \begin{bmatrix} e^{i\theta} & 0 \\ 0 & e^{-i\theta} \end{bmatrix} \in SU(2).$$

To determine $f(B) \in O(3)$, we make the computation

$$B \begin{bmatrix} x_3 & x_1 - ix_2 \\ x_1 + ix_2 & -x_3 \end{bmatrix} B^{-1} = \begin{bmatrix} x_3 & e^{2i\theta}(x_1 - ix_2) \\ e^{-2i\theta}(x_1 + ix_2) & -x_3 \end{bmatrix}.$$

Using Euler's formula to rewrite the upper right-hand corner yields:

$$\begin{aligned} e^{2i\theta}(x_1 - ix_2) &= (\cos(2\theta) + i\sin(2\theta))(x_1 - ix_2) \\ &= (\cos(2\theta)x_1 + \sin(2\theta)x_2) - i(\cos(2\theta)x_2 - \sin(2\theta)x_1) \\ &= y_1 - iy_2. \end{aligned}$$

Together with the observation that $y_3 = x_3$, we see that

$$f(B) = \begin{bmatrix} \cos(2\theta) & \sin(2\theta) & 0 \\ -\sin(2\theta) & \cos(2\theta) & 0 \\ 0 & 0 & 1 \end{bmatrix}.$$

Note the curious fact that $f(B)$ describes a clockwise rotation in the $x_1 x_2$-plane through the angle 2θ—twice the angle θ appearing in the matrix B (see example 1.15 and Figure 1.2).

As mentioned earlier, $SU(2)$ is path connected, so by continuity, the image of f must be contained in the path connected component of the identity of $O(3)$. This is the set of matrices in $O(3)$ that may be joined to the identity by a continuous path in $O(3)$.

Exercise 2.36 (♣). *Show that the set of 3×3 orthogonal matrices that may be joined to the identity matrix by a continuous path in $O(3)$ is the subgroup of rotations $SO(3)$. (Hint: the determinant is a polynomial in the entries of a matrix, hence continuous.)*

Exercise 2.37. *Prove that the kernel of $f\colon SU(2) \to SO(3)$ is the set of matrices in $SU(2)$ that commute with all of the Pauli matrices $\sigma_1, \sigma_2, \sigma_3$. Then prove that the only such matrices in $GL(2, \mathbb{C})$ are scalar matrices (i.e., multiples of the identity). Conclude that the kernel of f is $\{\pm I\}$.*

Spinor Space 43

The previous two exercises demonstrate that f induces an isomorphism between $SU(2)/\{\pm I\}$ and a subgroup of $SO(3)$, namely the image of f. In fact, f is a surjection, so that $SU(2)/\{\pm I\}$ is isomorphic to $SO(3)$. To see this, let $A \in SO(3)$ be arbitrary. Then by exercise 1.19, there exists a unit vector $\mathbf{u} = (u_1, u_2, u_3) \in \mathbb{R}^3$ and an angle θ such that A is the rotation about the axis spanned by $\mathbf{u}$ through the angle 2θ. Consider the following 2×2 complex matrix:

$$
\begin{aligned}
B \quad &:= \quad \cos(\theta)I + 2\sin(\theta)F^{-1}(\mathbf{u}) \\
&= \quad \cos(\theta)I - i\sin(\theta)(u_1\sigma_1 + u_2\sigma_2 + u_3\sigma_3) \\
&= \quad \begin{bmatrix} \cos(\theta) - u_3\sin(\theta)i & -\sin(\theta)(u_2 + u_1 i) \\ \sin(\theta)(u_2 - u_1 i) & \cos(\theta) + u_3\sin(\theta)i \end{bmatrix}.
\end{aligned}
$$

Exercise 2.38 (♣). *Show that B is an element of $SU(2)$, and that $f(B) = A$. (Hints: extend $\{\mathbf{u}\}$ to a positively oriented orthonormal basis $\{\mathbf{u}, \mathbf{v}, \mathbf{w}\}$ for $(\mathbb{R}^3, \cdot)$. Then show that left-multiplication by the matrix $f(B) \in SO(3)$ fixes $\mathbf{u}$ and rotates the plane spanned by $\mathbf{v}$ and $\mathbf{w}$ through the angle 2θ.)*

The results in this section combine to establish the following theorem.

Theorem 2.39. *Consider the mapping $f \colon SU(2) \to SO(3)$ defined by sending $B \in SU(2)$ to the matrix $f(B) \in SO(3)$ taking (x_1, x_2, x_3) to (y_1, y_2, y_3), where*

$$
B \begin{bmatrix} x_3 & x_1 - ix_2 \\ x_1 + ix_2 & -x_3 \end{bmatrix} B^{-1} = \begin{bmatrix} y_3 & y_1 - iy_2 \\ y_1 + iy_2 & -y_3 \end{bmatrix}.
$$

Then $f \colon SU(2) \to SO(3)$ is a continuous, surjective, two-to-one homomorphism with kernel $\{\pm I\}$, so that $SU(2)$ is a double cover of $SO(3)$.

The group $SU(2)$ is called the *universal double cover* of $SO(3)$ because it is simply connected—a theorem from topology ensures that $SU(2)$ is the unique simply connected group that covers $SO(3)$. For a nice exposition of the relationship between $SO(3)$ and $SU(2)$ that makes use of the geometry of quaternions, see [18], which goes on to study the "generalized rotation groups" $SO(n), U(n), SU(n),$ and $Sp(n)$.

2.5 Back to spinor space

Recall the question facing M and P at the end of Section 2.3: suppose that M's basis for physical space is obtained from P's through a rotation $\mathcal{A} \colon V \to V$ represented by a matrix $A \in SO(3)$. The columns of A are P's description of M's basis. What matrix in $SU(2)$ represents the transformation that sends P's z-basis for spinor space to M's z'-basis? Well, by theorem 2.39,

44   *Symmetry and Quantum Mechanics*

there are exactly two matrices $\pm B \in SU(2)$ such that $f(\pm B) = A$, where $f: SU(2) \to SO(3)$ is the double cover. These matrices represent (in the z-basis) transformations $\pm \mathcal{B}: W \to W$. In particular, the columns of B are P's description of a pair of orthogonal spin-states. (Note that since $-1 \in U(1)$ is a phase, the choice of sign does not affect the spin-states.)

P is convinced: the spin-states corresponding to the basis $\mathcal{B}|\pm z\rangle$ must correspond to M's third basis vector $\mathbf{u}_3'$ for V. That is, her SGz' will produce exactly the spin-states represented by $\mathcal{B}|z\rangle, \mathcal{B}|-z\rangle$ upon measurement. M agrees, but emphasizes that they have not derived or proved this fact, but rather discovered a desirable feature to add to their model. That is, just as physical space turned out to be $(\mathbb{R}^3, \cdot)$ together with the rotation action of $SO(3)$, spinor space has turned out to be $(\mathbb{C}^2, \cdot)$ with its $SU(2)$-action. But the model is not complete until the connection between physical space and spinor space has been specified. Just as the model for physical space was motivated by physical intuition and the model for spinor space was motivated by experiment, the proposed connection between the two spaces is motivated by the purely mathematical fact that $SU(2)$ is the universal double cover of $SO(3)$.

M summarizes the story so far: when P sets up his right-handed coordinate system by choosing the orthonormal basis $\{\mathbf{u}_1, \mathbf{u}_2, \mathbf{u}_3\}$ for physical space $(V, \langle, \rangle)$, his model of physical space becomes the rotation action of $SO(3)$ on $(\mathbb{R}^3, \cdot)$. He interprets a particular matrix $A \in SO(3)$ as telling him how to rotate his coordinate axes to obtain M's right-handed coordinate system corresponding to her basis $\{\mathbf{u}_1', \mathbf{u}_2', \mathbf{u}_3'\}$. Furthermore, he decides to identify his copy of $\mathbb{R}^3$ with the space $iH_0(2)$ of traceless 2×2 skew-Hermitian matrices by sending the standard basis vector $\mathbf{e}_j$ to $\frac{1}{2i}\sigma_j$, where σ_j is the jth Pauli matrix. Via this identification, $SO(3)$ acts on $iH_0(2)$. In fact, this action comes from the conjugation action of $SU(2)$ on $iH_0(2)$ via the universal double cover $f: SU(2) \to SO(3)$.

When M and P start observing electrons, they model the electron spin using spinor space $(W, \langle | \rangle)$. When P sets up his SGz, he identifies spinor space with the $SU(2)$-action on $(\mathbb{C}^2, \cdot)$. When M sets up her SGz', she makes a different identification of spinor space with $(\mathbb{C}^2, \cdot)$ by sending her basis $|\pm z'\rangle$ to the standard basis. In particular, her SGz' defines an ordered pair of orthogonal spin-states, and these are obtained from P's basis by applying either of the elements $\pm B = f^{-1}(A) \in SU(2)$.

P decides to check whether this connection between physical space and spinor space accords with their beliefs about electron spin from the beginning of Section 2.2. In particular, he sets out to check whether belief 4 holds in the model. Recall that belief 4 states that if $\mathbf{u}$ and $\mathbf{u}'$ are unit vectors making an angle α in physical space, then an electron that exits an $SG\mathbf{u}$ spin up will measure spin up in an $SG\mathbf{u}'$ with probability $\cos^2(\frac{\alpha}{2})$. To investigate, P takes $\mathbf{u} = \mathbf{u}_3$ and $\mathbf{u}' = \sin(\alpha)\mathbf{u}_1 + \cos(\alpha)\mathbf{u}_3$, so $\mathbf{u}'$ is obtained from $\mathbf{u}$ by a rotation in the xz-plane through an angle α (see Figure 2.7). The question is: what is

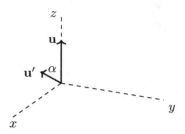

FIGURE 2.7: A rotation by α in the xz-plane. The dashed lines show P's coordinate axes.

the probability that $|+z\rangle$ will be found to be spin up when measured by an $SG\mathbf{u}'$? This probability is given in the model by the quantity $|\langle +\mathbf{u}'|+z\rangle|^2$.

The matrix describing the rotation in the xz-plane through an angle α is:

$$A = \begin{bmatrix} \cos(\alpha) & 0 & \sin(\alpha) \\ 0 & 1 & 0 \\ -\sin(\alpha) & 0 & \cos(\alpha) \end{bmatrix}.$$

Exercise 2.40. *Check that left multiplication by A on $\mathbb{R}^3$ corresponds under the isomorphism F to the conjugation on $iH_0(2)$ by the matrix*

$$B = \begin{bmatrix} \cos(\frac{\alpha}{2}) & -\sin(\frac{\alpha}{2}) \\ \sin(\frac{\alpha}{2}) & \cos(\frac{\alpha}{2}) \end{bmatrix}.$$

That is, verify explicitly that $f(B) = A$.

The matrix B represents in the z-basis the automorphism of spinor space obtained by sending the z-basis to the basis $|\pm \mathbf{u}'\rangle$. Thus,

$$|+\mathbf{u}'\rangle = \cos\left(\frac{\alpha}{2}\right)|+z\rangle + \sin\left(\frac{\alpha}{2}\right)|-z\rangle.$$

The probability that $|+z\rangle$ will be measured spin up by an $SG\mathbf{u}'$ is then

$$\begin{aligned} |\langle +\mathbf{u}'|+z\rangle|^2 &= |\langle +z|+\mathbf{u}'\rangle|^2 \\ &= \left|\langle +z|\left(\cos\left(\frac{\alpha}{2}\right)|+z\rangle + \sin\left(\frac{\alpha}{2}\right)|-z\rangle\right)\right|^2 \\ &= \left|\cos\left(\frac{\alpha}{2}\right)\langle +z|+z\rangle + \sin\left(\frac{\alpha}{2}\right)\langle +z|-z\rangle\right|^2 \\ &= \cos^2\left(\frac{\alpha}{2}\right). \end{aligned}$$

Chapter 3

Observables and Uncertainty

In which M and P discover the Lie algebra $\mathfrak{su}(2)$ and its complexification $\mathfrak{sl}_2(\mathbb{C})$.

3.1 Spin observables

In classical mechanics, observables are modeled by functions on the phase space of the physical system under observation. As a simple example, consider P's representation of physical space as $(\mathbb{R}^3, \cdot)$. Then the phase space for a free particle is the six-dimensional space $\mathbb{R}^3 \times \mathbb{R}^3$ describing all of the particle's possible locations and momenta: $(\mathbf{x}, \mathbf{p}) = (x, y, z, p_x, p_y, p_z)$. The act of P making an observation of the particle corresponds to the evaluation of a function $g \colon \mathbb{R}^3 \times \mathbb{R}^3 \to \mathbb{R}$ at the state of the particle. For instance, if P wants to observe the z-coordinate of the particle, then he evaluates the function $g(\mathbf{x}, \mathbf{p}) = z$ at the state of the particle. If P wants to observe the z-coordinate of the particle's angular momentum, then he evaluates the function $l(\mathbf{x}, \mathbf{p}) = xp_y - yp_x$ at the state of the particle.

Notice that in the preceding description of classical observables, the act of evaluation is the measurement, while the function itself is the observable. In practice, P doesn't think about this distinction too much, because he is confident that making a measurement doesn't affect the state of the particle. However, he knows that measuring the spin of an electron *does* affect the state of the electron, which raises the question: how are spin observables modeled in this theory?

The function corresponding to a classical observable is a gadget that incorporates all of the possible observed values into a single entity. For instance, the function $l(\mathbf{x}, \mathbf{p}) = xp_y - yp_z$ packages together all the possible measurements of the z-coordinate of the particle's angular momentum. Thus, the mathematical object corresponding to a spin observable should package together all of the possible observed values. In the case of electron spin, belief 1 from Section 2.2 says that the observable "spin in the $\mathbf{u}$-direction" has only two possible observed values, $\pm\frac{\hbar}{2}$, corresponding to orthogonal spin-states represented by the kets $|\pm\mathbf{u}\rangle$ in spinor space. These kets are only determined up to phases in $U(1)$, but by proposition 2.6, the orthogonal projections $|\pm\mathbf{u}\rangle\langle\pm\mathbf{u}| \colon W \to W$ are uniquely determined by the spin-states. Thus, we can package the possible

47

48 *Symmetry and Quantum Mechanics*

observed values together by defining the operator

$$S_{\mathbf{u}} := \frac{\hbar}{2}|+\mathbf{u}\rangle\langle+\mathbf{u}| - \frac{\hbar}{2}|-\mathbf{u}\rangle\langle-\mathbf{u}|.$$

The operator $S_{\mathbf{u}}$ is a sum of the orthogonal projections corresponding to the states of definite "spin in the $\mathbf{u}$-direction," each weighted by the observed value for that spin-state. The important point is that $S_{\mathbf{u}}$ contains all of the information about the original spin-states and observed values. Indeed, $S_{\mathbf{u}}$ is a Hermitian operator on $(W, \langle | \rangle)$ with eigenvectors $|\pm\mathbf{u}\rangle$ and associated eigenvalues $\pm\frac{\hbar}{2}$. For example, the operator S_z is represented by the Hermitian matrix $\frac{\hbar}{2}\sigma_3$ with respect to the z-basis, where σ_3 is the third Pauli matrix introduced just before exercise 2.32.

Conversely, the Spectral Theorem A.32 may be phrased in terms of projections as follows: if H is an arbitrary Hermitian operator on $(W, \langle | \rangle)$, then its eigenspaces are orthogonal and H has a *spectral decomposition* as the weighted sum of the orthogonal projections, P_1, P_2, onto the eigenspaces, the weights being the real eigenvalues, $\lambda_1 \geq \lambda_2$:

$$H = \lambda_1 P_1 + \lambda_2 P_2.$$

If $P_1 = I$ and $P_2 = 0$, then $H = \lambda I$ is a scalar operator, corresponding to an observable with only one possible observed value for all spin-states. For an example, consider the operator $S_z^2 = \frac{\hbar^2}{4}\sigma_3^2 = \frac{\hbar^2}{4}I$, which corresponds to the observable "square of the z-component of spin angular momentum." Otherwise, P_1 and P_2 have orthogonal, one-dimensional ranges corresponding to a pair of orthogonal spin-states ϕ_1, ϕ_2. The operator H corresponds to the observable characterized as follows: the state ϕ_j has the definite value of λ_j in the sense that measurement of ϕ_j with an "H-device" will yield the value λ_j with probability 1. For a general spin-state, ψ, we write $|\psi\rangle = c_1|\phi_1\rangle + c_2|\phi_2\rangle$, where $|c_1|^2 + |c_2|^2 = 1$. Then each act of measuring ψ with an "H-device" is modeled by the following procedure: choose one of the two H-eigenstates ϕ_1, ϕ_2 according to the probabilities $|c_1|^2, |c_2|^2$ respectively. Then the measurement of ψ with H changes the spin-state to the randomly chosen state ψ_j and yields the observed value λ_j.

In light of this discussion, M proposes the following definition.

Definition 3.1. *A* quantum observable *on spinor space* $(W, \langle | \rangle)$ *is a Hermitian operator* $H: W \to W$.

Note that, if H is a quantum observable, then the act of measuring ψ with an "H-device" *does not* corresponding to applying H to the ket $|\psi\rangle$. Rather, the probabilistic procedure outlined above generalizes beliefs 1–3 from Section 2.2 to statements about arbitrary quantum observables on spinor space. At the same time, it successfully implements the idea (announced by M and P in Section 2.2) that quantum measurements should involve "operating" on spin-states: a quantum measurement projects the state onto a randomly chosen eigenstate of the corresponding observable.

Observables and Uncertainty

Of course, some functions on $\mathbb{R}^3 \times \mathbb{R}^3$ represent more interesting classical observables than others. Similarly, not all quantum observables will be physically interesting, and in particular, not all of them will correspond to Stern-Gerlach devices in physical space. The observables $S_{\mathbf{u}}$ that do correspond to Stern-Gerlach devices will be called the *spin observables*.

To get a better sense of these spin observables, consider their matrices in the z-basis, $|\pm z\rangle$, for W. It is easy to write down the matrix of $S_{\mathbf{u}}$ in the $\mathbf{u}$-basis:

$$[S_{\mathbf{u}}]_{\mathbf{u}} = \frac{\hbar}{2}\begin{bmatrix} 1 & 0 \\ 0 & -1 \end{bmatrix} = \frac{\hbar}{2}\sigma_3 \in H_0(2).$$

To obtain the matrix in the z-basis, write

$$[S_{\mathbf{u}}]_z = B[S_{\mathbf{u}}]_{\mathbf{u}}B^{-1} = B\frac{\hbar}{2}\sigma_3 B^{-1},$$

where $B \in SU(2)$ is the change of basis matrix from the $\mathbf{u}$-basis to the z-basis. We have been led (once again) to consider the conjugation action of $SU(2)$ on $H_0(2)$, the real vector space of 2×2 Hermitian matrices with trace 0.

Example 3.2. *Spin in the x-direction corresponds to the operator*

$$S_x = \frac{\hbar}{2}|+x\rangle\langle+x| - \frac{\hbar}{2}|-x\rangle\langle-x|.$$

To find the change of basis matrix, B, first note that the following matrix describes a $\frac{\pi}{2}$-rotation of the xz-plane, mapping P's z-axis onto his x-axis.

$$A = \begin{bmatrix} 0 & 0 & 1 \\ 0 & 1 & 0 \\ -1 & 0 & 0 \end{bmatrix}.$$

By exercise 2.38, the following matrix B maps to A via the double cover $f: SU(2) \to SO(3)$:

$$B = \frac{1}{\sqrt{2}}\begin{bmatrix} 1 & -1 \\ 1 & 1 \end{bmatrix},$$

and hence is the change of basis matrix from the x-basis to the z-basis. We compute

$$[S_x]_z = \frac{\hbar}{4}\begin{bmatrix} 1 & -1 \\ 1 & 1 \end{bmatrix}\begin{bmatrix} 1 & 0 \\ 0 & -1 \end{bmatrix}\begin{bmatrix} 1 & 1 \\ -1 & 1 \end{bmatrix} = \frac{\hbar}{2}\begin{bmatrix} 0 & 1 \\ 1 & 0 \end{bmatrix} = \frac{\hbar}{2}\sigma_1.$$

A similar computation reveals that $[S_y]_z = \frac{\hbar}{2}\sigma_2$, using the change of basis matrix

$$B' = \frac{1}{\sqrt{2}}\begin{bmatrix} 1 & i \\ i & 1 \end{bmatrix}.$$

Thus, we find that the matrix representations of the spin observables S_x, S_y, S_z with respect to the z-basis are $\frac{\hbar}{2}$ times the Pauli matrices σ_j.

50   *Symmetry and Quantum Mechanics*

In fact, we could have discovered this based on our previous work, without any explicit computation. Indeed, multiplication by $-\frac{i}{\hbar}$ leads to the $SU(2)$-action on $iH_0(2)$ that we considered in Section 2.4.4. There, we identified $iH_0(2)$ with $\mathbb{R}^3$ according to the isomorphism F defined by $F(\frac{1}{2i}\sigma_j) = \mathbf{e}_j$. By doing so, the conjugation action of $SU(2)$ on $iH_0(2)$ became identified with the rotation action of $SO(3)$ on $\mathbb{R}^3$, thereby yielding the double cover $f\colon SU(2) \to SO(3)$. This means that if $\mathbf{u} = (u_1, u_2, u_3)$ is an arbitrary unit vector, and $f(B) = A \in SO(3)$ describes a rotation of the z-axis onto the line spanned by $\mathbf{u}$, then

$$B\frac{1}{2i}\sigma_3 B^{-1} = F^{-1}(A\mathbf{e}_3) = F^{-1}(\mathbf{u}) = \frac{1}{2i}\left(u_1\sigma_1 + u_2\sigma_2 + u_3\sigma_3\right).$$

Multiplying this equation by $-\frac{\hbar}{i}$ yields

$$[S_{\mathbf{u}}]_z = B\frac{\hbar}{2}\sigma_3 B^{-1} = \frac{\hbar}{2}\left(u_1\sigma_1 + u_2\sigma_2 + u_3\sigma_3\right) \in H_0(2).$$

The vector space $iH_0(2)$ has now come to our attention twice, once as the tangent space to $SU(2) = S^3$ at the identity, and once as a space of quantum observables (scaled by $-\frac{i}{\hbar}$). At this point it seems wise to take a closer look.

3.2   The Lie algebra $\mathfrak{su}(2)$

$SU(2)$ is both a geometric and an algebraic object, and these two structures interact nicely. More precisely, we have seen that $SU(2) = S^3$ is the 3-dimensional unit sphere, which is a *smooth manifold* in $\mathbb{R}^4$. Roughly speaking, this means that near each point, the sphere S^3 looks like $\mathbb{R}^3$, with a well-defined tangent space. (For a one-dimensional example, think about the unit circle S^1 contained in $\mathbb{R}^2$; see Figure 2.6.) This allows us to speak of differentiable functions $S^3 \to \mathbb{R}$, and more generally of differentiable mappings $S^3 \to \mathcal{M}$ and $\mathcal{M} \to S^3$ to or from other smooth manifolds. In particular, since the group operations of multiplication and inversion on $SU(2)$ are given by polynomials in the matrix entries, they define differentiable mappings mult$\colon S^3 \times S^3 \to S^3$ and inv$\colon S^3 \to S^3$. These facts make $SU(2)$ an example of a *Lie group*: a group that is also a smooth manifold such that the group operations are differentiable mappings.

In Section 2.4.2, we found that the tangent space to $SU(2) = S^3$ at the identity is given by the real vector space of 2×2 skew-Hermitian matrices with trace 0:

$$T_I S^3 = iH_0(2).$$

Moreover, we saw that $SU(2)$ acts by conjugation on itself, as well as on its tangent space $iH_0(2)$. At the time, the action on the tangent space was verified

Observables and Uncertainty

in a completely algebraic fashion: the reader checked in exercise 2.33 that if iY is a 2×2 skew-Hermitian matrix with trace zero, then so is $B(iY)B^{-1}$ for all $B \in SU(2)$. We would now like to reveal the geometric content of this action on the tangent space, showing that it is the derivative of the conjugation action of $SU(2)$ on itself.

For a fixed $B \in SU(2)$, consider the mapping $g_B \colon SU(2) \to SU(2)$ defined by conjugation: $g_B(M) := BMB^{-1}$. Given a tangent vector $iX \in iH_0(2)$, define $c \colon (-\epsilon, \epsilon) \to SU(2)$ by $c(t) := \exp(itX)$. Then by propositions 2.21 and 2.31, c is a differentiable curve in $SU(2)$ satisfying $c(0) = I$ and $\dot{c}(0) = iX$. Applying g_B to c yields another differentiable curve in $SU(2)$, given by $g_B(c(t)) = Bc(t)B^{-1}$. The derivative of this transformed curve at $t = 0$ is also a tangent vector to $SU(2)$ at the identity $I = g_B(c(0))$:

$$
\begin{aligned}
\frac{d}{dt}\left(g_B(c(t))\right)|_{t=0} &= \frac{d}{dt}(Bc(t)B^{-1})|_{t=0} \\
&= B\dot{c}(0)B^{-1} \\
&= B(iX)B^{-1}.
\end{aligned}
$$

The process described in the previous paragraph associates to each tangent vector $iX \in T_I S^3$ a new tangent vector via differentiation. As explained more fully in Appendix A.2, this association of tangent vectors to tangent vectors is the *derivative* of the map $g_B \colon S^3 \to S^3$, which is a linear operator denoted[1] by $Dg_B \colon T_I S^3 \to T_I S^3$. Indeed, since g_B is built out of matrix multiplication and inversion, it is continuously differentiable. Moreover, since g_B maps the identity of $SU(2)$ to itself, its derivative maps the tangent space at the identity to itself. We record the results of this discussion in the following proposition.

Proposition 3.3. *The conjugation action of $SU(2)$ on its tangent space $T_I S^3 = iH_0(2)$ arises via differentiation from the conjugation action of $SU(2)$ on itself. That is, for $B \in SU(2)$*

$$
Dg_B(iX) = B(iX)B^{-1} \qquad \text{for all } iX \in iH_0(2),
$$

where $g_B(M) := BMB^{-1}$ for $M \in SU(2)$. This $SU(2)$-action on its tangent space is called the adjoint action of $SU(2)$.

The vector space structure on $T_I S^3$ comes entirely from the geometric aspect of the Lie group $SU(2)$: tangent spaces to manifolds are *always* vector spaces. We now want to ask whether $T_I S^3$ has any extra algebraic structure, coming from the algebraic aspect of the Lie group $SU(2)$. We have just seen that $T_I S^3$ carries an action of $SU(2)$, but is there any structure that makes reference only to the tangent space itself?

To investigate this question, we employ a similar line of analysis as

---

[1] In order to simplify the notation, we will generally write Dg_B instead of $(Dg_B)_I$ as in Appendix A.2, since we will almost always be working with the derivative at the identity I of our Lie groups.

52   *Symmetry and Quantum Mechanics*

above. For each tangent vector $iY \in iH_0(2)$, we have the mapping $h_Y \colon SU(2) \to iH_0(2)$ defined by the adjoint action:

$$h_Y(B) := B(iY)B^{-1}.$$

The mapping h_Y from $SU(2) = S^3$ to $T_I S^3$ is differentiable, taking the identity of $SU(2)$ to the tangent vector iY. Hence, its derivative at the identity is a linear transformation $Dh_Y \colon T_I S^3 \to T_I S^3$. Here, as usual, we identify the tangent space to the vector space $T_I S^3$ at the point iY with $T_I S^3$ itself.

Proposition 3.4. *The derivative of the adjoint action of $SU(2)$ on its tangent space $iH_0(2)$ is given by the following formula: for $iY \in iH_0(2)$,*

$$Dh_Y(iX) = (iX)(iY) - (iY)(iX) \qquad \text{for all } iX \in iH_0(2),$$

where $h_Y(B) := B(iY)B^{-1}$ for $B \in SU(2)$. The matrix

$$[iX, iY] := (iX)(iY) - (iY)(iX)$$

is called the commutator *of iX and iY.*

Proof. The derivative of h_Y is computed just as described in the paragraphs before proposition 3.3: given a tangent vector $iX \in iH_0(2)$, consider the differentiable curve $c(t) := \exp(itX)$ satisfying $c(0) = I$ and $\dot{c}(0) = iX$. Then applying h_Y yields a curve in $iH_0(2)$, and the derivative of this transformed curve at $t = 0$ is Dh_Y evaluated at iX:

$$
\begin{aligned}
Dh_Y(iX) \quad :=\ & \frac{d}{dt}\left(g_Y(c(t))\right)|_{t=0} \\
=\ & \frac{d}{dt}(c(t)(iY)c(t)^{-1})|_{t=0} \\
=\ & \frac{d}{dt}(c(t)(iY)c(t)^{\dagger})|_{t=0} \\
=\ & \dot{c}(0)(iY)c(0)^{\dagger} + c(0)(iY)\dot{c}(0)^{\dagger} \\
=\ & (iX)(iY) - (iY)(iX) \\
=:\ & [iX, iY].
\end{aligned}
$$

$\square$

The next exercise reveals some properties of the commutator.

Exercise 3.5. *Show that the commutator*

$$[iX, iY] := (iX)(iY) - (iY)(iX)$$

satisfies the following properties for all $iX, iY, iZ \in iH_0(2)$ and all $a \in \mathbb{R}$:

i) $[iX, iY] \in iH_0(2)$ (closure);

Observables and Uncertainty 53

ii) $[aiX + iY, iZ] = a[iX, iZ] + [iY, iZ]$ *(linearity in first slot);*

iii) $[iX, iY] = -[iY, iX]$ *(skew-symmetry);*

iv) $[iX, [iY, iZ]] + [iY, [iZ, iX]] + [iZ, [iX, iY]] = 0$ *(Jacobi identity).*

Properties ii) and iii) together imply linearity in the second slot as well.

Definition 3.6. *A* Lie algebra *is a vector space* $\mathfrak{g}$ *together with a* Lie bracket *operation* $[,]: \mathfrak{g} \times \mathfrak{g} \to \mathfrak{g}$ *satisfying the properties listed in exercise 3.5. (Note: if* $\mathfrak{g}$ *is a vector space over the field* $\mathbb{F}$*, then condition ii) means that the Lie bracket must be* $\mathbb{F}$*-linear.)*

Exercise 3.7. *Check that the real vector space* $i\mathbb{R}$ *with the trivial bracket* $[ix, iy] = 0$ *for all* $x, y \in \mathbb{R}$ *is a Lie algebra. Since* $i\mathbb{R}$ *is the tangent space at the identity to the Lie group* $U(1)$ *(see Section 2.4.1), we denote this Lie algebra by* $\mathfrak{u}(1)$*.*

We see that the tangent space of $SU(2)$ at the identity, endowed with the commutator as Lie bracket, is a real Lie algebra, denoted $\mathfrak{su}(2)$. Since the Lie bracket is bilinear and skew-symmetric, it is completely determined by its values on pairs of distinct basis elements. Using the basis $\left\{ \frac{1}{2i}\sigma_j \right\}$ for $iH_0(2)$, these values are called the *commutation relations* of $\mathfrak{su}(2)$:

$$\left[\frac{1}{2i}\sigma_1, \frac{1}{2i}\sigma_2 \right] = \frac{1}{2i}\sigma_3$$

$$\left[\frac{1}{2i}\sigma_2, \frac{1}{2i}\sigma_3 \right] = \frac{1}{2i}\sigma_1$$

$$\left[\frac{1}{2i}\sigma_3, \frac{1}{2i}\sigma_1 \right] = \frac{1}{2i}\sigma_2.$$

Exercise 3.8. *Verify the commutation relations for* $\mathfrak{su}(2)$ *listed above.*

Now recall that we have identified $iH_0(2)$ with $\mathbb{R}^3$ via the isomorphism F defined by $F(\frac{1}{2i}\sigma_j) = \mathbf{e}_j$. Transferring the Lie bracket on $\mathfrak{su}(2)$ to $\mathbb{R}^3$ via this isomorphism turns $\mathbb{R}^3$ into a Lie algebra with the same commutation relations:

$$\mathbf{e}_1, \mathbf{e}_2] = \mathbf{e}_3$$
$$[\mathbf{e}_2, \mathbf{e}_3] = \mathbf{e}_1$$
$$[\mathbf{e}_3, \mathbf{e}_1] = \mathbf{e}_2.$$

P immediately recognizes these relations as defining the familiar cross product of vectors in physical space! It is as if physical space (together with the cross product) has "spilled out" from the Lie group $SU(2)$ in the form of its Lie algebra $\mathfrak{su}(2)$. Even more, we have seen that $SU(2)$ acts on $\mathfrak{su}(2)$, and sending $B \in SU(2)$ to the matrix representing its conjugation action with respect to the basis $\left\{ \frac{1}{2i}\sigma_j \right\}$ defines the double cover $f: SU(2) \to SO(3)$ connecting

54 *Symmetry and Quantum Mechanics*

rotations of physical space to the corresponding change of basis matrices for spinor space.

The exponential map introduced in Section 2.4.3 furnishes the connection between elements of the Lie algebra and the Lie group: if $iX \in \mathfrak{su}(2) = iH_0(2)$, then $\exp(iX) \in SU(2)$. For example, if θ is a real number, then

$$\exp(i\theta\sigma_1) = \exp \begin{bmatrix} 0 & i\theta \\ i\theta & 0 \end{bmatrix} = \begin{bmatrix} \cos(\theta) & i\sin(\theta) \\ i\sin(\theta) & \cos(\theta) \end{bmatrix}$$

$$\exp(i\theta\sigma_2) = \exp \begin{bmatrix} 0 & \theta \\ -\theta & 0 \end{bmatrix} = \begin{bmatrix} \cos(\theta) & \sin(\theta) \\ -\sin(\theta) & \cos(\theta) \end{bmatrix}$$

$$\exp(i\theta\sigma_3) = \exp \begin{bmatrix} i\theta & 0 \\ 0 & -i\theta \end{bmatrix} = \begin{bmatrix} e^{i\theta} & 0 \\ 0 & e^{-i\theta} \end{bmatrix}.$$

Exercise 3.9. *Verify the preceding computations.*

Applying the universal covering map $f: SU(2) \to SO(3)$ to the matrices computed above, we find that (see example 2.35 and exercise 2.38)

$$f(\exp(i\theta\sigma_1)) = \begin{bmatrix} 1 & 0 & 0 \\ 0 & \cos(2\theta) & \sin(2\theta) \\ 0 & -\sin(2\theta) & \cos(2\theta) \end{bmatrix},$$

$$f(\exp(i\theta\sigma_2)) = \begin{bmatrix} \cos(2\theta) & 0 & -\sin(2\theta) \\ 0 & 1 & 0 \\ \sin(2\theta) & 0 & \cos(2\theta) \end{bmatrix},$$

$$f(\exp(i\theta\sigma_3)) = \begin{bmatrix} \cos(2\theta) & \sin(2\theta) & 0 \\ -\sin(2\theta) & \cos(2\theta) & 0 \\ 0 & 0 & 1 \end{bmatrix}.$$

Note that $f(\exp(i\theta\sigma_j))$ gives a *clockwise* rotation through the angle 2θ around the jth coordinate axis in $\mathbb{R}^3$. That is, the "phase angle" θ gets doubled in the passage from spinor space to physical space. It is in order to account for this doubling, and for the *counter-clockwise* orientation of planes in physical space, that we use the basis $\{\frac{1}{2i}\sigma_j\}$ for $\mathfrak{su}(2)$ rather than $\{i\sigma_j\}$. Indeed, the computations above show that $f(\exp(\frac{\theta}{2i}\sigma_j))$ is a counter-clockwise rotation through the angle θ around the jth coordinate axis, for which reason the matrices $\frac{1}{2i}\sigma_j$ are called *generators of rotations*.

In Section 3.1, we discovered a relationship between spin observables and generators of rotations: the observable $S_z = \frac{\hbar}{2}\sigma_3$ corresponds to the generator $-\frac{i}{\hbar}S_z = \frac{1}{2i}\sigma_3$ of rotation around the z-axis. More generally, multiplication by $-\frac{i}{\hbar}$ yields a correspondence between spin observables and generators of rotations about axes in physical space: if $\mathbf{u} = (u_1, u_2, u_3)$ is a unit vector in physical space, then the observable "spin in the $\mathbf{u}$-direction" is represented by the operator $S_\mathbf{u} = \frac{\hbar}{2}(u_1\sigma_1 + u_2\sigma_2 + u_3\sigma_3)$, and $-\frac{i}{\hbar}S_\mathbf{u} = \frac{1}{2i}(u_1\sigma_1 + u_2\sigma_2 + u_3\sigma_3)$ is the generator of rotation about the $\mathbf{u}$-axis.

M summarizes as follows: the adjoint action of $SU(2)$ on its Lie algebra

Observables and Uncertainty 55

$\mathfrak{su}(2)$ (given by conjugation) is P's model of physical space, with the Lie bracket corresponding to the cross product. P's model of spinor space is the action of $SU(2)$ on $\mathbb{C}^2$. The quantum observables on spinor space are given by the Hermitian operators, and the spin observables correspond to the orbit of $\frac{\hbar}{2}\sigma_3$ under the conjugation action of $SU(2)$ on $H_0(2)$. The connection between spinor space and physical space is effected by means of a correspondence between spin observables and infinitesimal generators of rotations: if $S_{\mathbf{u}}$ is the spin observable corresponding to a Stern-Gerlach device in the direction $\mathbf{u}$, then $-\frac{i}{\hbar}S_{\mathbf{u}}$ is the generator of rotation about the $\mathbf{u}$-axis.

3.3 Commutation relations and uncertainty

While P is quite pleased to see his old friend the cross product emerge from all this formalism, M wonders about the meaning of those commutation relations for the spin observables. Strictly speaking, the spin observables live in the space $H_0(2) = i\mathfrak{su}(2)$ of 2-by-2 Hermitian matrices, rather than in $iH_0(2) = \mathfrak{su}(2)$ itself, and this space is not closed under the commutator. For instance, consider the commutator of the observables S_x and S_y, represented by the Hermitian matrices $\frac{\hbar}{2}\sigma_1$ and $\frac{\hbar}{2}\sigma_2$ respectively:

$$[S_x, S_y] = \left[\frac{\hbar}{2}\sigma_1, \frac{\hbar}{2}\sigma_2\right] = \hbar^2 \left[\frac{1}{2}\sigma_1, \frac{1}{2}\sigma_2\right] = i\hbar\frac{\hbar}{2}\sigma_3 = i\hbar S_z \in iH_0(2).$$

For this reason, we introduce a new, larger Lie algebra, called the *complexification* of $\mathfrak{su}(2)$.

Definition 3.10. *If* $(\mathfrak{g}, [,])$ *is a real Lie algebra, then the* complexification *of* $\mathfrak{g}$ *is the complex Lie algebra*

$$\mathfrak{g}_{\mathbb{C}} := \mathfrak{g} \oplus i\mathfrak{g}$$

with complex scalar multiplication

$$(a + ib)(X + iY) := aX - bY + i(bX + aY)$$

for all $(a + ib) \in \mathbb{C}$ *and* $X + iY \in \mathfrak{g}_{\mathbb{C}}$. *The Lie bracket is defined by*

$$[X + iY, X' + iY'] := [X, X'] - [Y, Y'] + i([X, Y'] + [Y, X'])$$

for all $X + iY, X' + iY' \in \mathfrak{g}_{\mathbb{C}}$.

Exercise 3.11. *Check that* $\mathfrak{g}_{\mathbb{C}}$ *is a complex Lie algebra.*

In the case of $\mathfrak{su}(2)$, we find that

$$\mathfrak{su}(2)_{\mathbb{C}} := \mathfrak{su}(2) \oplus i\mathfrak{su}(2) = iH_0(2) \oplus H_0(2) = M_0(2, \mathbb{C}),$$

56   *Symmetry and Quantum Mechanics*

where $M_0(2, \mathbb{C})$ denotes the vector space of 2-by-2 complex matrices with trace 0.

Exercise 3.12 (♣). *Show that if M is an arbitrary 2-by-2 complex matrix, then there exist unique Hermitian matrices H, K such that $M = H + iK$. This is called the* Cartesian decomposition *of M. Moreover, show that M has trace 0 if and only if H and K both have trace 0.*

The complexified Lie algebra $\mathfrak{su}(2)_\mathbb{C}$ is a 3-dimensional *complex* Lie algebra, with the same commutation relations as $\mathfrak{su}(2)$ when expressed using the basis $\{\frac{1}{2i}\sigma_j\}$. But the Hermitian operators $\{\frac{\hbar}{2}\sigma_j\} = \{S_x, S_y, S_z\}$ also provide a basis for $\mathfrak{su}(2)_\mathbb{C}$, for which the commutation relations are

$$
\begin{aligned}
[S_x, S_y] &= i\hbar S_z, \\
[S_y, S_z] &= i\hbar S_x, \\
[S_z, S_x] &= i\hbar S_y,
\end{aligned}
$$

and these are the ones that M is interested in at the moment. More generally, the next exercise provides an even larger Lie algebra where the commutators of arbitrary quantum observables may be computed.

Exercise 3.13. *Show that the real vector space $iH(2)$ consisting of 2-by-2 skew-Hermitian matrices is closed under the commutator $[,]$, and that $(iH(2), [,])$ is a real Lie algebra. Furthermore, show that the complexification of this Lie algebra is $(M(2, \mathbb{C}), [,])$, consisting of all 2-by-2 complex matrices. Finally, show that $\{\frac{\hbar}{2}I, S_x, S_y, S_z\}$ forms a basis for $M(2, \mathbb{C})$.*

Since the observables S_x and S_y do not commute, they are *not* simultaneously diagonalizable, i.e., they have *different* eigenvectors. Of course, we already knew this because we know the eigenvectors explicitly (they are $|\pm x\rangle$ and $|\pm y\rangle$ respectively). In physical terms: states with a definite value of S_x do not have a definite value of S_y. But again, this is just a weak form of belief 4 from Section 2.2 which states that an electron in the spin-state $|+x\rangle$ has probability $\frac{1}{2}$ of being observed in each of the states $|\pm y\rangle$. As described below, in order to understand what the commutation relation means in this context, we need to imagine performing the S_y-measurement on $|+x\rangle$ not just once, but many times.

More generally, suppose that $|\psi\rangle = c_+|+z\rangle + c_-|-z\rangle$ is a ket in W representing the spin-state of an electron. Recall that ψ has a probability $|c_+|^2$ of being measured spin up by an SGz, and a probability $|c_-|^2$ of being measured spin down. We define the *expectation value* of S_z for the state ψ to be

$$
|c_+|^2 \frac{\hbar}{2} + |c_-|^2 \left(-\frac{\hbar}{2}\right) = \langle\psi|S_z|\psi\rangle =: \langle S_z\rangle_\psi.
$$

This is the average value P would find if he made many measurements of S_z on identical copies of the state ψ. We extend this definition to an arbitrary quantum observable as follows.

Observables and Uncertainty

Definition 3.14. *If $H\colon W \to W$ is an observable (i.e., Hermitian operator), then the* expectation value *of H in the spin-state ψ is the inner product $\langle\psi|H|\psi\rangle$, which we denote by $\langle H\rangle_\psi$.*

Proposition 3.15. *If $H\colon W \to W$ is an observable and ψ is a spin-state, then the expectation value $\langle H\rangle_\psi$ is a real number. We interpret the expectation value as the average value we would find if we made a large number of H-measurements on identical copies of the state ψ.*

Proof. Since H is Hermitian, the Spectral Theorem A.32 guarantees that there is an orthonormal basis of eigenkets for W, say $\{|\phi_1\rangle, |\phi_2\rangle\}$. Thus, we may express the state ψ as a linear combination of H-eigenstates:

$$|\psi\rangle = c_1|\phi_1\rangle + c_2|\phi_2\rangle.$$

Applying H, we find that

$$H|\psi\rangle = c_1 H|\phi_1\rangle + c_2 H|\phi_2\rangle = c_1\lambda_1|\phi_1\rangle + c_2\lambda_2|\phi_2\rangle,$$

where λ_1 and λ_2 are the real eigenvalues of H. Then the expectation value $\langle H\rangle_\psi = \langle\psi|H|\psi\rangle$ is obtained by applying the bra $\langle\psi| = c_1^\star\langle\phi_1| + c_2^\star\langle\phi_2|$ (see exercise 2.5):

$$\begin{aligned}
\langle\psi|H|\psi\rangle &= (c_1^\star\langle\phi_1| + c_2^\star\langle\phi_2|)(c_1\lambda_1|\phi_1\rangle + c_2\lambda_2|\phi_2\rangle) \\
&= |c_1|^2\lambda_1 + |c_2|^2\lambda_2,
\end{aligned}$$

which is a real number since the eigenvalues λ_j are real. The expectation value is thus a weighted average of the possible observed values λ_j, where the weights are the probabilities of obtaining these values on any given measurement of the state ψ. $\qquad\square$

Definition 3.16. *The* uncertainty *of an observable H in the spin-state ψ, denoted $\Delta_\psi H$, is the standard deviation, i.e., the square root of the expectation value of the operator $(H - \langle\psi|H|\psi\rangle\, I)^2$ in the state ψ:*

$$\begin{aligned}
(\Delta_\psi H)^2 &:= \left\langle\psi|(H - \langle\psi|H|\psi\rangle\, I)^2|\psi\right\rangle \\
&= \left\langle\psi|(H^2 - 2\langle\psi|H|\psi\rangle\, H + \langle\psi|H|\psi\rangle^2\, I)|\psi\right\rangle \\
&= \langle\psi|H^2|\psi\rangle - \langle\psi|H|\psi\rangle^2.
\end{aligned}$$

The uncertainty provides a measure of the "spread" of the observed H-values for ψ around the expectation value $\langle\psi|H|\psi\rangle$. In particular, if ϕ is an eigenstate for H with eigenvalue $\lambda \in \mathbb{R}$, then it is a state of zero uncertainty for H:

$$(\Delta_\phi H)^2 = \langle\phi|H^2|\phi\rangle - \langle\phi|H|\phi\rangle^2 = \lambda^2\langle\phi|\phi\rangle^2 - (\lambda\langle\phi|\phi\rangle)^2 = 0.$$

This makes sense, because the state ϕ will always yield the value λ when measured by an "H-device."

58 *Symmetry and Quantum Mechanics*

Exercise 3.17 (♣). *Show, conversely, that if ϕ is a state of zero uncertainty for an observable H, then ϕ is an eigenstate for H.*

In the following proposition, we show that the Cauchy-Schwarz inequality provides some information about the product of uncertainties.

Proposition 3.18. *Suppose that $H, K: W \to W$ are observables on spinor space. Fix a spin-state ψ and define new observables by subtracting the expectation values:*

$$\bar{H} := H - \langle\psi|H|\psi\rangle I, \qquad \bar{K} := K - \langle\psi|K|\psi\rangle I.$$

Then

$$(\Delta_\psi H)(\Delta_\psi K) \geq |\langle\psi|\bar{H}\bar{K}|\psi\rangle|.$$

Proof. For the purposes of this proof, we think ψ as an actual unit vector in W rather than as an equivalence class, so that we may use the defining property of the adjoint operator (see proposition A.28):

$$(\Delta_\psi H)^2 = \langle\psi|\bar{H}^2\psi\rangle = \langle\psi|\bar{H}\bar{H}\psi\rangle = \langle\bar{H}^\dagger\psi|\bar{H}\psi\rangle = \langle\bar{H}\psi|\bar{H}\psi\rangle = \|\bar{H}\psi\|^2,$$

using the fact that $\bar{H}^\dagger = \bar{H}$ because $\bar{H}$ is Hermitian. Similarly, we have $(\Delta_\psi K)^2 = \|\bar{K}\psi\|^2$.

The Cauchy-Schwarz inequality then says that the product of these norms is greater than or equal to the modulus of the inner product:

$$(\Delta_\psi H)(\Delta_\psi K) = \|\bar{H}\psi\|\|\bar{K}\psi\| \geq |\langle\bar{H}\psi|\bar{K}\psi\rangle| = |\langle\psi|\bar{H}\bar{K}\psi\rangle|.$$

In the final equality, we have again used the defining property of the adjoint and the fact that $\bar{H}$ is Hermitian. □

Note that the product $\bar{H}\bar{K} \in M(2, \mathbb{C})$ is not Hermitian unless H and K commute, so it does not represent an observable in general. Nevertheless, by exercise 3.12 we can write it uniquely as a sum $\bar{H}\bar{K} = X + iY$ where X and Y are Hermitian.

Exercise 3.19. *Show that if $\bar{H}\bar{K} := (H - \langle H\rangle_\psi I)(K - \langle K\rangle_\psi I) = X + iY$ is the Cartesian decomposition of $\bar{H}\bar{K}$, then $Y = \frac{1}{2i}[H, K]$.*

Putting all of these results together, we obtain the uncertainty principle for quantum observables on spinor space:

Theorem 3.20 (Uncertainty Principle). *Suppose that $H, K: W \to W$ are observables on spinor space, and fix a spin-state ψ. Then*

$$(\Delta_\psi H)(\Delta_\psi K) \geq \frac{1}{2}|\langle\psi|[H, K]|\psi\rangle|.$$

Observables and Uncertainty

Proof. Using the notation of proposition 3.18 and exercise 3.19, we have

$$
\begin{aligned}
(\Delta_\psi H)(\Delta_\psi K) &\geq \left|\langle\psi|\bar{H}\bar{K}|\psi\rangle\right| \\
&= \left|\langle\psi|(X+iY)|\psi\rangle\right| \\
&= \left|\langle\psi|X|\psi\rangle + i\langle\psi|Y|\psi\rangle\right| \\
&\geq \left|\langle\psi|Y|\psi\rangle\right| \\
&= \left|\left\langle\psi\left|\frac{1}{2i}[H,K]\right|\psi\right\rangle\right| \\
&= \frac{1}{2}\left|\langle\psi|[H,K]|\psi\rangle\right|.
\end{aligned}
$$

$\square$

Applying the uncertainty principle to the spin observables S_x and S_y with the commutation relation $[S_x, S_y] = i\hbar S_z$, we obtain:

$$
(\Delta_\psi S_x)(\Delta_\psi S_y) \geq \frac{1}{2}\left|\langle\psi|i\hbar S_z|\psi\rangle\right| = \frac{\hbar}{2}\left|\langle\psi|S_z|\psi\rangle\right|.
$$

In words: for a given spin-state ψ, the product of the uncertainties of S_x and S_y must be greater than or equal to $\frac{\hbar}{2}$ times the absolute value of the expectation value of S_z.

The uncertainty principle quantifies the sense in which there is a tradeoff between the simultaneous determination of non-commuting observables for a single spin-state. For instance, if ψ is a state with a nonzero expectation value for S_z, then the less uncertainty there is in its values for S_x, the more uncertainty there must be in its values for S_y. The inequality even holds when ψ is a state of zero uncertainty for S_x, since the expectation value $\langle\psi|S_z|\psi\rangle$ on the right-hand side is also zero in that case. Indeed, we know from exercise 3.17 that if $\Delta_\psi S_x = 0$, then ψ must be one of the S_x-eigenstates $|\pm x\rangle$. Thus, the absolute certainty about the value of S_x means that we have no information about the values of S_y or S_z: each of the S_x-eigenstates has equal probability of being measured spin up or spin down by an SGy or an SGz. Hence, we may compute the expectation value and uncertainty as:

$$
\begin{aligned}
\langle\psi|S_z|\psi\rangle &= \frac{1}{2}\left(\frac{\hbar}{2}\right) + \frac{1}{2}\left(-\frac{\hbar}{2}\right) = 0, \\
(\Delta_\psi S_y)^2 &= \langle\psi|S_y^2|\psi\rangle - \langle\psi|S_y|\psi\rangle^2 = \left\langle\psi\left|\frac{\hbar^2}{4}I\right|\psi\right\rangle - 0 = \frac{\hbar^2}{4}.
\end{aligned}
$$

We see that the uncertainty inequality holds in the trivial form $0 \cdot \frac{\hbar}{2} \geq 0$.

Thus, the commutation relations for $\mathfrak{su}(2)_{\mathbb{C}}$ lead to the uncertainty principle expressing a numerical inequality that goes further than anything explicitly stated in beliefs 1–4 from Section 2.2. Both P and M are delighted to see that they have gotten more out of their model than they explicitly put in.

3.4 Some related Lie algebras

We have now encountered several different Lie algebras, all with the commutator as Lie bracket:

$$
\begin{aligned}
\mathfrak{u}(1) &= i\mathbb{R} & \text{Lie algebra of the Lie group } U(1), \\
\mathfrak{su}(2) &= iH_0(2) & \text{Lie algebra of the Lie group } SU(2), \\
\mathfrak{su}(2)_{\mathbb{C}} &= M_0(2,\mathbb{C}) & \text{complexification of the Lie algebra } \mathfrak{su}(2), \\
& iH(2) & \text{Lie algebra of 2-by-2 skew-Hermitian matrices}, \\
& M(2,\mathbb{C}) & \text{complexification of } iH(2).
\end{aligned}
$$

Note that in the case of $\mathfrak{u}(1) = i\mathbb{R}$, the Lie bracket is trivial: $[ix, iy] = 0$ for all $x, y \in \mathbb{R}$. This is a reflection of the fact that the Lie group $U(1)$ is abelian. It is natural to ask whether the final three Lie algebras come from Lie groups, in the same way that $\mathfrak{u}(1)$ and $\mathfrak{su}(2)$ come from $U(1)$ and $SU(2)$ respectively. We will take them one at a time, and in each case, show that the answer is "yes."

3.4.1 Warmup: The Lie algebra $\mathfrak{u}(1)$

In order to present the outline of the general argument, we begin with the case of the one-dimensional Lie algebra $i\mathbb{R}$ with trivial bracket. Starting with this Lie algebra, how might we discover the group $U(1)$, together with the fact that it is actually a smooth curve? Following a hunch from our previous work, we consider the exponential mapping

$$
\exp \colon i\mathbb{R} \to \mathbb{C}
$$

and observe that for any $it \in i\mathbb{R}$, we have $\exp(it) \in U(1)$, the group of complex numbers of modulus 1. In Section 2.4.1 we simply noted that $U(1)$ is the unit circle in the complex plane, and thus clearly a smooth curve. We now make a more careful argument using the Implicit Function Theorem A.40, organizing things so that the generalization to higher-dimensional Lie algebras will be clear.

To this end (see Figure 3.1), we think of $\mathbb{C}$ as $\mathbb{R}^2$ by identifying a complex number $x + iy$ with the point (x, y). Consider the function $\phi \colon \mathbb{R}^2 \to \mathbb{R}$ defined by $\phi(x, y) = x^2 + y^2$. Then $(x, y) \in U(1)$ if and only if $\phi(x, y) = 1$; that is, $U(1) = \phi^{-1}(1)$. Now the derivative of ϕ at any point (a, b) is represented by the Jacobian matrix of partial derivatives:

$$
(D\phi)_{(a,b)} = \begin{bmatrix} 2a & 2b \end{bmatrix}.
$$

If $(a, b) \neq (0, 0)$, then this matrix has rank 1. Let's assume that $b \neq 0$, although a similar argument goes through under the assumption that $a \neq 0$. If we make

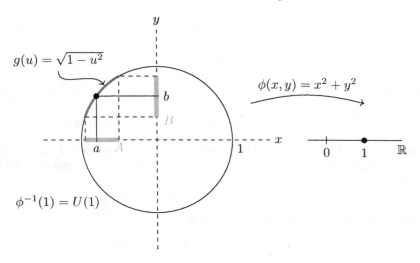

FIGURE 3.1: The Implicit Function Theorem for the circle $U(1)$.

the further assumption that $(a,b) \in U(1)$ (i.e., that $\phi(a,b) = a^2 + b^2 = 1$), then we may apply the Implicit Function Theorem to obtain open sets $a \in A \subset \mathbb{R}$ and $b \in B \subset \mathbb{R}$ and a differentiable function $g \colon A \to B$ such that

$$\begin{aligned}\text{graph}(g) := \{(u, g(u)) \mid u \in A\} &= \{(u,v) \in A \times B \mid u^2 + v^2 = 1\} \\ &= U(1) \cap (A \times B).\end{aligned}$$

Thus, the portion of $U(1)$ near the point (a,b) is given by the graph of a differentiable function, and is thus a smooth curve in $\mathbb{R}^2$. The tangent space to $U(1)$ at (a,b) is simply the tangent line to the graph of g at that point.

Of course, in this simple case, the use of the Implicit Function Theorem is unnecessary, because we can write down a formula for the function g explicitly:

$$g(u) = \pm\sqrt{1 - u^2},$$

where the sign is chosen to match the sign of b. However, in the examples below, such explicit formulas are difficult to obtain, and the Implicit Function Theorem comes to the rescue.

3.4.2 The Lie algebra $\mathfrak{sl}_2(\mathbb{C})$

The complex Lie algebra $\mathfrak{su}(2)_\mathbb{C}$ may also be considered as a six-dimensional *real* Lie algebra, with basis $\{\frac{1}{2i}\sigma_j; \frac{1}{2}\sigma_j\}$. Viewed this way, we want to know whether $\mathfrak{su}(2)_\mathbb{C}$ is the Lie algebra of some (six-dimensional) Lie group, in the same way as $\mathfrak{su}(2)$ is the Lie algebra of the (three-dimensional) Lie group $SU(2)$. To figure this out, let's follow the pattern of our analysis of $\mathfrak{u}(1)$, and consider the exponential of matrices in $M_0(2, \mathbb{C})$.

Note that every traceless complex matrix $M \in M_0(2, \mathbb{C})$ has the form

$$M = \begin{bmatrix} \alpha & \beta \\ \gamma & -\alpha \end{bmatrix} \qquad \alpha, \beta, \gamma \in \mathbb{C}.$$

The eigenvalues of M are given by the roots of the characteristic polynomial

$$\det(M - \lambda I) = \lambda^2 - (\alpha^2 + \beta\gamma) = \lambda^2 + \det(M).$$

Hence, if $\det(M) \neq 0$, then M has two distinct eigenvalues $\pm\lambda$, and hence is diagonalizable: $QMQ^{-1} = \operatorname{diag}(\lambda, -\lambda)$ for some invertible matrix Q. We then have

$$\exp(M) = \exp(Q^{-1}\operatorname{diag}(\lambda, -\lambda)Q) = Q^{-1}\operatorname{diag}(e^\lambda, e^{-\lambda})Q.$$

Taking the determinant now reveals that

$$\det(\exp(M)) = \det(Q^{-1}\operatorname{diag}(e^\lambda, e^{-\lambda})Q) = 1,$$

so $\exp(M)$ is an element of the *special linear group* $SL(2, \mathbb{C})$ of 2-by-2 complex matrices with determinant 1.

Now suppose that $\det(M) = 0$. Then compute the square of M:

$$M^2 = (\alpha^2 + \beta\gamma)I = -\det(M)I = 0.$$

Hence, the infinite series defining the matrix exponential (see Section 2.4.3) truncates at the second term, and

$$\det(\exp(M)) = \det(I + M) = \begin{vmatrix} 1 + \alpha & \beta \\ \gamma & 1 - \alpha \end{vmatrix} = 1 + \det(M) = 1.$$

Hence, in all cases we have $\exp(M) \in SL(2, \mathbb{C})$ for $M \in M_0(2, \mathbb{C})$.

$SL(2, \mathbb{C})$ is certainly a group, but in order to be a *Lie* group, it must have a nice geometric structure as a smooth manifold. To prove that this is the case, consider $SL(2, \mathbb{C}) \subset M(2, \mathbb{C}) = \mathbb{R}^8$, where we identify the space of 2-by-2 complex matrices with $\mathbb{R}^8$ via the map

$$(x_1, \ldots, x_8) \mapsto \begin{bmatrix} x_1 + ix_2 & x_3 + ix_4 \\ x_5 + ix_6 & x_7 + ix_8 \end{bmatrix}.$$

Identifying $\mathbb{C}$ with $\mathbb{R}^2$ via $(y_1, y_2) \mapsto y_1 + iy_2$, the determinant becomes a function $\det \colon \mathbb{R}^8 \to \mathbb{R}^2$ given by

$$\det(\mathbf{x}) = (x_1x_7 - x_2x_8 - x_5x_3 + x_6x_4, x_2x_7 + x_1x_8 - x_6x_3 - x_5x_4).$$

At any point $\mathbf{x}$, the derivative of $\det$ is a linear transformation $(D\det)_{\mathbf{x}} \colon \mathbb{R}^8 \to \mathbb{R}^2$ with Jacobian matrix of partial derivatives

$$\begin{bmatrix} x_7 & -x_8 & -x_5 & x_6 & -x_3 & x_4 & x_1 & -x_2 \\ x_8 & x_7 & -x_6 & -x_5 & -x_4 & -x_3 & x_2 & x_1 \end{bmatrix}.$$

Observables and Uncertainty 63

If $\mathbf{x} \neq 0$, then this matrix has rank 2. For the purposes of this computation, we will assume that $x_1 + ix_2 \neq 0$, although a similar argument works for any other nonzero matrix element. Under this assumption, the final 2-by-2 block of the Jacobian matrix is invertible. Now suppose further that $\mathbf{x}$ corresponds to a matrix in $SL(2, \mathbb{C})$, so that $\det(\mathbf{x}) = (1, 0)$. Then writing $\mathbf{x} = (\mathbf{a}, \mathbf{b})$ with $\mathbf{a} = (x_1, \ldots, x_6) \in \mathbb{R}^6$ and $\mathbf{b} = (x_7, x_8) \in \mathbb{R}^2$, we have a continuously differentiable function $\det \colon \mathbb{R}^6 \times \mathbb{R}^2 \to \mathbb{R}^2$ with $\det(\mathbf{a}, \mathbf{b}) = (1, 0)$. Moreover, we have seen that the final 2-by-2 block of the Jacobian matrix at $(\mathbf{a}, \mathbf{b})$ is invertible. By the Implicit Function Theorem, there exist open sets $\mathbf{a} \in U \subset \mathbb{R}^6$ and $\mathbf{b} \in V \subset \mathbb{R}^2$ and a differentiable map $g : U \to V$ such that

$$
\begin{aligned}
\mathrm{graph}(g) := \{(\mathbf{u}, g(\mathbf{u})) \mid \mathbf{u} \in U\} \ &= \ \{\mathbf{w} \in U \times V \mid \det(\mathbf{w}) = (1, 0)\} \\
&= \ SL(2, \mathbb{C}) \cap (U \times V).
\end{aligned}
$$

Since $U \times V$ is an open set in $\mathbb{R}^8$ containing $\mathbf{x} = (\mathbf{a}, \mathbf{b})$, this means that the portion of $SL(2, \mathbb{C})$ near $\mathbf{x}$ is given by the graph of the differentiable $g \colon U \to \mathbb{R}^2$. The inverse of the differentiable map $G \colon \mathbf{u} \mapsto (\mathbf{u}, g(\mathbf{u}))$ is given by the projection $(\mathbf{u}, \mathbf{v}) \mapsto \mathbf{u}$, showing that $SL(2, \mathbb{C})$ looks like (is *diffeomorphic* to) $\mathbb{R}^6$ near $\mathbf{x}$. The tangent space to $SL(2, \mathbb{C})$ at $\mathbf{x}$ is given by the image of the linear transformation $(DG)_{\mathbf{a}} : \mathbb{R}^6 \to \mathbb{R}^8$.

So, $SL(2, \mathbb{C})$ is both a group and a six-dimensional manifold. Moreover, since matrix multiplication and inversion are given by polynomials in the matrix entries, it follows that these operations are continuously differentiable, so $SL(2, \mathbb{C})$ is a Lie group. Moreover, we have shown that the matrix exponential defines a map $\exp \colon M_0(2, \mathbb{C}) \to SL(2, \mathbb{C})$. All that remains is to show that $M_0(2)$ is actually the tangent space to $SL(2, \mathbb{C})$ at the identity.

For this, we make the argument which should be familiar from our work with $SU(2)$ in Section 2.4.2: suppose that $c \colon (-\epsilon, \epsilon) \to SL(2, \mathbb{C})$ is a smooth parametrized curve with $c(0) = I$. Then $\dot{c}(0)$ is tangent to $SL(2, \mathbb{C})$ at the identity. But $\det(c(t)) = 1$ for all t, so by the chain rule (see theorem A.37 and proposition A.38), taking the derivative at time $t = 0$ yields

$$
0 = \frac{d}{dt} \det(c(t))|_{t=0} = (D \det)_I(\dot{c}(0)).
$$

The identity matrix corresponds to the vector $(1, 0, 0, 0, 0, 0, 1, 0) \in \mathbb{R}^8$, so the Jacobian matrix at the identity is

$$
(D \det)_I = \begin{bmatrix} 1 & 0 & 0 & 0 & 0 & 0 & 1 & 0 \\ 0 & 1 & 0 & 0 & 0 & 0 & 0 & 1 \end{bmatrix}.
$$

Recall that we are identifying the matrix $\dot{c}(0)$ with a vector $(x_1, \ldots, x_8)$ as follows:

$$
(x_1, \ldots, x_8) \mapsto \begin{bmatrix} x_1 + ix_2 & x_3 + ix_4 \\ x_5 + ix_6 & x_7 + ix_8 \end{bmatrix} = \dot{c}(0).
$$

Viewing $\dot{c}(0)$ as a column vector in $\mathbb{R}^8$, the derivative condition becomes

$$
(0, 0) = (D \det)_I(\dot{c}(0)) = (x_1 + x_7, x_2 + x_8).
$$

64 *Symmetry and Quantum Mechanics*

As a matrix condition, this is equivalent to demanding that

$$0 = x_1 + ix_2 + x_7 + ix_8 = \text{tr}(\dot{c}(0)).$$

Hence, every tangent vector to $SL(2, \mathbb{C})$ has trace zero. On the other hand, if $M \in M_0(2, \mathbb{C})$ is arbitrary, then defining $c(t) := \exp(tM)$ defines a curve in $SL(2, \mathbb{C})$ satisfying $c(0) = I$ and $\dot{c}(0) = M$. This shows that $T_I SL(2, \mathbb{C}) = M_0(2, \mathbb{C})$ as claimed.

From here, the argument is exactly like the one for $SU(2)$ from Section 3.2: the group $SL(2, \mathbb{C})$ acts by conjugation on itself, hence on its tangent space $M_0(2, \mathbb{C})$. This defines the adjoint action of $SL(2, \mathbb{C})$. By differentiation, we discover the commutator bracket on $M_0(2, \mathbb{C})$, which endows $M_0(2, \mathbb{C})$ with the structure of a Lie algebra, denoted $\mathfrak{sl}_2(\mathbb{C})$. Putting all of this together, we have established the following proposition:

Proposition 3.21. *The complexification of the Lie algebra of $SU(2)$ is the Lie algebra of $SL(2, \mathbb{C})$:*

$$\mathfrak{su}(2)_\mathbb{C} = \mathfrak{sl}_2(\mathbb{C}).$$

While the Implicit Function Theorem allows us to see that $SL(2, \mathbb{C})$ looks *locally* like $\mathbb{R}^6$, it tells us nothing about the *global* topology of the group. For that, we will make an algebraic argument. We begin by recalling the *polar decomposition* of a matrix.

If $M \in GL(n, \mathbb{C})$ is an invertible matrix, then we may write $M = UP$, where $U \in U(n)$ is unitary and $P \in H(n)$ is Hermitian with positive eigenvalues (i.e., *positive definite*). Moreover, U and P are uniquely determined by M. To justify these assertions and describe U and P explicitly, we begin by considering the matrix $M^\dagger M$.

Exercise 3.22 (♣). *Show that $M^\dagger M$ is Hermitian with positive eigenvalues, so it has a positive definite square root P such that $P^2 = M^\dagger M$.*

Now define $U := MP^{-1}$, so that $M = UP$. Note that U is unitary as required:

$$U^\dagger U = P^{-1} M^\dagger M P^{-1} = P^{-1} P^2 P^{-1} = I.$$

Exercise 3.23. *Show that the polar decomposition $M = UP$ of an invertible matrix M is unique.*

Now specialize to the case of a special linear matrix $M \in SL(n, \mathbb{C})$, i.e., a matrix with determinant 1. Then we have

$$1 = \det(M) = \det(UP) = \det(U)\det(P).$$

Since P is positive definite and U unitary, we have $\det(P) = r > 0$ and $\det(U) = e^{i\theta} \in U(1)$. It follows that $r = 1 = e^{i\theta}$, so that in fact $U \in SU(n)$ and P is positive definite with determinant 1.

Finally, we specialize further to the case $C \in SL(2, \mathbb{C})$. Then we have

Observables and Uncertainty 65

the polar decomposition $C = BP$ where $B \in SU(2)$ and P is positive definite with determinant 1. We may diagonalize P via a unitary matrix: $P = Q\mathrm{diag}(\lambda, \lambda^{-1})Q^{-1}$ for some $\lambda > 0$. Setting $a := \log(\lambda)$, it follows that

$$P = Q\exp(\mathrm{diag}(a, -a))Q^{-1} = \exp(Q\mathrm{diag}(a, -a)Q^{-1}).$$

Hence, P may be expressed (uniquely) as the exponential of a Hermitian matrix of trace zero. Since $H_0(2)$ is a 3-dimensional real vector space and $SU(2) = S^3$, it follows that as a topological space we have

$$SL(2, \mathbb{C}) = S^3 \times \mathbb{R}^3.$$

In particular, $SL(2, \mathbb{C})$ is connected and simply connected, but not compact.

We will study the Lie group $SL(2, \mathbb{C})$ further in Section 9.3, where we will develop a physical interpretation for it as a symmetry group in the context of special relativity.

3.4.3 The Lie algebra $\mathfrak{u}(2)$

The real Lie algebra $(iH(2), [,])$ is four-dimensional, with basis $\{\frac{1}{2i}\sigma_j\}$, where we have added the identity $\sigma_0 := I$ to the list of Pauli matrices. Our question is thus whether $iH(2)$ is the Lie algebra of some four-dimensional Lie group. Following the outline of our investigation of $\mathfrak{su}(2)_\mathbb{C}$, the following exercise gives a proof of the fact that $iH(2)$ is the Lie algebra of the unitary group $U(2)$.

Exercise 3.24. *Recall the definition of the 2-by-2 unitary group:*

$$U(2) := \{U \in M(2, \mathbb{C}) \mid UU^\dagger = I\}.$$

i) Show that if $iX \in iH(2)$ is an arbitrary skew-Hermitian matrix, then $\exp(iX) \in U(2)$ is a unitary matrix. (See proposition 2.31.)

ii) Identify $M(2, \mathbb{C})$ with $\mathbb{R}^8$ as in the previous section, and identify $H(2)$ with $\mathbb{R}^4$ via

$$(y_1, y_2, y_3, y_4) \mapsto \begin{bmatrix} y_1 & y_2 + iy_3 \\ y_2 - iy_3 & y_4 \end{bmatrix}.$$

Show that, via these identifications, the map $M \mapsto MM^\dagger$ becomes the continuously differentiable function $\Phi \colon \mathbb{R}^8 \to \mathbb{R}^4$ defined by

$$\Phi(\mathbf{x}) = \begin{bmatrix} x_1^2 + x_2^2 + x_3^2 + x_4^2 \\ x_1 x_5 + x_2 x_6 + x_3 x_7 + x_4 x_8 \\ -x_1 x_6 + x_2 x_5 - x_3 x_8 + x_4 x_7 \\ x_5^2 + x_6^2 + x_7^2 + x_8^2 \end{bmatrix},$$

and that $U(2) = \Phi^{-1}(1, 0, 0, 1)$.

66 *Symmetry and Quantum Mechanics*

iii) Compute the Jacobian matrix of partial derivatives representing the derivative $(D\Phi)_{\mathbf{x}}$ at any point $\mathbf{x} \in \mathbb{R}^8$. Show that if $\mathbf{x}$ corresponds to an invertible matrix, then $(D\Phi)_{\mathbf{x}}$ has rank 4.

iv) Now suppose that $\mathbf{x} \in \mathbb{R}^8$ corresponds to a unitary matrix. Use the Implicit Function Theorem to show that $U(2)$ is diffeomorphic to $\mathbb{R}^4$ near $\mathbf{x}$, and conclude that $U(2)$ is a Lie group.

v) Argue that the tangent space to $U(2)$ at the identity is given by the space of skew-Hermitian matrices, $iH(2)$.

vi) Conclude that $iH(2)$, endowed with the commutator $[,]$, is the Lie algebra of $U(2)$, denoted $\mathfrak{u}(2)$.

Once again, to determine the global topology of $U(2)$, we make an algebraic argument. Given a unitary matrix $M \in U(2)$, we know that $\det(M) = e^{i\theta} \in U(1)$. It follows that the matrix $M' := \operatorname{diag}(e^{-i\theta}, 1)M$ has determinant 1 and hence is in $SU(2)$. Thus, we may define a mapping from $U(2)$ to the product $U(1) \times SU(2) = S^1 \times S^3$ by

$$M \mapsto (\det(M), M').$$

The inverse mapping from $U(1) \times SU(2)$ to $U(2)$ is clearly given by $(e^{i\theta}, B) \mapsto \operatorname{diag}(e^{i\theta}, 1)B$. Since both of these mappings are continuous, we see that, as a topological space, $U(2)$ is the product of the circle and the 3-sphere:

$$U(2) = S^1 \times S^3.$$

In particular, $U(2)$ is compact and connected but not simply connected, due to the "hole" coming from the S^1-factor.

3.4.4 The Lie algebra $\mathfrak{gl}_2(\mathbb{C})$

We have saved the easiest for last: consider the general linear group $GL(2, \mathbb{C}) \subset M(2, \mathbb{C})$, consisting of all invertible 2-by-2 complex matrices. This group is defined by the non-vanishing of the determinant function. In terms of our usual identification of $M(2, \mathbb{C})$ with $\mathbb{R}^8$, Section 3.4.2 shows that a point $\mathbf{x}$ corresponds to an invertible matrix if and only if

$$x_1 x_7 - x_2 x_8 - x_5 x_3 + x_6 x_4 \neq 0 \quad \text{or} \quad x_2 x_7 + x_1 x_8 - x_6 x_3 - x_5 x_4 \neq 0.$$

That is, the subset $GL(2, \mathbb{C})$ is obtained by removing the intersection of two quadric hyper-surfaces from $\mathbb{R}^8$. In particular, $GL(2, \mathbb{C})$ is an open subset of $\mathbb{R}^8$, and hence is an 8-dimensional submanifold. Thus, $GL(2, \mathbb{C})$ is a Lie group, and its tangent space at every point (in particular the identity) is $\mathbb{R}^8 = M(2, \mathbb{C})$. The Lie algebra of $GL(2, \mathbb{C})$ is $(M(2, \mathbb{C}), [,])$, which we denote by $\mathfrak{gl}_2(\mathbb{C})$. By the Cartesian decomposition of an arbitrary complex matrix (exercise 3.12), it follows that $\mathfrak{gl}_2(\mathbb{C})$ is the complexification of the Lie algebra $\mathfrak{u}(2)$:

$$\mathfrak{u}(2)_{\mathbb{C}} = \mathfrak{gl}_2(\mathbb{C}).$$

Observables and Uncertainty 67

Exercise 3.25. *Extend the argument above to show that $GL(n, \mathbb{C})$ is a Lie group for any $n \geq 1$, and that its Lie algebra $\mathfrak{gl}_n(\mathbb{C}) = (M(n, \mathbb{C}), [,])$ consists of all n-by-n complex matrices.*

More generally, by [11, Corollary 3.45], any closed subgroup $G \subset GL(n, \mathbb{C})$ is a Lie group (called a *matrix Lie group*), and the Lie bracket on the associated Lie algebra, $\mathfrak{g}$, is given by the commutator, as in all of our examples above. Moreover, the matrix exponential function maps $\mathfrak{g}$ into G. As a particular example that will be important in Section 5.5, the unitary groups $U(n)$ are all matrix Lie groups, with Lie algebras $iH(n)$ consisting of skew-Hermitian matrices. As we showed explicitly in proposition 2.31, the exponential mapping $\exp: iH(n) \to U(n)$ takes skew-Hermitian matrices to unitary matrices.

Chapter 4

Dynamics

In which M and P discover the Schrödinger equation.

P wants to actually do some physics. That is, he would like to see what happens to the electron under various external forces, due to the presence of electric or magnetic fields for instance. These external fields will presumably have some effect on the spin-state of the electron, which will therefore evolve in time. So M and P are led to the question: how should they model time-evolution?

Their thinking goes as follows: as time elapses, the spin-states change. That is, if the electron is initially in the spin-state ψ, then at time t it will be in a possibly different spin-state ψ_t. Thus, for every $t \in \mathbb{R}$, we have a function $\mathcal{U}_t$ taking spin-states to spin-states defined by $\mathcal{U}_t(\psi) = \psi_t$. Recall that a spin-state is actually an equivalence class of unit vectors (kets) in spinor space W, where two kets are equivalent if they differ by a phase in the group $U(1)$. In order to obtain a simple model, M and P assume[1]

1. Each function $\mathcal{U}_t$ is actually defined on all of spinor space W, taking unit vectors to unit vectors. Hence, if $|\psi\rangle$ is a ket representing the spin-state ψ, then the ket $|\psi\rangle_t := \mathcal{U}_t|\psi\rangle$ represents the spin-state ψ_t. For ease of notation, we will write $|\psi_t\rangle = |\psi\rangle_t$ to indicate that (unless explicitly stated otherwise), we use the time-evolution of the initial ket $|\psi\rangle$ to represent the time-evolution of the spin-state ψ.

2. Each $\mathcal{U}_t$ preserves the superposition of spin-states. That is, if $|\psi\rangle = c_1|\phi_1\rangle + c_2|\phi_2\rangle$ is the initial ket, then at time t the ket will have evolved to $|\psi_t\rangle = c_1|\phi_1\rangle_t + c_2|\phi_2\rangle_t$. That is, $\mathcal{U}_t\colon W \to W$ is a linear map.

Since $\mathcal{U}_t$ is linear and preserves unit vectors, it actually preserves the norm of all vectors, and so must be a unitary linear operator on W by exercise 2.11.

In terms of P's identification of W with $\mathbb{C}^2$ via the z-basis, we see that $\mathcal{U}_t \in U(2)$, the group of 2×2 unitary matrices. Explicitly, $\mathcal{U}_t$ is invertible and $\mathcal{U}_t^{-1} = \mathcal{U}_t^\dagger$. Thus, the time evolution of the electron is modeled by a function $\mathcal{U}\colon \mathbb{R} \to U(2)$ defined by $\mathcal{U}(t) := \mathcal{U}_t$. Observer M notes that $\mathcal{U}(0) = \mathcal{U}_0 = I$ is the identity transformation, and wonders if $\mathcal{U}$ might be a group homomorphism. That is, should we expect that $\mathcal{U}(s + t) = \mathcal{U}(s)\mathcal{U}(t)$ for all

---

[1]In fact, a fundamental theorem of E. Wigner implies that any bijection from the space of spin-states to itself that preserves the modulus of the inner product between spin-states comes from either a unitary or anti-unitary operator on W, uniquely defined up to a phase. See [23, Chapter 2, Appendix A] for a precise statement and proof of Wigner's theorem.

69

70   *Symmetry and Quantum Mechanics*

$s, t \in \mathbb{R}$? Observer P doesn't think so: if this condition holds, then an initial spin-state ψ would evolve in time $s + t$ to the state

$$\psi_{s+t} = \mathcal{U}_{s+t}\psi = \mathcal{U}_s\left(\mathcal{U}_t\psi_t\right) = \mathcal{U}_s(\psi_t) = (\psi_t)_s.$$

In words, an electron initially in the spin-state ψ would evolve in time $s + t$ to the *same* state as that achieved in time s by an electron initially in the spin-state ψ_t. While this seems reasonable when the external fields are time-independent, it seems quite unlikely to occur when the fields are changing with time.

4.1   Time-independent external fields

M and P decide to investigate the time-independent case first. Thus, they imagine that the electron is in the presence of some external fields that are constant in time, and that the time-evolution is modeled by a group homomorphism $\mathcal{U}\colon \mathbb{R} \to U(2)$. They make the further assumption that $\mathcal{U}$ is continuous, which captures their intuition that small changes in time should correspond to small changes in spin-states. P is surprised when M tells him that the combination of $\mathcal{U}$ being continuous *and* a homomorphism forces $\mathcal{U}$ to be C^∞, i.e., to possess derivatives of all orders (see [17, Chapter 10, corollary 11].)

Thus, we can think of $\mathcal{U}\colon \mathbb{R} \to U(2)$ as defining a smooth curve in the Lie group $U(2)$, passing through the identity at time $t = 0$. Hence, the derivative of $\mathcal{U}$ at $t = 0$ yields an element of the tangent space, i.e., the Lie algebra $\mathfrak{u}(2)$ consisting of 2×2 skew-Hermitian matrices (see Section 3.4.3). Remembering the correspondence $S_{\mathbf{u}} \leftrightarrow -\frac{i}{\hbar}S_{\mathbf{u}}$ between spin observables and generators of rotations in $\mathfrak{su}(2)$, we define a Hermitian operator $\mathcal{H}$ by the formula

$$\mathcal{H} = -\frac{\hbar}{i}\dot{\mathcal{U}}(0) \in i\mathfrak{u}(2) = H(2).$$

Now fix a time, t, and compute the derivative of $\mathcal{U}$ while using the group homomorphism condition $\mathcal{U}(s + t) = \mathcal{U}(s)\mathcal{U}(t)$:

$$\begin{aligned}
\dot{\mathcal{U}}(t) &= \lim_{s \to 0} \frac{\mathcal{U}(s + t) - \mathcal{U}(t)}{s} \\
&= \lim_{s \to 0} \frac{\mathcal{U}(s)\mathcal{U}(t) - \mathcal{U}(t)}{s} \\
&= \left(\lim_{s \to 0} \frac{\mathcal{U}(s) - I}{s}\right)\mathcal{U}(t) \\
&= \dot{\mathcal{U}}(0)\mathcal{U}(t) \\
&= -\frac{i}{\hbar}\mathcal{H}\mathcal{U}(t).
\end{aligned}$$

Dynamics

By the Existence and Uniqueness Theorem [13, theorem 6.2.3], $\mathcal{U}$ is the unique solution to this equation passing through the identity at time $t = 0$.

Conversely, given *any* Hermitian operator $\mathcal{H}$ on W, multiplication by $-\frac{i}{\hbar}$ yields an element $-\frac{i}{\hbar}\mathcal{H} \in \mathfrak{u}(2)$. The function $\mathcal{U} \colon \mathbb{R} \to U(2)$ defined by $\mathcal{U}(t) := \exp(-\frac{i}{\hbar}t\mathcal{H})$ is a smooth group homomorphism with derivative $-\frac{i}{\hbar}\mathcal{H}$ at $t = 0$. Moreover, by proposition 2.21e), it satisfies the differential equation

$$\dot{\mathcal{U}}(t) = -\frac{i}{\hbar}\mathcal{H}\mathcal{U}(t) \qquad \text{for all } t.$$

Thus, there is a one-to-one correspondence between continuous homomorphisms $\mathcal{U} \colon \mathbb{R} \to U(2)$ and Hermitian operators $\mathcal{H}$ on spinor space $\mathbb{C}^2$. The operator $\mathcal{H}$ is called the *Hamiltonian* of the physical system, and the *Schrödinger equation* specifies the relationship between the Hamiltonian $\mathcal{H}$ and the time-evolution $\mathcal{U}$:

$$i\hbar\dot{\mathcal{U}}(t) = \mathcal{H}\mathcal{U}(t).$$

Getting back to the physics, P wants to *discover* the time-evolution $\mathcal{U}$, so he must *choose* a particular Hamiltonian $\mathcal{H}$ to model the specific external fields that impinge upon the electron. This is real physics, not abstract mathematics: P must use his knowledge of *classical* physics to cook up a Hamiltonian operator to input into his *quantum* model. A good choice of Hamiltonian should lead to a model of the physical system with predictive and explanatory power.

As M wonders how to select a Hamiltonian, P points out that the operator $\mathcal{H}$, being Hermitian, must correspond to a quantum observable. Moreover, the Schrödinger equation reveals $\mathcal{H}$ to have the units of energy (kg $\cdot$ m^2/s^2), so with any luck the Hamiltonian $\mathcal{H}$ will correspond to the energy of the electron in the presence of the given (time-independent) fields. So, P decides on the following strategy: he will write down the classical expression for the energy of the electron in the given external fields, then *quantize*[2] the expression by replacing classical observables with appropriate quantum observables, thus obtaining his Hamiltonian $\mathcal{H}$. The time evolution $\mathcal{U}(t) = \exp(-\frac{i}{\hbar}t\mathcal{H})$ will then be obtained by exponentiation.

As we have seen several times now, the easiest way to exponentiate a skew-Hermitian matrix is to diagonalize it. That is, to compute $\mathcal{U}$, we should first find the eigenvalues and eigenvectors for the Hamiltonian $\mathcal{H}$. This amounts to finding the real eigenvalues $E_2 \geq E_1$ for $\mathcal{H}$, and then solving the corresponding eigenvector equations for nonzero kets $|\phi_j\rangle \in \mathbb{C}^2$:

$$\mathcal{H}|\phi_j\rangle = E_j|\phi_j\rangle.$$

In the physical context, the preceding equation is called the *time-independent Schrödinger equation*, and the spin-states ϕ_j are called *stationary states*, because they remain constant in time, although their representing kets evolve

---

[2]This is a highly non-trivial process in general, see [10, Chapter 13] and [20, Chapter 2, Section 2].

72 *Symmetry and Quantum Mechanics*

via a phase:

$$\mathcal{U}(t)|\phi_j\rangle = \exp\left(-\frac{i}{\hbar}t\mathcal{H}\right)|\phi_j\rangle = e^{-\frac{iE_jt}{\hbar}}|\phi_j\rangle.$$

These two stationary states provide an $\mathcal{H}$-eigenbasis for spinor space, and the associated eigenvalues E_j are their definite energies. An arbitrary spin-state, ψ, may be expressed as a superposition of the two stationary states:

$$|\psi\rangle = c_1|\phi_1\rangle + c_2|\phi_2\rangle.$$

It follows that the time-evolution of ψ is given by

$$|\psi_t\rangle = \mathcal{U}(t)|\psi\rangle = c_1\mathcal{U}(t)|\phi_1\rangle + c_2\mathcal{U}(t)|\phi_2\rangle = c_1 e^{-\frac{iE_1t}{\hbar}}|\phi_1\rangle + c_2 e^{-\frac{iE_2t}{\hbar}}|\phi_2\rangle.$$

Note that if $E_1 \neq E_2$, then the spin-state $|\psi_t\rangle$ will generally differ from $|\psi\rangle$ by more than just a phase, so that $\psi_t \neq \psi$ and we really do have a non-trivial time-evolution of spin-states.

Example 4.1. *The classical expression for the energy of an electron in a magnetic field* $\mathbf{B}\colon \mathbb{R}^3 \to \mathbb{R}^3$ *is*

$$E = -\boldsymbol{\mu} \cdot \mathbf{B},$$

where $\boldsymbol{\mu}$ is the magnetic dipole moment of the electron. Recall from Section 2.1 that $\boldsymbol{\mu}$ is proportional to the angular momentum of the electron, which we now interpret as the electron's spin. So, replacing $\boldsymbol{\mu}$ by the vector operator $\gamma\mathbf{S} := \gamma(S_x, S_y, S_z)$, we obtain the Hamiltonian

$$\mathcal{H} := -\gamma\mathbf{S} \cdot \mathbf{B}.$$

For the electron, the proportionality constant is $\gamma = -\frac{ge}{2m}$, where m is the mass of the electron, $-e$ is the (negative) charge of the electron, and g is an experimentally determined dimensionless constant with a value very close to 2.

Now suppose that $\mathbf{B} = (0, 0, B_z)$ is a time-independent field pointing along the z-axis. Then the Hamiltonian becomes

$$\mathcal{H} = \frac{ge}{2m}B_z S_z = \omega\frac{\hbar}{2}\sigma_3,$$

where we have introduced the quantity $\omega := \frac{ge}{2m}B_z$. The stationary states for $\mathcal{H}$ are simply the eigenkets for S_z, namely the basis vectors $|\pm z\rangle$, with corresponding energies $E_\pm = \pm\omega\frac{\hbar}{2}$. In this basis, the time-evolution is given by

$$\mathcal{U}(t) = \exp\left(-\frac{i}{\hbar}t\mathcal{H}\right) = \exp\left(-\frac{i\omega t}{2}\sigma_3\right) = \begin{bmatrix} e^{-\frac{i\omega t}{2}} & 0 \\ 0 & e^{\frac{i\omega t}{2}} \end{bmatrix}.$$

Consider a general spin-state, ψ, expressed in the z-basis as $|\psi\rangle = c_1|+z\rangle + c_2$ Its time evolution is given by

$$\begin{aligned} |\psi_t\rangle &= c_1 e^{-\frac{i\omega t}{2}}|+z\rangle + c_2 e^{\frac{i\omega t}{2}}|-z\rangle \\ &= e^{-\frac{i\omega t}{2}}\left(c_1|+z\rangle + c_2 e^{i\omega t}|-z\rangle\right). \end{aligned}$$

Dynamics
73

Thus (ignoring the phase out front), the spin-state ψ_t is represented by the ket $c_1|{+}z\rangle + c_2 e^{i\omega t}|{-}z\rangle$, from which we see that the time-evolution is periodic, with period $T = \frac{2\pi}{\omega}$, inversely proportional to the magnetic field strength B_z. Note that $\mathcal{U}(t)$ is actually in $SU(2)$ for all t, and that it maps to a rotation around the z-axis under the double cover $f \colon SU(2) \to SO(3)$. For this reason, physicists often describe this time-evolution as "spin precession around the z-axis." Of course, this isn't quite right, since the kets $|\psi_t\rangle$ live in $\mathbb{C}^2$ while the z-axis is in $\mathbb{R}^3$, but the intuitive physical picture is appealing.

However, there is an experimentally accessible quantity that actually does precess around the z-axis, namely the expectation value of the vector observable $\mathbf{S}$:

$$\langle \mathbf{S}\rangle_{\psi_t} := \left(\langle S_x\rangle_{\psi_t}, \langle S_y\rangle_{\psi_t}, \langle S_z\rangle_{\psi_t}\right).$$

This is the average (time-dependent) vector that we would obtain, were we to make many separate measurements of S_x, S_y, and S_z on identical copies of the time-dependent spin-state $|\psi_t\rangle = c_1|{+}z\rangle + c_2 e^{i\omega t}|{-}z\rangle$. After adjusting by an overall phase and perhaps shifting the time parameter, we may assume that the coefficients, c_i, are real, with $c_1 = \cos(\alpha)$ and $c_2 = \sin(\alpha)$ for some $\alpha \in \mathbb{R}$. Then we compute

$$
\begin{aligned}
\langle S_x\rangle_{\psi_t} &= \langle \psi_t | S_x | \psi_t\rangle \\
&= \frac{\hbar}{2}\left[\ \cos(\alpha)\quad \sin(\alpha)e^{-i\omega t}\ \right]\begin{bmatrix} 0 & 1 \\ 1 & 0 \end{bmatrix}\begin{bmatrix} \cos(\alpha) \\ \sin(\alpha)e^{i\omega t} \end{bmatrix} \\
&= \frac{\hbar}{2}\sin(\alpha)\cos(\alpha)\left(e^{i\omega t} + e^{-i\omega t}\right) \\
&= \frac{\hbar}{2}\sin(2\alpha)\cos(\omega t).
\end{aligned}
$$

Similar computations reveal that

$$
\begin{aligned}
\langle S_y\rangle_{\psi_t} &= \frac{\hbar}{2}\sin(2\alpha)\sin(\omega t) \\
\langle S_z\rangle_{\psi_t} &= \frac{\hbar}{2}\cos(2\alpha).
\end{aligned}
$$

Hence, as a vector in $\mathbb{R}^3$, we have

$$\langle \mathbf{S}\rangle_{\psi_t} = \frac{\hbar}{2}\left(\sin(2\alpha)\cos(\omega t), \sin(2\alpha)\sin(\omega t), \cos(2\alpha)\right),$$

which is a time-dependent vector of constant length $\frac{\hbar}{2}$, making a constant angle of 2α with the z-axis, and precessing counterclockwise around the z-axis with an angular speed of ω.

Exercise 4.2. *Verify the expressions for the expectation values in the previous example.*

4.2 Time-dependent external fields

Now M and P turn to the case of external fields that are time-dependent. In this case, the time-evolution is modeled by a function $\mathcal{U}\colon \mathbb{R} \to U(2)$ that is *not* a homomorphism. Nevertheless, M and P make the assumption that $\mathcal{U}$ is continuously differentiable. Thus, $\mathcal{U}$ still defines a differentiable curve in the Lie group $U(2)$, passing through the identity at time $t = 0$. At an arbitrary time t, the derivative of $\mathcal{U}$ is an element of the tangent space to $U(2)$ at the element $\mathcal{U}_t$:

$$\dot{\mathcal{U}}(t) \in T_{\mathcal{U}_t} U(2).$$

In order to obtain an element of the Lie algebra $\mathfrak{u}(2) = T_I U(2)$, we need a way to identify these two tangent spaces. Define $R_t\colon U(2) \to U(2)$ to be the map "right multiplication by $\mathcal{U}_t$." That is, $R_t(A) := A\mathcal{U}_t$ for all $A \in U(2)$. The map R_t is simply the restriction to $U(2) \subset M(2, \mathbb{C}) = \mathbb{R}^8$ of the linear operator on $M(2, \mathbb{C})$ given by right-multiplication by $\mathcal{U}_t$. Since linear operators are their own derivatives (see definition A.34), it follows that $(DR_t)_I\colon \mathfrak{u}(2) \to T_{\mathcal{U}_t} U(2)$ is given by $(DR_t)_I(iX) = iX\mathcal{U}_t$ for all skew-Hermitian $iX \in \mathfrak{u}(2)$. Thus, at every time t, we can define a Hermitian operator $\mathcal{H}(t)$ via the formula

$$-\frac{i}{\hbar}\mathcal{H}(t) := \dot{\mathcal{U}}(t)\mathcal{U}_t^{-1} \in \mathfrak{u}(2).$$

Hence, the continuously differentiable time-evolution function $\mathcal{U}\colon \mathbb{R} \to U(2)$ yields a continuous *Hamiltonian function* $\mathcal{H}\colon \mathbb{R} \to i\mathfrak{u}(2)$, where $i\mathfrak{u}(2)$ denotes the space of Hermitian matrices. Multiplying both sides of the forgoing equation on the right by $i\hbar\mathcal{U}_t$ yields the Schrödinger equation once again, this time with a time-dependent Hamiltonian:

$$i\hbar\dot{\mathcal{U}}(t) = \mathcal{H}(t)\mathcal{U}(t).$$

Again, by the Existence and Uniqueness Theorem [13, theorem 6.2.3], $\mathcal{U}$ is the unique solution to this differential equation satisfying $\mathcal{U}(0) = I$.

So, if P wants to model the dynamics of electron spin in the presence of external time-dependent fields, he must choose an appropriate Hamiltonian *function*, starting with his knowledge of energy in classical physics and then quantizing the result as before. But now the passage from the Hamiltonian to the time-evolution is more difficult: instead of simply exponentiating, P must solve the Schrödinger equation for $\mathcal{U}(t)$. Depending on the particular structure of the Hamiltonian $\mathcal{H}(t)$, this may be difficult, and approximation techniques may be necessary in practice.

Note that the time-evolution $\mathcal{U}(t)$ acts on kets via $\mathcal{U}(t)|\psi\rangle = |\psi_t\rangle$, so we may rewrite the Schrödinger equation in terms of kets as

$$i\hbar\frac{d|\psi_t\rangle}{dt} = \mathcal{H}(t)|\psi_t\rangle, \tag{4.1}$$

Dynamics

and solve directly for the ket-valued function $|\psi_t\rangle$ rather than the matrix-valued function $\mathcal{U}(t)$.

Example 4.3. *Let's modify the previous example by considering an oscillating magnetic field aligned with the z-axis:* $\mathbf{B} = (0, 0, B_z \cos(\beta t))$. *The Hamiltonian[3] for the electron in this field is time-dependent:*

$$\mathcal{H}(t) = \frac{ge}{2m} \mathbf{S} \cdot \mathbf{B} = \frac{ge}{2m} B_z \cos(\beta t) S_z = \omega \frac{\hbar}{2} \cos(\beta t) \sigma_3.$$

To find the time-evolution of a spin-state $|\psi\rangle = a_1|+z\rangle + a_2|-z\rangle$, *we write* $|\psi_t\rangle = c_1(t)|+z\rangle + c_2(t)|-z\rangle$, *and we must solve the Schrödinger equation:*

$$\begin{bmatrix} \dot{c}_1(t) \\ \dot{c}_2(t) \end{bmatrix} = \begin{bmatrix} -\frac{i\omega}{2} \cos(\beta t) & 0 \\ 0 & \frac{i\omega}{2} \cos(\beta t) \end{bmatrix} \begin{bmatrix} c_1(t) \\ c_2(t) \end{bmatrix},$$

with initial conditions $c_j(0) = a_j$. *This is an uncoupled system of linear, first-order equations, which are easily solved by separation of variables. We find that*

$$\begin{bmatrix} c_1(t) \\ c_2(t) \end{bmatrix} = \begin{bmatrix} a_1 \exp\left(-\frac{i\omega}{2\beta} \sin(\beta t) \right) \\ a_2 \exp\left(\frac{i\omega}{2\beta} \sin(\beta t) \right) \end{bmatrix},$$

so that the time-evolution of ψ is given by

$$\begin{aligned} |\psi_t\rangle &= a_1 \exp\left(-\frac{i\omega}{2\beta} \sin(\beta t) \right) |+z\rangle + a_2 \exp\left(\frac{i\omega}{2\beta} \sin(\beta t) \right) |-z\rangle \\ &= \exp\left(-\frac{i\omega}{2\beta} \sin(\beta t) \right) \left(a_1|+z\rangle + a_2 \exp\left(\frac{i\omega}{\beta} \sin(\beta t) \right) |-z\rangle \right). \end{aligned}$$

Ignoring the phase out front, we see that ψ_t is represented by the ket $a_1|+z\rangle + a_2 \exp\left(\frac{i\omega}{\beta} \sin(\beta t) \right) |-z\rangle$. *Hence, the time-evolution is periodic, with period* $T = \frac{1}{\beta} \arcsin\left(\frac{2\pi\beta}{\omega} \right)$.

Exercise 4.4. *Show that the expectation value of the vector observable $\mathbf{S}$ is*

$$\langle \mathbf{S} \rangle_{\psi_t} = \frac{\hbar}{2} \left(\sin(2\alpha) \cos\left(\frac{\omega}{\beta} \sin(\beta t) \right), \sin(2\alpha) \sin\left(\frac{\omega}{\beta} \sin(\beta t) \right), \cos(2\alpha) \right).$$

4.3 The energy-time uncertainty principle

Viewing kets as column vectors, we may take the conjugate transpose of equation (4.1) to obtain the Schrödinger equation for bras, viewed as row

---

[3]We are ignoring the electric field that would be generated by the time-varying magnetic field.

76   *Symmetry and Quantum Mechanics*

vectors:

$$-i\hbar \frac{d\langle \psi_t|}{dt} = \langle \psi_t|\mathcal{H}(t).$$

Suppose that $O(t)$ is a (possibly time-dependent) Hermitian operator on spinor space, corresponding to a quantum observable. If ψ is an initial spin-state, then the expectation value of the observable at any time t is given by the inner product

$$\langle O(t)\rangle_{\psi_t} = \langle \psi_t|O(t)|\psi_t\rangle.$$

Note that the time-dependence enters via both copies of the evolving state ψ_t, as well as through the explicit time-dependence of $O(t)$. Using the Leibniz rule twice, we compute:

$$
\begin{aligned}
\frac{d}{dt}\langle \psi_t|O(t)|\psi_t\rangle &= \frac{d}{dt}\left(\langle \psi_t|\right)O(t)|\psi_t\rangle + \langle \psi_t|\frac{d}{dt}\left(O(t)|\psi_t\rangle\right) \\
&= \frac{i}{\hbar}\langle \psi_t|\mathcal{H}(t)O(t)|\psi_t\rangle + \langle \psi_t|\dot{O}(t)|\psi_t\rangle \\
&\quad + \left\langle \psi_t|-\frac{i}{\hbar}O(t)\mathcal{H}(t)|\psi_t\right\rangle \\
&= \frac{i}{\hbar}\langle \psi_t|\left(\mathcal{H}(t)O(t)-O(t)\mathcal{H}(t)\right)|\psi_t\rangle + \langle \psi_t|\dot{O}(t)|\psi_t\rangle \\
&= \frac{i}{\hbar}\langle \psi_t|[\mathcal{H}(t),O(t)]|\psi_t\rangle + \langle \psi_t|\dot{O}(t)|\psi_t\rangle.
\end{aligned}
$$

In particular, if the operator O is time-independent and commutes with the Hamiltonian $\mathcal{H}(t)$ for all t, then the expectation value $\langle \psi_t|O|\psi_t\rangle$ is independent of time, i.e., a constant of the dynamics. For instance, in examples 4.1 and 4.3 from this chapter, the Hamiltonian is a multiple of the spin observable S_z. Hence, S_z commutes with the Hamiltonians, but S_x and S_y do not. As a consequence, the expectation value of S_z is constant in both examples, while the expectation values of S_x and S_y vary with time.

Now suppose that O is an arbitrary time-independent observable, so that

$$\frac{d}{dt}\langle \psi_t|O|\psi_t\rangle = \frac{i}{\hbar}\langle \psi_t|[\mathcal{H}(t),O]|\psi_t\rangle.$$

The appearance of the commutator on the right-hand side makes observer P think of uncertainty (see Section 3.3). He reminds M that at any particular time t, Theorem 3.20 asserts the inequality

$$\Delta_{\psi_t}\mathcal{H}(t)\Delta_{\psi_t}O \geq \frac{1}{2}|\langle \psi_t|[\mathcal{H}(t),O]|\psi_t\rangle| = \frac{\hbar}{2}\left|\frac{d}{dt}\langle \psi_t|O|\psi_t\rangle\right|.$$

If the right-hand side of the previous inequality is non-zero, we define the quantity

$$\Delta_{\psi_t}^O \tau := \frac{\Delta_{\psi_t}O}{\left|\frac{d}{dt}\langle \psi_t|O|\psi_t\rangle\right|},$$

Dynamics 77

and obtain the inequality

$$\Delta_{\psi_t}\mathcal{H}(t)\Delta_{\psi_t}^O\tau \geq \frac{\hbar}{2}$$

which is reminiscent of the uncertainty relations for the spin operators derived in Section 3.3. But M and P wonder: what is the physical interpretation of this inequality?

First of all, note that the left-hand side is a function of t, which enters in two ways: through the spin-state ψ_t and through the Hamiltonian $\mathcal{H}(t)$. The first factor on the left-hand side gives the uncertainty in the energy of the state ψ_t. The second factor may be interpreted as follows: it is the time necessary for the expectation value of O in the state ψ_t to change by an amount equal to the uncertainty of O in the state ψ_t. The inequality thus expresses a precise tradeoff between the energy uncertainty and some sort of characteristic time for the observable O as the state ψ_t evolves.

If we set $\Delta_{\psi_t}\tau := \inf_O \Delta_{\psi_t}^O\tau$, then this quantity represents the time required for the state ψ_t to change in any appreciable way[4]. Moreover, since the preceding inequality holds for all observables O, we must have

$$\Delta_{\psi_t}\mathcal{H}(t)\Delta_{\psi_t}\tau \geq \frac{\hbar}{2}.$$

This *energy-time uncertainty principle* expresses a tradeoff between the energy uncertainty and the evolutionary time for the state ψ_t.

4.3.1   Conserved quantities

In the previous section, we saw that if O is a time-independent observable that commutes with the Hamiltonian $\mathcal{H}(t)$ for all times t, then for any initial spin-state ψ, the expectation value $\langle O \rangle_{\psi_t}$ is independent of t. We say that the expectation value $\langle O \rangle_{\psi_t}$ is a *conserved quantity* for the dynamics specified by $\mathcal{H}(t)$.

Going further, suppose that ψ is an eigenstate for the observable O, so that $O|\psi\rangle = \lambda|\psi\rangle$ for some $\lambda \in \mathbb{R}$. Hence, ψ is a state of zero uncertainty for O: it will return the value λ with probability 1 when measured with an "O-device." Still assuming that O commutes with $\mathcal{H}(t)$ for all t, we find that ψ_t is also an eigenstate for O with the same eigenvalue λ:

$$O|\psi_t\rangle = O\mathcal{H}(t)|\psi\rangle = \mathcal{H}(t)O|\psi\rangle = \mathcal{H}(t)(\lambda|\psi\rangle) = \lambda\mathcal{H}(t)|\psi\rangle = \lambda|\psi_t\rangle.$$

Thus, for all times t, the state ψ_t has definite O-value λ, so that λ is a conserved quantity for the time-evolution of the state ψ. Since this holds for all eigenvalues of the operator O, we say that O-values are conserved quantities for the dynamics specified by $\mathcal{H}(t)$.

---

[4]I am indebted to [5, p. 52] for this line of analysis. Here, $\inf_O$ denotes the *greatest lower bound* as O ranges over the space of time-independent quantum observables.

78   Symmetry and Quantum Mechanics

The Hamiltonians in examples 4.1 and 4.3 are multiples of the operator S_z, and hence commute with S_z. It follows that the z-component of spin is a conserved quantity for the dynamics in both cases.

Chapter 5

Higher Spin

In which M and P classify the representations of $SU(2)$.

While P is still coming to grips with the "naturalness" of this model, M marvels at its tidiness: starting with some intuitions about physical space and the Stern-Gerlach experiments, they have arrived at a model based entirely on the Lie group $SU(2)$. Physical space is the adjoint action of $SU(2)$ on its Lie algebra $\mathfrak{su}(2)$, and spinor space is the multiplication action of $SU(2)$ on $\mathbb{C}^2$. Moreover, these two spaces are connected by the correspondence between spin observables and generators of rotations: if $S_{\mathbf{u}}$ is the observable "spin in the $\mathbf{u}$-direction," then $-\frac{i}{\hbar}S_{\mathbf{u}}$ generates rotation about the $\mathbf{u}$-axis in physical space. Moreover, the commutation relations for the complexified Lie algebra $\mathfrak{su}(2)_{\mathbb{C}} = \mathfrak{sl}_2(\mathbb{C})$ yield uncertainty inequalities for the spin observables. Finally, the dynamics of spin-states are determined by a Hermitian Hamiltonian function $\mathcal{H}(t) \in i\mathfrak{u}(2)$, which determines the unitary time-evolution $\mathcal{U}(t) \in U(2)$ via the Schrödinger equation

$$i\hbar\dot{\mathcal{U}}(t) = \mathcal{H}(t)\mathcal{U}(t).$$

Just as P starts to get out some laboratory equipment, M interrupts him with an intriguing question. Since the multiplication action of $SU(2)$ on $\mathbb{C}^2$ models electron spin and the adjoint action models physical space, what might other actions of $SU(2)$ be good for? Note that $SU(2)$ acts linearly on both spinor space and its Lie algebra $\mathfrak{su}(2)$, so they are *representations* of $SU(2)$ in the sense of definition 1.20. In this language, the multiplication action of $SU(2)$ on $\mathbb{C}^2$ is called the *defining representation* of $SU(2)$. Observer M would like to know whether a different representation of $SU(2)$ might describe particles that behave differently than the electron in Stern-Gerlach experiments? Can an analysis of the representations of $SU(2)$ predict the possibilities for the outcomes of such Stern-Gerlach experiments? Quite reasonably, P wonders what other representations exist, and if they can be classified. Before answering this question, we need to introduce some general concepts and terminology about group representations. This will lead into a lengthy and purely mathematical analysis of the representation theory of $SU(2)$ and its Lie algebra $\mathfrak{su}(2)$. But in Section 5.5 we will return to the physics and provide an interpretation of this representation theory in terms of spin angular momentum.

80 Symmetry and Quantum Mechanics

5.1 Group representations

Suppose that G is a group and V is a vector space. Recall from definition 1.20 that V is a *representation* of G if G acts on V as linear transformations. That is, we have a function $G \times V \to V$ (denoted $(g, \mathbf{v}) \mapsto g \star \mathbf{v}$) such that for all $g_1, g_2 \in G$ and $\mathbf{v} \in V$:

$$(g_1 g_2) \star \mathbf{v} = g_1 \star (g_2 \star \mathbf{v})$$
$$e \star \mathbf{v} = \mathbf{v},$$

and for each $g \in G$,

$$(g, \mathbf{v}) \mapsto g \star \mathbf{v} \qquad \text{is a linear operator on } V.$$

Before we can classify the representations of G, we need to specify when two representations should be considered the same.

Definition 5.1. *Suppose that $G \times V_1 \to V_1$ and $G \times V_2 \to V_2$ are two representations of the same group G. Then a linear transformation $\phi \colon V_1 \to V_2$ is a* morphism *of G-representations if the diagram below commutes, so that $\phi(g \star_1 \mathbf{v}) = g \star_2 \phi(\mathbf{v})$ for all $g \in G$ and $\mathbf{v} \in V_1$:*

$$
\begin{array}{ccc}
G \times V_1 & \xrightarrow{\ \star_1\ } & V_1 \\
{\scriptstyle \mathrm{id} \times \phi} \downarrow & & \downarrow {\scriptstyle \phi} \\
G \times V_2 & \xrightarrow{\ \star_2\ } & V_2.
\end{array}
$$

If, in addition, ϕ is an isomorphism of vector spaces, then ϕ is called an isomorphism *of G-representations.*

We are mainly interested in the case where V is a finite-dimensional vector space over the complex numbers.

Exercise 5.2. *Suppose that V is a finite-dimensional complex vector space, and $G \times V \to V$ is a representation. Choose an isomorphism $\phi \colon V \xrightarrow{\sim} \mathbb{C}^n$, and consider the function $\Phi \colon G \times \mathbb{C}^n \to \mathbb{C}^n$ defined by $\Phi(g, \mathbf{x}) := \phi(g \star \phi^{-1}(\mathbf{x}))$. Show that Φ is a representation of G on $\mathbb{C}^n$, and that ϕ is an isomorphism of G-representations.*

Hence, in order to classify finite-dimensional complex representations of G up to isomorphism, it suffices to consider representations on the spaces $\mathbb{C}^n$. So suppose that $G \times \mathbb{C}^n \to \mathbb{C}^n$ is a representation. This action determines a map $\rho \colon G \to GL(n, \mathbb{C})$ defined by

$$\rho(g) := \text{matrix of } \mathbf{x} \mapsto g \star \mathbf{x} \text{ with respect to the standard basis of } \mathbb{C}^n.$$

The group action condition $(g_1 g_2) \star \mathbf{x} = g_1 \star (g_2 \star \mathbf{x})$ translates into the fact that

Higher Spin

ρ is a group homomorphism. Moreover, we have $g \star \mathbf{x} = \rho(g)\mathbf{x}$ for all $g \in G$ and $\mathbf{x} \in \mathbb{C}^n$, so the homomorphism ρ determines the original representation of G. Making use of the fact that $GL(n, \mathbb{C})$ is a Lie group, we will require the mapping ρ to be differentiable in the case where G is also a Lie group.

Definition 5.3. *An n-dimensional complex matrix representation of a Lie group, G, is a differentiable homomorphism $\rho\colon G \to GL(n, \mathbb{C})$.*

Given two representations of a group, G, we can build a larger representation that contains them both.

Definition 5.4. *Suppose that V_1 and V_2 are representations of G. Then the direct sum $V_1 \oplus V_2$ is the G-representation defined by*

$$(g, (\mathbf{v}_1, \mathbf{v}_2)) \mapsto (g \star \mathbf{v}_1, g \star \mathbf{v}_2).$$

Example 5.5. *Consider the defining representation of $SU(2)$ on $\mathbb{C}^2$ and the complexification of the adjoint representation of $SU(2)$ on $\mathbb{C}^3$. In terms of group homomorphisms, the fundamental representation corresponds to the identity mapping $\mathrm{id}\colon SU(2) \to SU(2) \subset GL(2, \mathbb{C})$, while the complexified adjoint representation corresponds to the universal double cover $f\colon SU(2) \to SO(3) \subset GL(3, \mathbb{R}) \subset GL(3, \mathbb{C})$. Then the direct sum is the representation on $\mathbb{C}^2 \oplus \mathbb{C}^3 = \mathbb{C}^5$ defined by the homomorphism $\rho\colon SU(2) \to SU(2) \times SO(3) \subset GL(5, \mathbb{C})$ given by*

$$\rho(B) = \begin{bmatrix} B & 0_{2\times 3} \\ 0_{3\times 2} & f(B) \end{bmatrix}.$$

Definition 5.6. *Suppose that $G \times V \to V$ is a G-representation, and $W \subset V$ is a vector subspace of V. Then W is* invariant *if $g \star \mathbf{w} \in W$ for all $g \in G$ and $\mathbf{w} \in W$. In this case we say that $G \times W \to W$ is a* subrepresentation *of V.*

Clearly, if $V = V_1 \oplus V_2$ is the direct sum of two G-representations, then $V_1 = V_1 \oplus \{\mathbf{0}\}$ and $V_2 = \{\mathbf{0}\} \oplus V_2$ are each invariant subspaces of V. This motivates the following definition.

Definition 5.7. *The representation V is* irreducible *if its only invariant subspaces are $\{\mathbf{0}\}$ and V.*

So, if V is an irreducible representation of G, then V is not the direct sum of positive-dimensional G-representations. For general groups, G, the converse is not true. That is, a G-representation may fail to be irreducible, even if it cannot be written as a direct sum.

Exercise 5.8. *Consider the* Heisenberg group *$H_1 \subset GL(3, \mathbb{C})$ defined as*

$$H_1 := \left\{ \begin{bmatrix} 1 & a & c \\ 0 & 1 & b \\ 0 & 0 & 1 \end{bmatrix} \mid a, b, c \in \mathbb{R} \right\},$$

82   *Symmetry and Quantum Mechanics*

and its representation on $\mathbb{C}^3$ by left-multiplication. Show that the only invariant subspaces are

$$\{\mathbf{0}\}, \quad \mathrm{span}\{\mathbf{e}_1\}, \quad \mathrm{span}\{\mathbf{e}_1, \mathbf{e}_2\}, \quad \mathbb{C}^3,$$

and conclude that this representation cannot be written as a direct sum of positive-dimensional representations.

Unlike the Heisenberg group, some groups[1] (including $SU(2)$) have the *complete reducibility property*, meaning that every finite-dimensional complex representation may be written as a direct sum of irreducible representations. For such groups, the problem of classifying their complex, finite-dimensional representations amounts to providing a complete description of the isomorphism classes of their irreducible complex matrix representations. This is our goal for $SU(2)$ in the remainder of this chapter.

Proposition 5.9. *The Lie group $SU(2)$ has the complete reducibility property: if V is a finite-dimensional complex representation of $SU(2)$, then V is isomorphic to a direct sum of irreducible representations.*

Proof. Choose a complex inner product, $\langle | \rangle$, on V. Then construct a new inner product by averaging over the group $SU(2) = S^3 \subset \mathbb{R}^4$:

$$\langle \mathbf{v} | \mathbf{w} \rangle' := \frac{1}{\mathrm{vol}(S^3)} \int_{SU(2)} \langle B \star \mathbf{v} | B \star \mathbf{w} \rangle dS^3,$$

where dS^3 denotes the measure on S^3 induced by Lebesgue measure on $\mathbb{R}^4$. This inner product is invariant under the action of $SU(2)$ on V:

$$\langle B \star \mathbf{v} | B \star \mathbf{w} \rangle' = \langle \mathbf{v} | \mathbf{w} \rangle' \qquad \text{for all } B \in SU(2).$$

Now suppose that $W_1 \subset V$ is a nonzero invariant and irreducible subspace, which must exist by the finite-dimensionality of V. If $W_1 = V$ then we are done, so suppose that $W_1 \neq V$, and consider the orthogonal complement $V_1 := W_1^\perp$. Then $V = W_1 \oplus V_1$, and V_1 is an invariant subspace: if $\mathbf{v} \in V_1$, then for all $B \in SU(2)$ and $\mathbf{w} \in W_1$ we have

$$\langle B \star \mathbf{v} | \mathbf{w} \rangle' = \langle B^{-1} \star (B \star \mathbf{v}) | B^{-1} \star \mathbf{w} \rangle' = \langle \mathbf{v} | B^{-1} \star \mathbf{w} \rangle' = 0,$$

since $B^{-1} \star \mathbf{w} \in W_1$. This shows that $B \star \mathbf{v} \in W_1^\perp = V_1$ as claimed. If V_1 is irreducible, then we are done. If not, then choose a nonzero invariant and irreducible subspace $W_2 \subset V_1$ and repeat the above argument to find the invariant subspace $V_2 = W_2^\perp$ satisfying $V = W_1 \oplus W_2 \oplus V_2$. By the finite-dimensionality of V, this procedure must eventually terminate with a direct sum decomposition of V into irreducible representations. $\qquad\square$

---

[1]Compactness is a sufficient but not necessary condition. A necessary and sufficient condition is that the Lie group be *semi-simple* [7, Appendix C.2].

Higher Spin 83

Corollary 5.10. *If V is a finite-dimensional complex representation of $SU(2)$, then there exists[2] a complex inner product on V that is $SU(2)$-invariant. The group $SU(2)$ acts via unitary transformations on the resulting complex inner product space.*

5.2 Representations of $SU(2)$

Consider the ring of polynomials in two variables over the complex numbers:

$$\mathbb{C}[w_1, w_2] = \left\{ p(w_1, w_2) = \sum_{j,k=0}^{N} c_{jk} w_1^j w_2^k \mid c_{jk} \in \mathbb{C} \right\}.$$

As a complex vector space, this ring is countably infinite dimensional, with a basis consisting of the monomials $w_1^j w_2^k$ for $j, k \geq 0$. The polynomials $p(w_1, w_2)$ may be viewed as functions on the space $\mathbb{C}^2$, and the defining representation of $SU(2)$ on $\mathbb{C}^2$ then yields a representation of $SU(2)$ on $\mathbb{C}[w_1, w_2]$ defined by composition:

$$B \star p := p \circ B^{-1}.$$

The appearance of the inverse is related to the fact, already noticed at the end of Section 2.3, that if $B \in SU(2)$ describes the automorphism that moves P's z-basis onto M's z'-basis for spinor space, then B^{-1} is the change of basis matrix that describes the translation from P's coordinates to M's. Moreover, as the following computation shows, the inverse is required in order that $\star$ defines an action of $SU(2)$:

$$
\begin{aligned}
(AB) \star p &= p \circ (AB)^{-1} \\
&= p \circ (B^{-1} A^{-1}) \\
&= (p \circ B^{-1}) \circ A^{-1} \\
&= A \star (p \circ B^{-1}) \\
&= A \star (B \star p).
\end{aligned}
$$

Example 5.11. *Let's compute the action of $SU(2)$ on the polynomial $w_1^2 + w_1 w_2$. Consider a general matrix $B \in SU(2)$:*

$$B = \begin{bmatrix} \alpha & -\beta^* \\ \beta & \alpha^* \end{bmatrix}.$$

---

[2]In Section 5.5 we will see that this inner product is unique up to scalar multiples if V is irreducible.

84   *Symmetry and Quantum Mechanics*

Then

$$
\begin{aligned}
B \star (w_1^2 + w_1 w_2) &= (w_1^2 + w_1 w_2) \circ B^{-1} \\
&= (w_1^2 + w_1 w_2) \circ \begin{bmatrix} \alpha^* & \beta^* \\ -\beta & \alpha \end{bmatrix} \\
&= (\alpha^* w_1 + \beta^* w_2)^2 + (\alpha^* w_1 + \beta^* w_2)(-\beta w_1 + \alpha w_2) \\
&= ((\alpha^*)^2 - \alpha^* \beta) w_1^2 + (\alpha^* \beta^* + |\alpha|^2 - |\beta|^2) w_1 w_2 \\
&\quad + ((\beta^*)^2 + \alpha\beta) w_2^2.
\end{aligned}
$$

Note that this polynomial is also homogeneous of degree 2, just like the original polynomial $w_1^2 + w_1 w_2$. As the next exercise shows, this preservation of degree is a general feature of the action.

Exercise 5.12. *Show that the $SU(2)$-action on $\mathbb{C}[w_1, w_2]$ preserves degrees, in the sense that if $p(w_1, w_2)$ is homogeneous of degree $m \geq 0$, then so is $B \star p$ for every $B \in SU(2)$.*

Since this action preserves the degree of polynomials, the representation is not irreducible, and in fact we have the following direct sum decomposition:

$$
\mathbb{C}[w_1, w_2] = \bigoplus_{m \geq 0} W_m,
$$

where $W_m :=$ {homogeneous polynomials of degree m}, an invariant subspace of dimension $m+1$ with basis $\{w_1^m, w_1^{m-1} w_2, \ldots, w_2^m\}$. Let's examine the first few of these finite-dimensional representations.

- For $m = 0$, we have the constant polynomials, $W_0 = \mathbb{C}$, on which every element of $SU(2)$ acts as the identity. This is the *trivial representation* of $SU(2)$, and is clearly irreducible.

- For $m = 1$, we have the linear polynomials, $W_1 \simeq \mathbb{C}^2$, where we have used the isomorphism $c_1 w_1 + c_2 w_2 \mapsto (c_1, c_2) \in \mathbb{C}^2$, thought of as a row-vector. Then the $SU(2)$ action on W_1 becomes $B \star (c_1, c_2) = (c_1, c_2) B^{-1}$. By inspection, this representation is irreducible. Moreover, it is actually isomorphic to the defining representation of $SU(2)$ on $\mathbb{C}^2$, defined by $(B, \mathbf{c}) \mapsto B\mathbf{c}$, where $\mathbf{c} \in \mathbb{C}^2$ are column-vectors. Indeed, define $\phi \colon \mathbb{C}^2 \to \mathbb{C}^2$ to be the following map from the space of row-vectors to the space of column-vectors:

$$
\phi(c_1, c_2) = \begin{bmatrix} -c_2 \\ c_1 \end{bmatrix}.
$$

To see that ϕ is an isomorphism of representations, consider any element B of SU(2):

$$
B = \begin{bmatrix} \alpha & -\beta^* \\ \beta & \alpha^* \end{bmatrix}.
$$

Then we compute:

$$
\begin{aligned}
\phi(B \star \phi^{-1}(\mathbf{c})) &= \phi((c_2, -c_1)B^{-1}) \\
&= \phi\left((c_2, -c_1)\begin{bmatrix} \alpha* & \beta^* \\ -\beta & \alpha \end{bmatrix}\right) \\
&= \phi(c_2\alpha^* + c_1\beta, c_2\beta^* - c_1\alpha) \\
&= \begin{bmatrix} c_1\alpha - c_2\beta^* \\ c_1\beta + c_2\alpha^* \end{bmatrix} \\
&= \begin{bmatrix} \alpha & -\beta^* \\ \beta & \alpha^* \end{bmatrix}\begin{bmatrix} c_1 \\ c_2 \end{bmatrix} \\
&= B\mathbf{c}.
\end{aligned}
$$

Note that we have made essential use of the fact that $B^{-1} = B^\dagger$ for $B \in SU(2)$.

- For $m = 2$ we have the quadratic polynomials, $W_2 \simeq \mathbb{C}^3$. In fact, this representation is isomorphic to the 3-dimensional representation of $SU(2)$ that we have already encountered: the complexified adjoint representation $\mathfrak{su}(2)_{\mathbb{C}}$. Recall that this is the conjugation-action of $SU(2)$ on the space $M_0(2, \mathbb{C})$ of 2×2 traceless complex matrices. To discover an isomorphism between the representations W_2 and $\mathfrak{su}(2)_{\mathbb{C}}$, first note that the monomials in W_2 are eigenvectors for the elements $B_\theta := \mathrm{diag}(e^{i\theta}, e^{-i\theta}) \in SU(2)$:

$$
\begin{aligned}
B_\theta \star w_1^2 &= e^{-2i\theta} w_1^2 \\
B_\theta \star w_1 w_2 &= w_1 w_2 \\
B_\theta \star w_2^2 &= e^{2i\theta} w_2^2.
\end{aligned}
$$

Hence, we should look for matrices in $M_0(2, \mathbb{C})$ that are eigenvectors for the conjugation-action of B_θ. A glance at example 2.35 from Section 2.4.2 and a little playing around with the Pauli matrices reveals that $\sigma_1 \pm i\sigma_2$ and σ_3 do the job:

$$
\begin{aligned}
B_\theta(\sigma_1 - i\sigma_2)B_\theta^{-1} &= e^{-2i\theta}(\sigma_1 - i\sigma_2) \\
B_\theta\sigma_3 B_\theta^{-1} &= \sigma_3 \\
B_\theta(\sigma_1 + i\sigma_2)B_\theta^{-1} &= e^{2i\theta}(\sigma_1 + i\sigma_2).
\end{aligned}
$$

Since $\{\sigma_1 \pm i\sigma_2, \sigma_3\}$ is a basis of $\mathfrak{su}(2)_{\mathbb{C}}$ over the complex numbers, we may define an isomorphism of vector spaces, $\phi\colon W_2 \to \mathfrak{su}(2)_{\mathbb{C}}$ by

$$
\phi(-w_1^2) = \sigma_1 - i\sigma_2, \quad \phi(w_1 w_2) = \sigma_3, \quad \phi(w_2^2) = \sigma_1 + i\sigma_2.
$$

We wish to show that ϕ is actually an isomorphism of $SU(2)$-representations. To that end, suppose that

$$
B = \begin{bmatrix} \alpha & -\beta^* \\ \beta & \alpha^* \end{bmatrix}, \qquad |\alpha|^2 + |\beta|^2 = 1
$$

86   *Symmetry and Quantum Mechanics*

is an arbitrary element of $SU(2)$. Then by explicit computation we see that

$$
\begin{aligned}
B\sigma_3 B^{-1} &= \begin{bmatrix} \alpha & -\beta^* \\ \beta & \alpha^* \end{bmatrix} \begin{bmatrix} 1 & 0 \\ 0 & -1 \end{bmatrix} \begin{bmatrix} \alpha^* & \beta^* \\ -\beta & \alpha \end{bmatrix} \\
&= \begin{bmatrix} \alpha & -\beta^* \\ \beta & \alpha^* \end{bmatrix} \begin{bmatrix} \alpha^* & \beta^* \\ \beta & -\alpha \end{bmatrix} \\
&= \begin{bmatrix} |\alpha|^2 - |\beta|^2 & 2\alpha\beta^* \\ 2\alpha^*\beta & |\beta|^2 - |\alpha|^2 \end{bmatrix} \\
&= \alpha^*\beta(\sigma_1 - i\sigma_2) + (|\alpha|^2 - |\beta|^2)\sigma_3 + \alpha\beta^*(\sigma_1 + i\sigma_2).
\end{aligned}
$$

On the other hand, we have

$$
\begin{aligned}
B \star w_1 w_2 &= w_1 w_2 \circ B^{-1} \\
&= (w_1\alpha^* + w_2\beta^*)(-w_1\beta + w_2\alpha) \\
&= -\alpha^*\beta w_1^2 + (|\alpha|^2 - |\beta|^2)w_1 w_2 + \alpha\beta^* w_2^2,
\end{aligned}
$$

which shows that $\phi(B \star \phi^{-1}(\sigma_3)) = B\sigma_3 B^{-1}$ as required. Similar computations for $\sigma_1 \pm i\sigma_2$ and w_j^2 establish that $\phi \colon W_2 \to \mathfrak{su}(2)_{\mathbb{C}}$ is an isomorphism of $SU(2)$ representations.

Exercise 5.13. *Finish the argument by showing that*

$$
\phi(B \star \phi^{-1}(\sigma_1 \pm i\sigma_2)) = B(\sigma_1 \pm i\sigma_2)B^{-1}.
$$

You should begin by showing that

$$
\sigma_1 - i\sigma_2 = \begin{bmatrix} 0 & 0 \\ 2 & 0 \end{bmatrix} \qquad and \qquad \sigma_1 + i\sigma_2 = \begin{bmatrix} 0 & 2 \\ 0 & 0 \end{bmatrix}.
$$

So, all of the representations of $SU(2)$ that we have seen in previous chapters occur at the beginning of the list of representations W_m. The remarkable fact (to be established in the remainder of this chapter) is that each W_m is irreducible, and together they account for all of the irreducible complex representations of $SU(2)$. Rather than showing directly that each W_m is irreducible, we will instead show the irreducibility of an associated representation of the Lie algebra $\mathfrak{su}(2)$. The main classification result is stated as theorem 5.27 toward the end of Section 5.4. But before we can prove that theorem (or even really understand its statement) we need some general information on representations of Lie algebras, and their relationship to representations of Lie groups.

5.3 Lie algebra representations

Definition 5.14. *Suppose that $\mathfrak{g}_1$ and $\mathfrak{g}_2$ are Lie algebras. Then a Lie algebra homomorphism is a linear map $\varphi\colon \mathfrak{g}_1 \to \mathfrak{g}_2$ that preserves the Lie brackets:*

$$\varphi([X,Y]) = [\varphi(X),\varphi(Y)] \quad \text{for all } X, Y \in \mathfrak{g}_1.$$

Recall from exercise 3.25 that $\mathfrak{gl}_n(\mathbb{C}) = M(n,\mathbb{C})$ is the vector space of $n \times n$ complex matrices, with the commutator as Lie bracket. As an analogue of definition 5.3, we have

Definition 5.15. *An n-dimensional complex matrix representation of a (real or complex) Lie algebra, $\mathfrak{g}$, is a Lie algebra homomorphism $\varphi\colon \mathfrak{g} \to \mathfrak{gl}_n(\mathbb{C})$. (Note that if $\mathfrak{g}$ is a real Lie algebra, then we only require φ to be $\mathbb{R}$-linear.)*

The key point of this definition is that every complex matrix representation of a Lie group determines a complex matrix representation of its Lie algebra via differentiation. Although the result is true in general, we will prove it only for matrix Lie groups (i.e., closed subgroups of some $GL(n,\mathbb{C})$), where the Lie bracket is the commutator (see comments at the end of Section 3.4.4).

Proposition 5.16. *Suppose that $\rho\colon G \to GL(n,\mathbb{C})$ is a matrix representation of a matrix Lie group G. Then the derivative of ρ at the identity[3] of G is a matrix representation of the corresponding real Lie algebra, $\mathfrak{g}$:*

$$D\rho\colon \mathfrak{g} \to \mathfrak{gl}_n(\mathbb{C}).$$

Proof. The fact that $D\rho$ is $\mathbb{R}$-linear is immediate from the definition of the derivative as a linear map (see Appendix A.2), so we just need to check that it preserves the Lie brackets. So suppose that $X, Y \in \mathfrak{g}$ are elements of the Lie algebra, and choose parametrized curves $c_X, c_Y\colon (-\epsilon, \epsilon) \to G$ satisfying $c_X(0) = c_Y(0) = I$ and $\dot{c}_X(0) = X, \dot{c}_Y(0) = Y$. Recall (see proposition 3.4) that the Lie bracket on $\mathfrak{g}$ is obtained by differentiating the conjugation action of G on $\mathfrak{g}$:

$$[X,Y] = \frac{d}{dt}\left(c_X(t)Y c_X(t)^{-1}\right)|_{t=0}.$$

So we must compute

$$\begin{aligned}
D\rho([X,Y]) &= D\rho\left(\frac{d}{dt}\left(c_X(t)Y c_X(t)^{-1}\right)|_{t=0}\right) \\
&= \frac{d}{dt} D\rho\left(c_X(t)Y c_X(t)^{-1}\right)|_{t=0},
\end{aligned}$$

---

[3]As mentioned earlier, we will generally write $D\rho$ instead of the more cumbersome $(D\rho)_I$ as in Appendix A.2, trusting the reader to understand that we are working with the derivative of ρ at the identity of the Lie group G.

Symmetry and Quantum Mechanics

where in the second step we have used the fact that the linear transformation $D\rho$ is continuous, and hence commutes with differentiation. Now fix t, and set $B = c_X(t)$. Then we have (using the chain rule in the form of proposition A.38 in the third and in the final line):

$$
\begin{aligned}
D\rho\left(BYB^{-1}\right) &= D\rho\left(B\dot{c}_Y(0)B^{-1}\right) \\
&= D\rho\left(\frac{d}{ds}\left(Bc_Y(s)B^{-1}\right)|_{s=0}\right) \\
&= \frac{d}{ds}\rho\left(Bc_Y(s)B^{-1}\right)|_{s=0} \\
&= \frac{d}{ds}\rho\left(B\right)\rho\left(c_Y(s)\right)\rho(B)^{-1}|_{s=0} \\
&= \rho\left(B\right)\frac{d}{ds}\rho\left(c_Y(s)\right)|_{s=0}\rho(B)^{-1} \\
&= \rho\left(B\right)D\rho\left(Y\right)\rho(B)^{-1}.
\end{aligned}
$$

Returning to our previous computation and using the chain and Leibniz rules, we get

$$
\begin{aligned}
D\rho([X,Y]) &= \frac{d}{dt}\rho\left(c_X(t)\right)D\rho\left(Y\right)\rho\left(c_X(t)\right)^{-1}|_{t=0} \\
&= D\rho(\dot{c}_X(0))D\rho(Y)\rho(c_X(0))^{-1} \\
&\qquad -\rho(c_X(0))D\rho(Y)D\rho(\dot{c}_X(0)) \\
&= D\rho(X)D\rho(Y) - D\rho(Y)D\rho(X) \\
&= [D\rho(X), D\rho(Y)].
\end{aligned}
$$

$\square$

Note that if $\varphi: \mathfrak{g} \to \mathfrak{gl}_n(\mathbb{C})$ is a matrix representation of a Lie algebra, then $\mathfrak{g}$ acts on $\mathbb{C}^n$ as a vector space of linear transformations via

$$
\begin{aligned}
\mathfrak{g} \times \mathbb{C}^n &\to \mathbb{C}^n \\
(X, \mathbf{c}) &\mapsto \varphi(X)\mathbf{c}.
\end{aligned}
$$

Just as for Lie groups, we may form the direct sum of Lie algebra representations, and we also have a notion of irreducibility.

Definition 5.17. *Suppose that $\varphi: \mathfrak{g} \to \mathfrak{gl}_n(\mathbb{C})$ is a Lie algebra representation, with associated action $\mathfrak{g} \times \mathbb{C}^n \to \mathbb{C}^n$. A vector subspace $W \subset \mathbb{C}^n$ is invariant if $\varphi(X)\mathbf{w} \in W$ for all $X \in \mathfrak{g}$ and $\mathbf{w} \in W$. The representation is irreducible if its only invariant subspaces are $\{\mathbf{0}\}$ and $\mathbb{C}^n$.*

Proposition 5.18. *Suppose that $\rho: G \to GL(n, \mathbb{C})$ is a matrix representation of a matrix Lie group G. Furthermore, suppose that the corresponding Lie algebra representation $D\rho: \mathfrak{g} \to \mathfrak{gl}_n(\mathbb{C})$ is irreducible. Then ρ is an irreducible representation of G.*

Higher Spin 89

Proof. Suppose that $W \subset \mathbb{C}^n$ is a non-zero invariant subspace for the representation ρ. We wish to show that $W = \mathbb{C}^n$. Choose a basis for W, which yields an identification with $\mathbb{C}^m$ for some $m \leq n$. Since W is invariant, restriction to W yields a differentiable homomorphism $\rho|_W \colon G \to GL(m, \mathbb{C})$. The derivative of this restriction defines a Lie algebra homomorphism $D(\rho|_W) \colon \mathfrak{g} \to \mathfrak{gl}_m(\mathbb{C})$, which defines an action of $\mathfrak{g}$ on $W \subset \mathbb{C}^n$. But this action of $\mathfrak{g}$ on W is simply the restriction to W of the action of $\mathfrak{g}$ on $\mathbb{C}^n$ defined by $D\rho$. This shows that W is a non-zero invariant subspace for the action of $\mathfrak{g}$ on $\mathbb{C}^n$. But the representation $D\rho$ is irreducible by assumption, so $W = \mathbb{C}^n$ as desired. $\qquad\square$

Hence, in order to show that a representation of a Lie group G is irreducible, it suffices to show the irreducibility of the associated representation of its Lie algebra, $\mathfrak{g}$. Moreover, since we are dealing with complex representations, the following exercise shows that it suffices to investigate the complexified Lie algebra, $\mathfrak{g}_{\mathbb{C}}$.

Exercise 5.19 (♣). *Suppose that $\varphi \colon \mathfrak{g} \to \mathfrak{gl}_n(\mathbb{C})$ is a representation of a real Lie algebra, $\mathfrak{g}$. Consider the complexified map $\varphi_{\mathbb{C}} \colon \mathfrak{g}_{\mathbb{C}} \to \mathfrak{gl}_n(\mathbb{C})$ defined by*

$$\varphi_{\mathbb{C}}(X + iY) := \varphi(X) + i\varphi(Y) \qquad \text{for all } X, Y \in \mathfrak{g}.$$

Show that $\varphi_{\mathbb{C}}$ is a representation of the complex Lie algebra $\mathfrak{g}_{\mathbb{C}}$, and that every $\mathbb{C}$-linear representation of $\mathfrak{g}_{\mathbb{C}}$ arises in this fashion from an $\mathbb{R}$-linear representation of $\mathfrak{g}$. Moreover, show that $W \subset \mathbb{C}^n$ is an invariant subspace for φ if and only if W is an invariant subspace for $\varphi_{\mathbb{C}}$. Conclude that the complex irreducible $\mathbb{R}$-linear representations of $\mathfrak{g}$ are in bijection with the complex irreducible $\mathbb{C}$-linear representations of $\mathfrak{g}_{\mathbb{C}}$.

Thus, the irreducibility of the $SU(2)$-representations W_m described in Section 5.2 will follow from their irreducibility as representations of the complexified Lie algebra $\mathfrak{su}(2)_{\mathbb{C}} = \mathfrak{sl}_2(\mathbb{C})$. Hence, we are led to the strategy illustrated in the diagram below for classifying the complex irreducible representations of $SU(2)$ by instead classifying the complex irreducible representations of $\mathfrak{sl}_2(\mathbb{C})$:

$$\{\text{Irreps of } SU(2)\} \xrightarrow{\quad D \quad} \{\text{Irreps of } \mathfrak{su}(2)\} \xrightarrow{\quad \mathbb{C} \quad} \{\text{Irreps of } \mathfrak{sl}_2(\mathbb{C})\}.$$

The second arrow is a bijection by exercise 5.19 above. The existence of the first arrow will follow from the converse of proposition 5.18 in the case of $SU(2)$, which we will prove as proposition 5.22 below. We will establish the injectivity of the first arrow in proposition 5.23, and its surjectivity in theorem 5.25.

The first step will be a further study of the matrix exponential, introduced in Section 2.4.3. Suppose that $\rho \colon G \to GL(n, \mathbb{C})$ is a representation of a matrix Lie group, with corresponding Lie algebra representation $D\rho \colon \mathfrak{g} \to \mathfrak{gl}_n(\mathbb{C})$.

90 *Symmetry and Quantum Mechanics*

Then consider the following diagram:

$$
\begin{array}{ccc}
G & \xrightarrow{\ \rho\ } & GL(n,\mathbb{C}) \\
{\scriptstyle\exp}\uparrow & & {\scriptstyle\exp}\uparrow \\
\mathfrak{g} & \xrightarrow[D\rho]{} & \mathfrak{gl}_n(\mathbb{C}).
\end{array}
\tag{5.1}
$$

Here, the vertical arrows are the matrix exponential, defined by the infinite series

$$
\exp(M) := \sum_{j=0}^{\infty} \frac{1}{j!} M^j.
$$

Our aim is to prove that the diagram (5.1) commutes (proposition 5.20 below). To do so, we will need a more abstract characterization of the exponential. Recall from proposition 2.21e) that for any $M \in M(n,\mathbb{C})$, the mapping $c(t) := \exp(tM)$ defines a homomorphism $c\colon \mathbb{R} \to GL(n,\mathbb{C})$ with the property that $\dot{c}(0) = M$. In fact, c satisfies the differential equation $\dot{c}(t) = Mc(t)$, with the initial condition $c(0) = I$. Now suppose that $b\colon \mathbb{R} \to GL(n,\mathbb{C})$ is another differentiable group homomorphism satisfying $\dot{b}(0) = M$. Then fixing $t \in \mathbb{R}$ and letting s vary yields:

$$
\dot{b}(t) = \frac{d}{ds}b(s+t)|_{s=0} = \frac{d}{ds}b(s)b(t)|_{s=0} = \dot{b}(0)b(t) = Mb(t),
$$

so that b satisfies the same differential equation as c, with the same initial condition $b(0) = I$. By the Existence and Uniqueness Theorem [13, theorem 6.2.3], $b(t) = c(t) = \exp(tM)$ for all t. Hence, rather than thinking of $\exp(M)$ in terms of its defining infinite series, we may instead think of it as the value at $t = 1$ of the unique group homomorphism $\mathbb{R} \to GL(n,\mathbb{C})$ with tangent vector M at $t = 0$.

Proposition 5.20. *The diagram (5.1) commutes. That is, for all $X \in \mathfrak{g}$, we have*

$$
\rho(\exp(X)) = \exp(D\rho(X)).
$$

Proof. Let $X \in \mathfrak{g}$ be arbitrary, and consider the mapping $b\colon \mathbb{R} \to GL(n,\mathbb{C})$ defined by $b(t) = \rho(\exp(tX))$. Then b is a differentiable group homomorphism satisfying $\dot{b}(0) = D\rho(X)$ by proposition A.38. By the comments above, it follows that $b(t) = \exp(tD\rho(X))$ for all $t \in \mathbb{R}$. Evaluating at $t = 1$ yields $\rho(\exp(X)) = b(1) = \exp(D\rho(X))$. $\qquad\square$

We now specialize to the case of the Lie group $SU(2)$, although a weaker form of the following result actually holds for all connected Lie groups (see [11, corollary 3.47]).

Proposition 5.21. *The exponential map, $\exp\colon \mathfrak{su}(2) \to SU(2)$, is surjective.*

Higher Spin 91

Proof. By the Spectral Theorem for Normal Operators A.33, every element $B \in SU(2)$ may be diagonalized by a unitary matrix $U \in U(2)$. Thus we have $U^{-1}BU = \text{diag}(\alpha, \beta)$, where the eigenvalues α, β have modulus 1. Taking the determinant yields $\alpha\beta = \det(U^{-1}BU) = \det(B) = 1$, so $\alpha = \beta^{-1} = e^{i\theta}$ for some $\theta \in \mathbb{R}$. Hence, $U^{-1}BU = \text{diag}(e^{i\theta}, e^{-i\theta}) = \exp(i\theta\sigma_3)$. It follows that $B = U\exp(i\theta\sigma_3)U^{-1} = \exp(Ui\theta\sigma_3 U^{-1})$. But $Ui\theta\sigma_3 U^{-1} \in \mathfrak{su}(2)$. $\square$

As promised, we can now prove the converse of proposition 5.18 in the case of the Lie group $SU(2)$.

Proposition 5.22. *Suppose that $\rho\colon SU(2) \to GL(n, \mathbb{C})$ is an irreducible representation. Then the representation $D\rho\colon \mathfrak{su}(2) \to \mathfrak{gl}_n(\mathbb{C})$ is also irreducible.*

Proof. Suppose that $W \subset \mathbb{C}^n$ is a non-zero $\mathfrak{su}(2)$-invariant subspace. We wish to show that $W = \mathbb{C}^n$. Since ρ is irreducible, it will suffice to show that W is $SU(2)$-invariant. So let $B \in SU(2)$ and $\mathbf{w} \in W$ be arbitrary. By proposition 5.21, we may choose an element $iX \in \mathfrak{su}(2)$ such that $\exp(iX) = B$. Then using proposition 5.20 we have

$$\begin{aligned} \rho(B)\mathbf{w} &= \rho(\exp(iX))\mathbf{w} \\ &= \exp(D\rho(iX))\mathbf{w} \\ &= \sum_{j=0}^{\infty} \frac{1}{j!}(D\rho(iX))^j\mathbf{w} \in W, \end{aligned}$$

since W is a closed subspace of $\mathbb{C}^n$ and each $D\rho(iX)^j\mathbf{w}$ is an element of W by $\mathfrak{su}(2)$-invariance. $\square$

Proposition 5.23. *Suppose that $\xi_1, \xi_2\colon SU(2) \to GL(n, \mathbb{C})$ are representations inducing isomorphic Lie algebra representations $D\xi_1$ and $D\xi_2$. Then ξ_1 and ξ_2 are isomorphic as representations of $SU(2)$.*

Proof. Since $D\xi_1$ and $D\xi_2$ are isomorphic as representations of $\mathfrak{su}(2)$, there exists an invertible matrix $M \in GL(n, \mathbb{C})$ such that for all $iX \in \mathfrak{su}(2)$, we have the following equality of matrices in $\mathfrak{gl}_n(\mathbb{C})$:

$$D\xi_1(iX) = M^{-1}D\xi_2(iX)M.$$

Now let $B \in SU(2)$ be arbitrary, and choose $iX \in \mathfrak{su}(2)$ such that $\exp(iX) = B$. Then by proposition 5.20,

$$\begin{aligned} \xi_1(B) &= \xi_1(\exp(iX)) \\ &= \exp(D\xi_1(iX)) \\ &= \exp(M^{-1}D\xi_2(iX)M) \\ &= M^{-1}\exp(D\xi_2(iX))M \\ &= M^{-1}\xi_2(B)M. \end{aligned}$$

This shows that ξ_1 and ξ_2 are isomorphic $SU(2)$-representations. $\square$

92   *Symmetry and Quantum Mechanics*

Hence, an $SU(2)$-representation is determined by the corresponding Lie algebra representation. Combining this result with proposition 5.22, we have shown that the map from irreducible representations of $SU(2)$ to irreducible representations of $\mathfrak{su}(2)$ is an injection:

$$\{\text{Irreps of } SU(2)\} \xrightarrow{\;\;D\;\;} \{\text{Irreps of } \mathfrak{su}(2)\}.$$

To establish surjectivity, we will explicitly determine (in theorem 5.25) all complex irreducible $\mathfrak{su}(2)$-representations up to isomorphism, showing that they all arise from the $SU(2)$-representations W_m introduced in Section 5.2.

5.4   Representations of $\mathfrak{su}(2)_{\mathbb{C}} = \mathfrak{sl}_2(\mathbb{C})$

We begin by describing the representations of $\mathfrak{su}(2)$ associated to the $SU(2)$-representations W_m. It will suffice to determine the action of $-i\sigma_1, -i\sigma_2, -i\sigma_3$ on the monomial basis of W_m. Recall from exercise 3.9 that these matrices are tangent vectors at the identity of the following curves in $SU(2)$:

$$c_1(\theta) = \exp(-i\theta\sigma_1) \;=\; \exp\begin{bmatrix} 0 & -i\theta \\ -i\theta & 0 \end{bmatrix} = \begin{bmatrix} \cos(\theta) & -i\sin(\theta) \\ -i\sin(\theta) & \cos(\theta) \end{bmatrix}$$

$$c_2(\theta) = \exp(-i\theta\sigma_2) \;=\; \exp\begin{bmatrix} 0 & -\theta \\ \theta & 0 \end{bmatrix} = \begin{bmatrix} \cos(\theta) & -\sin(\theta) \\ \sin(\theta) & \cos(\theta) \end{bmatrix}$$

$$c_3(\theta) = \exp(-i\theta\sigma_3) \;=\; \exp\begin{bmatrix} -i\theta & 0 \\ 0 & i\theta \end{bmatrix} = \begin{bmatrix} e^{-i\theta} & 0 \\ 0 & e^{i\theta} \end{bmatrix}.$$

To determine the action of $-i\sigma_3$ on W_m, we must compute the derivative of the action of c_3 on W_m under the representation $\rho_m \colon SU(2) \to GL(m+1, \mathbb{C})$:

$$\begin{aligned}
D\rho_m(-i\sigma_3)(w_1^{m-k}w_2^k) &= \frac{d}{d\theta}(\rho_m(c_3(\theta))(w_1^{m-k}w_2^k))|_{\theta=0} \\
&= \frac{d}{d\theta}(e^{i(m-k)\theta}w_1^{m-k}e^{-ik\theta}w_2^k)|_{\theta=0} \\
&= \frac{d}{d\theta}(e^{i(m-2k)\theta}w_1^{m-k}w_2^k)|_{\theta=0} \\
&= i(m-2k)w_1^{m-k}w_2^k \\
&= i\left(w_1\frac{\partial}{\partial w_1} - w_2\frac{\partial}{\partial w_2}\right)w_1^{m-k}w_2^k.
\end{aligned}$$

Similarly, to determine the action of $-i\sigma_2$ on W_m, we compute (interpret-

Higher Spin

ing terms involving a negative exponent as zero)

$$
\begin{aligned}
D\rho_m(-i\sigma_2)(w_1^{m-k}w_2^k) &= \frac{d}{d\theta}(\rho_m(c_2(\theta))(w_1^{m-k}w_2^k))|_{\theta=0} \\
&= \frac{d}{d\theta}\left((\cos(\theta)w_1 + \sin(\theta)w_2)^{m-k}\cdot \right. \\
&\qquad \left. (-\sin(\theta)w_1 + \cos(\theta)w_2)^k\right)|_{\theta=0} \\
&= (m-k)w_1^{m-k-1}w_2^{k+1} - kw_1^{m-k+1}w_2^{k-1} \\
&= \left(w_2\frac{\partial}{\partial w_1} - w_1\frac{\partial}{\partial w_2}\right)w_1^{m-k}w_2^k.
\end{aligned}
$$

Exercise 5.24. *Show by a similar computation that*

$$
D\rho_m(-i\sigma_1)(w_1^{m-k}w_2^k) = i\left(w_2\frac{\partial}{\partial w_1} + w_1\frac{\partial}{\partial w_2}\right)w_1^{m-k}w_2^k.
$$

Thus, we see that the $\mathfrak{su}(2)$-representation on W_m may be described in terms of differential operators. Complexifying, we obtain a description of the corresponding representation of $\mathfrak{sl}_2(\mathbb{C})$ on W_m, which we may describe in terms of the basis $\{\sigma_1 \pm i\sigma_2, \sigma_3\}$ as follows:

$$
\begin{aligned}
\sigma_1 + i\sigma_2 \quad &\text{acts as} \quad -2w_2\frac{\partial}{\partial w_1} \\[6pt]
\sigma_1 - i\sigma_2 \quad &\text{acts as} \quad -2w_1\frac{\partial}{\partial w_2} \\[6pt]
\sigma_3 \quad &\text{acts as} \quad w_2\frac{\partial}{\partial w_2} - w_1\frac{\partial}{\partial w_1}.
\end{aligned}
$$

Note in particular that the monomial $w_1^{m-k}w_2^k$ is an eigenvector for the action of σ_3 with eigenvalue $2k - m$. Hence, σ_3 acts on W_m as a diagonalizable linear transformation with one-dimensional eigenspaces labeled by the $m + 1$ eigenvalues

$$
-m\,,-m+2\,,\dots\,,m-2\,,m.
$$

Moreover, the element $\sigma_1 + i\sigma_2$ acts as a "raising operator" that sends the eigenspace for λ to the eigenspace for $\lambda + 2$, annihilating the eigenspace for m:

$$
-2w_2\frac{\partial}{\partial w_1}(w_1^{m-k}w_2^k) = -2(m-k)w_1^{m-k-1}w_2^{k+1}.
$$

Similarly, the element $\sigma_1 - i\sigma_2$ acts as a "lowering operator" that sends the eigenspace for λ to the eigenspace for $\lambda - 2$, annihilating the eigenspace for $-m$:

$$
-2w_1\frac{\partial}{\partial w_2}(w_1^{m-k}w_2^k) = -2kw_1^{m-k+1}w_2^{k-1}.
$$

These observations allow us to conclude that each W_m is irreducible as a

94   *Symmetry and Quantum Mechanics*

representation of $\mathfrak{sl}_2(\mathbb{C})$. Indeed, suppose that $W \subset W_m$ is a nonzero $\mathfrak{sl}_2(\mathbb{C})$-invariant subspace. In particular, W is invariant under the action of σ_3, and since we are working over the complex numbers, W must contain an eigenvector for the action of σ_3. It follows that W contains one of the eigenspaces described above. Repeated application of the raising and lowering operators $\sigma_1 \pm i\sigma_2$ then shows that W must contain all $m+1$ eigenspaces, which implies that $W = W_m$. From exercise 5.19, we see that each W_m is also irreducible as a representation of $\mathfrak{su}(2)$. By Proposition 5.18, W_m is irreducible as a representation of the Lie group $SU(2)$. In the next theorem, we show that these account for all of the irreducible representations of $\mathfrak{sl}_2(\mathbb{C})$.

Theorem 5.25. *Every finite-dimensional complex irreducible $\mathbb{C}$-linear representation of $\mathfrak{sl}_2(\mathbb{C})$ is isomorphic to W_m for some $m \geq 0$.*

Proof. Suppose that $\varphi \colon \mathfrak{sl}_2(\mathbb{C}) \to \mathfrak{gl}_n(\mathbb{C})$ is an irreducible representation. To simplify the notation, set

$$X := \varphi(\sigma_1 + i\sigma_2), \qquad Y := \varphi(\sigma_1 - i\sigma_2), \qquad Z := \varphi(\sigma_3).$$

Then X, Y, Z are linear operators on $\mathbb{C}^n$ satisfying the same commutation relations as $\sigma_1 \pm i\sigma_2, \sigma_3$:

$$[X, Y] = 4Z, \qquad [Z, X] = 2X, \qquad [Z, Y] = -2Y.$$

Let $\lambda \in \mathbb{C}$ be an eigenvalue for Z, with eigenvector $\mathbf{v} \in \mathbb{C}^n$. Then compute

$$ZX\mathbf{v} = XZ\mathbf{v} + [Z, X]\mathbf{v} = X(\lambda\mathbf{v}) + 2X\mathbf{v} = (\lambda + 2)X\mathbf{v},$$

which demonstrates that $X\mathbf{v}$ (if non-zero) is an eigenvector for Z with eigenvalue $\lambda + 2$. Since Z has only finitely many eigenvalues, it follows that there exists a least integer, $j \geq 0$, such that $X^{j+1}\mathbf{v} = 0$. Set $\mathbf{w} = X^j\mathbf{v}$, which is an eigenvector for Z with eigenvalue $\omega = \lambda + 2j$, satisfying $X\mathbf{w} = 0$.

Now make a similar computation for the operator Y:

$$ZY\mathbf{w} = YZ\mathbf{w} + [Z, Y]\mathbf{w} = Y(\omega\mathbf{w}) - 2Y\mathbf{w} = (\omega - 2)Y\mathbf{w},$$

showing that $Y\mathbf{w}$ (if non-zero) is an eigenvector for Z with eigenvalue $\omega - 2$. Again by finite-dimensionality, there exists a least integer, $m \geq 0$, such that $Y^{m+1}\mathbf{w} = 0$. I claim that the subspace, W, spanned by $\mathbf{w}, Y\mathbf{w}, \ldots, Y^m\mathbf{w}$ is invariant, hence $W = \mathbb{C}^n$ by irreducibility.

It is clear that W is invariant under Y, and since each $Y^k\mathbf{w}$ is an eigenvector for Z with eigenvalue $\omega - 2k$, it follows that W is also invariant under Z. For invariance under X, first note that $X\mathbf{w} = 0$. We will show by induction that for $k = 1, \ldots, m+1$,

$$XY^k\mathbf{w} = 4k(\omega - k + 1)Y^{k-1}\mathbf{w}.$$

For $k = 1$ we have

$$XY\mathbf{w} = YX\mathbf{w} + [X, Y]\mathbf{w} = 4Z\mathbf{w} = 4\omega\mathbf{w}.$$

Higher Spin 95

So suppose the claim holds for some $1 \le k \le m$, and compute

$$
\begin{aligned}
XY^{k+1}\mathbf{w} &= Y(XY^k\mathbf{w}) + [X,Y]Y^k\mathbf{w} \\
&= Y(XY^k\mathbf{w}) + 4ZY^k\mathbf{w} \\
&= Y(4k(\omega - k + 1)Y^{k-1}\mathbf{w}) + 4(\omega - 2k)Y^k\mathbf{w} \\
&= 4((k+1)\omega - k(k-1) - 2k)Y^k\mathbf{w} \\
&= 4((k+1)\omega - k(k+1))Y^k\mathbf{w} \\
&= 4(k+1)(\omega - k)Y^k\mathbf{w}.
\end{aligned}
$$

This establishes the claim. Moreover, taking $k = m$ and remembering that $Y^{m+1}\mathbf{w} = 0$ yields

$$
0 = XY^{m+1}\mathbf{w} = 4(m+1)(\omega - m)Y^m\mathbf{w}.
$$

Since $Y^m\mathbf{w} \ne 0$, it follows that we must have $\omega = m$.

Thus, X maps each $Y^k\mathbf{w}$ into a multiple of the previous vector $Y^{k-1}\mathbf{w}$, which shows that W is $\mathfrak{sl}_2(\mathbb{C})$-invariant, so $W = \mathbb{C}^n$ as claimed. Moreover, since the $Y^k\mathbf{w}$ are eigenvectors for Z with distinct eigenvalues, it follows that $n = m + 1$, each eigenspace for Z is one-dimensional, and the $Y^k\mathbf{w}$ form a basis for $\mathbb{C}^{m+1}$. Now define an isomorphism of vector spaces $\phi \colon \mathbb{C}^{m+1} \to W_m$ by

$$
\phi(Y^k\mathbf{w}) := \frac{(-1)^k 2^k m!}{(m-k)!} w_1^k w_2^{m-k}.
$$

By the next exercise, ϕ is an isomorphism of representations. $\qquad\square$

Exercise 5.26. *Check that $\phi \colon \mathbb{C}^{m+1} \to W_m$ is an isomorphism of $\mathfrak{sl}_2(\mathbb{C})$-representations.*

Let's take stock of what we have shown: for every $m \ge 0$, we have constructed an $SU(2)$-representation ρ_m on the vector space W_m of complex dimension $m + 1$. Each of these representations induces a representation $D\rho_m$ of the Lie algebra $\mathfrak{su}(2)$. These $\mathfrak{su}(2)$-representations are irreducible, since the corresponding representations of $\mathfrak{sl}_2(\mathbb{C})$ obtained by complexification are irreducible (exercise 5.19). It then follows from proposition 5.18 that the $SU(2)$-representations ρ_m are irreducible. Moreover, by theorem 5.25 (and exercise 5.19 again), we know that the $D\rho_m$ account for all of the finite-dimensional irreducible representations of $\mathfrak{su}(2)$. Propositions 5.22 and 5.23 then imply that the ρ_m account for all of the finite-dimensional irreducible representations of $SU(2)$.

Hence, we have shown that there is a bijection between irreducible representations of the Lie group $SU(2)$ and irreducible representations of its Lie algebra $\mathfrak{su}(2)$, and we have been able to describe them explicitly. We record this fact as a theorem:

Theorem 5.27. *The complex irreducible representations of the Lie group*

96 *Symmetry and Quantum Mechanics*

$SU(2)$ *are in bijection with the complex irreducible representations of its Lie algebra* $\mathfrak{su}(2)$. *Moreover, for each integer* $m \geq 1$, *there exists a unique complex irreducible representation of dimension* $m + 1$ *on the vector space of homogeneous polynomials of degree* m *in two variables:*

$$W_m = \mathrm{span}_{\mathbb{C}}\{w_1^m, w_1^{m-1}w_2, \ldots, w_2^m\}.$$

The group $SU(2)$ *acts on* W_m *via composition:* $B \star p = p \circ B^{-1}$ *for* $B \in SU(2)$ *and* $p(w_1, w_2) \in W_m$. *The corresponding action of the Lie algebra* $\mathfrak{su}(2)$ *is given by differential operators:*

$$-i\sigma_1 \star p(w_1, w_2) = i\left(w_2 \frac{\partial p}{\partial w_1} + w_1 \frac{\partial p}{\partial w_2}\right)$$

$$-i\sigma_2 \star p(w_1, w_2) = w_2 \frac{\partial p}{\partial w_1} - w_1 \frac{\partial p}{\partial w_2}$$

$$-i\sigma_3 \star p(w_1, w_2) = i\left(w_1 \frac{\partial p}{\partial w_1} - w_2 \frac{\partial p}{\partial w_2}\right).$$

This bijection is actually a reflection of some topological features of the group $SU(2)$. In particular, the fact that every irreducible Lie algebra representation actually arises from a group representation depends on the fact that $SU(2)$ is simply connected, and is not true in general. In particular, consider the non-simply connected Lie group $SO(3)$. As we will show below in proposition 5.29, the double cover $f \colon SU(2) \to SO(3)$ induces an isomorphism of Lie algebras $Df \colon \mathfrak{su}(2) \to \mathfrak{so}(3)$, so the W_m are exactly the irreducible Lie algebra representations of $\mathfrak{so}(3)$. However, as we now show, only half of these occur as representations of the Lie group $SO(3)$.

Proposition 5.28. *The matrix Lie group* $SO(3)$ *has the complete reducibility property, and its complex irreducible representations are exactly the representations* W_{2d} *of odd-dimension.*

Proof. Suppose that $\rho \colon SO(3) \to GL(n, \mathbb{C})$ is a complex representation of $SO(3)$. Then $\rho \circ f \colon SU(2) \to GL(n, \mathbb{C})$ is a representation of $SU(2)$ for which $-I$ acts trivially. Moreover, any such representation of $SU(2)$ yields a representation of $SO(3)$. It follows that $SO(3)$ also has the complete reducibility property, and that the irreducible representations of $SO(3)$ are the representations W_m for which $-I$ acts trivially. But

$$(-I) \star (w_1^{m-k}w_2^k) = (-w_1)^{m-k}(-w_2)^k = (-1)^m w_1^{m-k}w_2^k,$$

so $-I$ acts trivially if and only if m is even. Thus, only the *odd-dimensional* representations W_{2d} occur as representations of $SO(3)$, and these account for all of the irreducible $SO(3)$-representations. $\qquad\square$

Proposition 5.29. *The Lie algebra* $\mathfrak{so}(3)$ *consists of* 3×3 *real skew-symmetric matrices, and* $Df \colon \mathfrak{su}(2) \to \mathfrak{so}(3)$ *is an isomorphism of Lie algebras.*

Higher Spin

Proof. Identify the space of 3×3 real matrices with $\mathbb{R}^9$ via

$$(x_1, x_2, \ldots, x_9) \longleftrightarrow \begin{bmatrix} x_1 & x_2 & x_3 \\ x_4 & x_5 & x_6 \\ x_7 & x_8 & x_9 \end{bmatrix}.$$

To determine the tangent space to $SO(3)$ at the identity, consider a smooth curve $c: (-\epsilon, \epsilon) \to SO(3) \subset \mathbb{R}^9$ such that $c(0) = I$. Since $c(t) \in SO(3)$ for all t, it satisfies the equation $c(t)c(t)^T = I$. Differentiating this condition at $t = 0$ yields

$$0 = \dot{c}(0)c(0) + c(0)\dot{c}(0)^T = \dot{c}(0) + \dot{c}(0)^T,$$

which implies that $\dot{c}(0)^T = -\dot{c}(0)$ is a skew-symmetric matrix. Thus, tangent vectors to curves passing through the identity of $SO(3)$ are skew-symmetric matrices. Conversely, if S is any 3×3 skew-symmetric real matrix, then

$$\exp(tS)\exp(tS)^T = \exp(tS)\exp(tS^T) = \exp(tS)\exp(-tS) = I,$$

so $\exp(tS) \in SO(3)$ for all t. Taking the derivative at $t = 0$ displays S as an element of the tangent space $T_I SO(3)$:

$$\frac{d}{dt}(\exp(tS))|_{t=0} = S.$$

Thus, the Lie algebra $\mathfrak{so}(3)$ consists of the vector space of real 3×3 skew-symmetric matrices, with the commutator as Lie bracket.

Now consider the universal double cover $f: SU(2) \to SO(3)$. We will show that the derivative $Df: \mathfrak{su}(2) \to \mathfrak{so}(3)$ is an isomorphism by showing that Df sends the basis $\frac{1}{2i}\sigma_j$ for $\mathfrak{su}(2)$ to a basis for $\mathfrak{so}(3)$. To determine the image of $\frac{1}{2i}\sigma_3$, we compute (compare Section 3.2):

$$\begin{aligned} Df\left(\frac{1}{2i}\sigma_3\right) &= \frac{d}{d\theta}f\left(\exp\left(\frac{\theta}{2i}\sigma_3\right)\right)|_{\theta=0} \\ &= \frac{d}{d\theta}\begin{bmatrix} \cos(\theta) & -\sin(\theta) & 0 \\ \sin(\theta) & \cos(\theta) & 0 \\ 0 & 0 & 1 \end{bmatrix} \\ &= \begin{bmatrix} 0 & -1 & 0 \\ 1 & 0 & 0 \\ 0 & 0 & 0 \end{bmatrix}. \\ &=: L_3. \end{aligned}$$

Similar computations show that

$$Df\left(\frac{1}{2i}\sigma_1\right) = \begin{bmatrix} 0 & 0 & 0 \\ 0 & 0 & -1 \\ 0 & 1 & 0 \end{bmatrix} =: L_1$$

$$Df\left(\frac{1}{2i}\sigma_2\right) = \begin{bmatrix} 0 & 0 & 1 \\ 0 & 0 & 0 \\ -1 & 0 & 0 \end{bmatrix} =: L_2.$$

98   *Symmetry and Quantum Mechanics*

Since the matrices L_j form a basis for $\mathfrak{so}(3)$, the derivative Df is an isomorphism of Lie algebras as claimed. $\square$

5.5   Spin-s particles

Theorem 5.27 provides an explicit description of the irreducible representation of $SU(2)$, which are in one-to-one correspondence with the irreducible representations of the Lie algebra $\mathfrak{su}(2)$. But by exercise 5.19, the irreducible representations of $\mathfrak{su}(2)$ are in bijection with the irreducible representations of the complexified Lie algebra $\mathfrak{su}(2)_{\mathbb{C}} = \mathfrak{sl}_2(\mathbb{C})$. The following theorem, summarizing work from the previous section, presents an explicit description of the irreducible $\mathfrak{sl}_2(\mathbb{C})$-representations, which will be important for understanding their physical interpretation.

Theorem 5.30. *For each $m \geq 0$ there is a unique complex irreducible $\mathbb{C}$-linear $\mathfrak{sl}_2(\mathbb{C})$-representation of dimension $m + 1$, and its structure may be described explicitly in terms of the monomial basis for the space W_m of homogeneous degree-m polynomials in two variables w_1 and w_2:*

$$\begin{array}{lll} \sigma_3 & \text{acts as} & w_1^{m-k}w_2^k \mapsto (2k-m)w_1^{m-k}w_2^k \\[4pt] \sigma_1 + i\sigma_2 & \text{acts as} & w_1^{m-k}w_2^k \mapsto -2(m-k)w_1^{m-k-1}w_2^{k+1} \\[4pt] \sigma_1 - i\sigma_2 & \text{acts as} & w_1^{m-k}w_2^k \mapsto -2kw_1^{m-k+1}w_2^{k-1}. \end{array}$$

Let's take a closer look at the 2-dimensional representation W_1 to remind ourselves of the physical interpretation for electrons. In Section 5.2, we saw that the isomorphism $c_1 w_1 + c_2 w_2 \mapsto c_2|+z\rangle - c_1|-z\rangle$ identifies W_1 with the defining representation of $SU(2)$ on $(W, \langle | \rangle) \simeq (\mathbb{C}^2, \cdot)$ from Chapter 2. Via this identification, we interpret w_2 and w_1 as orthogonal spin-states, possessing the definite values of $\pm\frac{\hbar}{2}$ for their z-components of spin-angular momentum respectively. Note that here we are using the $SU(2)$-invariant inner product on W_1 determined by taking $\{w_1, w_2\}$ as an orthonormal basis. This inner product is essential to the Probability Interpretation, whereby $|\langle \psi | \phi \rangle|^2$ is the probability that the spin state ϕ will be found in the state ψ when measured by a "ψ-device."

In order to provide an analogous physical interpretation for the higher-dimensional representations W_m, we will likewise need an $SU(2)$-invariant inner product on W_m, so that $SU(2)$ will act on W_m via unitary transformations. From corollary 5.10 in Section 5.1, we know that there exists an $SU(2)$-invariant inner product on W_m. In fact, this invariant inner product is unique up to scalar multiples. To demonstrate this, let $\langle | \rangle$ denote any $SU(2)$-invariant inner product on W_m, normalized so that w_1^m has norm 1. Since the Lie algebra of the unitary group $U(m+1)$ is $\mathfrak{u}(m+1) = iH(m+1)$,

Higher Spin 99

the statement that $SU(2)$ acts via unitary transformations on $(W_m, \langle | \rangle)$ is equivalent to the statement that its Lie algebra $\mathfrak{su}(2)$ acts via skew-Hermitian transformations. Since the monomials $w_1^{m-k}w_2^k$ are eigenvectors for the action of $i\sigma_3 \in \mathfrak{su}(2)$ with distinct eigenvalues, it follows from the Spectral Theorem A.32 that they form an orthogonal basis for $(W_m, \langle | \rangle)$. The next exercise shows that the norms of these basis elements are completely determined by the normalization $\|w_1^m\| = 1$, so that any invariant inner product is a scalar multiple of $\langle | \rangle$ as claimed.

Exercise 5.31 (♣). *Show by induction on k that*

$$\|w_1^{m-k}w_2^k\|^2 = \binom{m}{k}^{-1} = \frac{k!(m-k)!}{m!}.$$

(Hint: Compute $\langle Xw_1^{m-k}w_2^k | Xw_2^{m-k}w_2^k \rangle$ in two ways, making use of the fact that $X^\dagger = Y$, where X, Y denote the linear transformations on W_m giving the action of $\sigma_1 \pm i\sigma_2$ as in theorem 5.25.)

From the previous exercise, we see that the scaled monomials

$$\binom{m}{k}^{\frac{1}{2}} w_1^{m-k}w_2^k$$

form an orthonormal basis for the complex inner product space $(W_m, \langle | \rangle)$.

For the physical interpretation, it will be convenient to use the ket notation. So denote the standard basis of $(\mathbb{C}^{m+1}, \cdot)$ by the kets

$$|-m\rangle \ , \ |-m+2\rangle \ , \ \dots \ , \ |m\rangle.$$

Then the map $\binom{m}{k}^{\frac{1}{2}} w_1^{m-k}w_2^k \mapsto (-1)^k |2k-m\rangle$ defines an isomorphism of inner product spaces. The $\mathfrak{sl}_2(\mathbb{C})$-action from theorem 5.30 becomes

$$
\begin{array}{lll}
\sigma_3 & \text{acts as} & |j\rangle \mapsto j|j\rangle \\
\sigma_1 + i\sigma_2 & \text{acts as} & |j\rangle \mapsto \sqrt{m(m+2) - j(j+2)}|j+2\rangle \\
\sigma_1 - i\sigma_2 & \text{acts as} & |j\rangle \mapsto \sqrt{m(m+2) - j(j-2)}|j-2\rangle.
\end{array}
$$

Exercise 5.32 (♣). *Verify the preceding formulas for the action of $\mathfrak{sl}_2(\mathbb{C})$.*

The spin observable $S_z = \frac{\hbar}{2}\sigma_3$ thus acts as $|j\rangle \mapsto \frac{j\hbar}{2}|j\rangle$. Hence, we may interpret the kets $|j\rangle$ as representing the orthogonal spin-states of a particle with $m+1$ possible observed values for the z-component of its spin angular momentum:

$$-\frac{m}{2}\hbar \ , \ \left(-\frac{m}{2}+1\right)\hbar \ , \ \dots \ , \ \frac{m}{2}\hbar.$$

Such a particle is said to be *spin-$\frac{m}{2}$*.

Because of this interpretation, we denote the unitary representations $(W_m, \langle | \rangle)$ by $\pi_s \colon SU(2) \to U(2s+1)$, where $s = \frac{m}{2} = 0, \frac{1}{2}, 1, \frac{3}{2}, 2, \dots$ refers

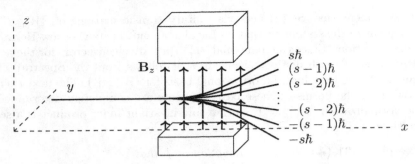

FIGURE 5.1: The behavior of a beam of spin-s particles in a Stern-Gerlach device.

to the highest possible observed spin of the associated particle. For instance, $\pi_{\frac{1}{2}}$ denotes the defining representation on $\mathbb{C}^2$ describing the electron, and we say that the electron is a spin-$\frac{1}{2}$ particle. For general s, the representation π_s has a basis of eigenvectors for the action of σ_3, representing spin-states with definite values for the observable S_z. The corresponding eigenvalues are the possible outcomes of measurements of a spin-s particle with a Stern-Gerlach device (see Figure 5.1). A general spin-state is represented by a unit vector in W_m, and may be expressed as a superposition of the S_z-eigenkets; two unit vectors describe the same spin-state if they differ by a phase $e^{i\theta} \in U(1)$.

Of course, just like in the case of spin-$\frac{1}{2}$, this description depends on P's choice of orthonormal basis for physical space, and his installation of a Stern-Gerlach device aligned with his positive z-axis. As in Chapter 2, suppose that M chooses a different orthonormal basis for physical space, and that $A \in SO(3)$ describes the rotation from P's coordinate system to M's. Then when M sets up a Stern-Gerlach device aligned with her positive z'-axis, she will obtain a different basis for the spinor space $\mathbb{C}^{2s+1}$ describing the possibilities for observed values of the z'-component of spin-angular momentum. If $B \in SU(2)$ maps to A under the double cover $f \colon SU(2) \to SO(3)$, then $\pi_s(B) \in U(m+1)$ sends P's S_z-eigenbasis for spinor space to M's $S_{z'}$-eigenbasis.

Just like in the case of the spin-$\frac{1}{2}$ electron, M and P make a list of shared beliefs about spin-s particles, phrased in terms of Stern-Gerlach experiments (here **u** denotes a unit vector in physical space):

1. A spin-s particle passing through an $SG\mathbf{u}$ will return an angular momentum measurement of one of the following $2s+1$ values:

$$-s\hbar,\ (-s+1)\hbar,\ \ldots,\ s\hbar,$$

   which we call the "spin in the **u**-direction."

2. Until we make a measurement with an $SG\mathbf{u}$, a particular spin-s particle may have no definite spin in the **u**-direction, but it *does* have a definite

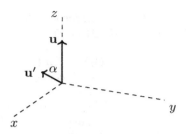

FIGURE 5.2: A rotation by α in the xz-plane. The dashed lines show P's coordinate axes.

probability of returning each of the possible values when measured by an $SG\mathbf{u}$;

3. If a spin-s particle exits an $SG\mathbf{u}$ with spin $\frac{j\hbar}{2}$ in the $\mathbf{u}$-direction, then it will measure spin $\frac{j\hbar}{2}$ in the $\mathbf{u}$-direction if measured immediately by a successive $SG\mathbf{u}$;

4. More generally, if the angle between $\mathbf{u}$ and $\mathbf{u}'$ is α, then a spin-s particle that exits an $SG\mathbf{u}$ with spin $\frac{j\hbar}{2}$ in the $\mathbf{u}$-direction will measure spin $\frac{j'\hbar}{2}$ in the direction $\mathbf{u}'$ with probability $\mathbb{P}_s(j, j', \alpha)$ if measured immediately by an $SG\mathbf{u}'$. This probability depends only on j, j' and the angle α.

The explicit formula for the probability $\mathbb{P}_s(j, j', \alpha)$ is complicated (see [21]), but we compute the specific case of the spin-1 particle in the next example.

Example 5.33. *Let's work out the probabilities $\mathbb{P}_1(j, j', \alpha)$ for a spin-1 particle. So suppose that $\mathbf{u}$ and $\mathbf{u}'$ are unit vectors making an angle α in physical space. To compute the probabilities, suppose that observer P chooses an orthonormal basis $\{\mathbf{u}_1, \mathbf{u}_2, \mathbf{u}_3\}$ for physical space such that $\mathbf{u}_3 = \mathbf{u}$ and $\mathbf{u}' = \sin(\alpha)\mathbf{u}_1 + \cos(\alpha)\mathbf{u}_3$. That is, P chooses his coordinates so that $\mathbf{u}$ points in his z-direction and $\mathbf{u}'$ is contained in his xz-plane (see Figure 5.2). Since $s = \frac{m}{2} = 1$, we have $m = 2$, so that P's basis for spinor-space is $\{|-2\rangle, |0\rangle, |+2\rangle\}$, representing states with the definite S_z-values $-\hbar, 0, \hbar$ respectively. Our question is: for a given pair of values j, j', what is the probability $\mathbb{P}_1(j, j', \alpha)$ that $|j\rangle$ will be found to be spin $\frac{j'\hbar}{2}$ when measured by an $SG\mathbf{u}'$? This probability is symmetric in j and j', so there are really six cases to consider: $\mathbb{P}_1(-2, -2, \alpha), \mathbb{P}_1(-2, 0, \alpha), \mathbb{P}_1(-2, 2, \alpha), \mathbb{P}_1(0, 0, \alpha), \mathbb{P}_1(0, 2, \alpha)$, and $\mathbb{P}_1(2, 2, \alpha)$.*

The matrix describing the rotation in the xz-plane through an angle α is:

$$A = \begin{bmatrix} \cos(\alpha) & 0 & \sin(\alpha) \\ 0 & 1 & 0 \\ -\sin(\alpha) & 0 & \cos(\alpha) \end{bmatrix}.$$

Symmetry and Quantum Mechanics

From exercise 2.40 in Chapter 2, we know that the following matrix, B, maps to A under the double cover $f: SU(2) \to SO(3)$.

$$B = \begin{bmatrix} \cos(\frac{\alpha}{2}) & -\sin(\frac{\alpha}{2}) \\ \sin(\frac{\alpha}{2}) & \cos(\frac{\alpha}{2}) \end{bmatrix}.$$

Applying the representation $\pi_1: SU(2) \to U(3) \subset GL(3, \mathbb{C})$ to the matrix B yields the automorphism of spinor space sending P's basis for spinor-space to the basis $\{|-2'\rangle, |0'\rangle, |+2'\rangle\}$ determined by an $SG\mathbf{u}'$. To determine this automorphism, we must compute the action of B on the scaled monomials $w_1^2, -\sqrt{2}w_1w_2, w_2^2$ under the representation π_1:

$$\begin{aligned}
B \star w_1^2 &= w_1^2 \circ B^{-1} \\
&= w_1^2 \circ \begin{bmatrix} \cos(\frac{\alpha}{2}) & \sin(\frac{\alpha}{2}) \\ -\sin(\frac{\alpha}{2}) & \cos(\frac{\alpha}{2}) \end{bmatrix} \\
&= \left(\cos\left(\frac{\alpha}{2}\right) w_1 + \sin\left(\frac{\alpha}{2}\right) w_2\right)^2 \\
&= \cos^2\left(\frac{\alpha}{2}\right) w_1^2 + \frac{\sqrt{2}}{2}\sin(\alpha)\sqrt{2}w_1w_2 + \sin^2\left(\frac{\alpha}{2}\right) w_2^2 \\
B \star -\sqrt{2}w_1w_2 &= -\sqrt{2}w_1w_2 \circ B^{-1} \\
&= -\sqrt{2}w_1w_2 \circ \begin{bmatrix} \cos(\frac{\alpha}{2}) & \sin(\frac{\alpha}{2}) \\ -\sin(\frac{\alpha}{2}) & \cos(\frac{\alpha}{2}) \end{bmatrix} \\
&= -\sqrt{2}(\cos\left(\frac{\alpha}{2}\right) w_1 + \sin\left(\frac{\alpha}{2}\right) w_2) \times \\
&\qquad (-\sin\left(\frac{\alpha}{2}\right) w_1 + \cos\left(\frac{\alpha}{2}\right) w_2) \\
&= \frac{\sqrt{2}}{2}\sin(\alpha) w_1^2 - \cos(\alpha)\sqrt{2}w_1w_2 - \frac{\sqrt{2}}{2}\sin(\alpha) w_2^2 \\
B \star w_2^2 &= w_2^2 \circ B^{-1} \\
&= w_2^2 \circ \begin{bmatrix} \cos(\frac{\alpha}{2}) & \sin(\frac{\alpha}{2}) \\ -\sin(\frac{\alpha}{2}) & \cos(\frac{\alpha}{2}) \end{bmatrix} \\
&= \left(-\sin\left(\frac{\alpha}{2}\right) w_1 + \cos\left(\frac{\alpha}{2}\right) w_2\right)^2 \\
&= \sin^2\left(\frac{\alpha}{2}\right) w_1^2 - \frac{\sqrt{2}}{2}\sin(\alpha)\sqrt{2}w_1w_2 + \cos^2\left(\frac{\alpha}{2}\right) w_2^2.
\end{aligned}$$

From these computations, we see that (expressed in terms of P's basis for spinor-space), we have

$$\pi_1(B) = \begin{bmatrix} \cos^2(\frac{\alpha}{2}) & \frac{\sqrt{2}}{2}\sin(\alpha) & \sin^2(\frac{\alpha}{2}) \\ -\frac{\sqrt{2}}{2}\sin(\alpha) & \cos(\alpha) & \frac{\sqrt{2}}{2}\sin(\alpha) \\ \sin^2(\frac{\alpha}{2}) & -\frac{\sqrt{2}}{2}\sin(\alpha) & \cos^2(\frac{\alpha}{2}) \end{bmatrix}.$$

Higher Spin 103

*The columns of this matrix are P's representation of the SG**u'**-basis:*

$$|-2'\rangle = \cos^2\left(\frac{\alpha}{2}\right)|-2\rangle - \frac{\sqrt{2}}{2}\sin(\alpha)|0\rangle + \sin^2\left(\frac{\alpha}{2}\right)|+2\rangle$$

$$|0'\rangle = \frac{\sqrt{2}}{2}\sin(\alpha)|-2\rangle + \cos(\alpha)|0\rangle - \frac{\sqrt{2}}{2}\sin(\alpha)|+2\rangle$$

$$|+2'\rangle = \sin^2\left(\frac{\alpha}{2}\right)|-2\rangle + \frac{\sqrt{2}}{2}\sin(\alpha)|0\rangle + \cos^2\left(\frac{\alpha}{2}\right)|+2\rangle.$$

We are now ready to compute the probabilities $\mathbb{P}_1(j, j', \alpha)$. *For instance, the probability that* $|-2\rangle$ *will be measured spin zero by an SG**u'** is given by the inner product*

$$\begin{aligned}
\mathbb{P}_1(-2, 0, \alpha) &= |\langle 0'|-2\rangle|^2 \\
&= |\langle -2|0'\rangle|^2 \\
&= \left|\langle -2|\left(\frac{\sqrt{2}}{2}\sin(\alpha)|-2\rangle + \cos(\alpha)|0\rangle - \frac{\sqrt{2}}{2}\sin(\alpha)|+2\rangle\right)\right|^2 \\
&= \frac{1}{2}\sin^2(\alpha),
\end{aligned}$$

where we have used the fact that $\{|-2\rangle, |0\rangle, |+2\rangle\}$ *is orthonormal. Similarly, we have*

$$\begin{aligned}
\mathbb{P}_1(0, 0, \alpha) &= |\langle 0'|0\rangle|^2 \\
&= |\langle 0|0'\rangle|^2 \\
&= \left|\langle 0|\left(\frac{\sqrt{2}}{2}\sin(\alpha)|-2\rangle + \cos(\alpha)|0\rangle - \frac{\sqrt{2}}{2}\sin(\alpha)|+2\rangle\right)\right|^2 \\
&= \cos^2(\alpha), \\
\mathbb{P}_1(2, 0, \alpha) &= |\langle 0'|+2\rangle|^2 \\
&= |\langle +2|0'\rangle|^2 \\
&= \left|\langle +2|\left(\frac{\sqrt{2}}{2}\sin(\alpha)|-2\rangle + \cos(\alpha)|0\rangle - \frac{\sqrt{2}}{2}\sin(\alpha)|+2\rangle\right)\right|^2 \\
&= \frac{1}{2}\sin^2(\alpha).
\end{aligned}$$

Analogous computations reveal that

$$\begin{aligned}
\mathbb{P}_1(-2, -2, \alpha) &= \cos^4\left(\frac{\alpha}{2}\right) \\
\mathbb{P}_1(-2, 2, \alpha) &= \sin^4\left(\frac{\alpha}{2}\right) \\
\mathbb{P}_1(2, 2, \alpha) &= \cos^4\left(\frac{\alpha}{2}\right).
\end{aligned}$$

104 *Symmetry and Quantum Mechanics*

Exercise 5.34. *Verify the last three probabilities displayed in the previous example by computing the relevant inner products.*

Note that in the general case of a spin-s particle, the probability $\mathbb{P}_s(j, j', \alpha)$ is simply the modulus squared of the matrix element in row j and column j' of the matrix $\pi_s(B)$ representing (in terms of P's orthonormal basis) the matrix that sends P's basis for spinor-space to the $SG\mathbf{u}'$-basis of spinor space. Hence, the task of computing these probabilities is equivalent to writing down explicit formulas for the matrix elements of the representations π_s.

Returning to the discussion at the beginning of this chapter, M provides a summary of their model of a spin-s particle: physical space is the adjoint representation of $SU(2)$ on its Lie algebra $\mathfrak{su}(2)$, and spinor space is the irreducible representation π_s of $SU(2)$ on $\mathbb{C}^{2s+1}$. Moreover, these two spaces are connected by the correspondence between generators of rotations and spin observables: if $-\frac{i}{\hbar}S_\mathbf{u} \in \mathfrak{su}(2)$ generates rotation about the $\mathbf{u}$-axis in physical space, then[4] $D\pi_s(S_\mathbf{u}) \in i\mathfrak{u}(2s+1) = H(2s+1)$ is the observable "spin in the $\mathbf{u}$-direction."

More generally, quantum observables on spinor space $\mathbb{C}^{2s+1}$ are given by Hermitian operators $O \in H(2s+1)$. Moreover, the discussion of Chapter 4 goes through in the higher-dimensional context: the energy of a spin-s particle in the presence of external fields is specified by a Hamiltonian function $\mathcal{H}(t) \in H(2s+1)$ which determines the unitary time-evolution $\mathcal{U}(t) \in U(2s+1)$ via the Schrödinger equation

$$i\hbar\dot{\mathcal{U}}(t) = \mathcal{H}(t)\mathcal{U}(t).$$

Finally, for any quantum observable O, we may define expectation values and uncertainties just as in the case of spin-$\frac{1}{2}$ (see definitions 3.14, 3.16). Moreover, the proof of theorem 3.20 extends verbatim to the higher-dimensional context, yielding the uncertainty principle for quantum observables on $\mathbb{C}^{2s+1}$. In particular, since the uncertainty inequalities for the spin observables S_x, S_y, and S_z depend only on the commutation relations for $\mathfrak{sl}_2(\mathbb{C})$ and not on the particular representation, we see that they are common to particles of all spins. For instance, if $|\psi\rangle \in \mathbb{C}^{2s+1}$ represents the spin-state of a spin-s particle, then we have

$$(\Delta_\psi S_x)(\Delta_\psi S_y) \geq \frac{1}{2}|\langle\psi|i\hbar S_z|\psi\rangle| = \frac{\hbar}{2}|\langle\psi|S_z|\psi\rangle|,$$

just like for the spin-$\frac{1}{2}$ electron discussed in Section 3.3.

Example 5.35. *As an example of an important observable, consider the* squared total spin operator

$$\mathbf{S}^2 := S_x^2 + S_y^2 + S_z^2.$$

---

[4]Technically speaking, we should write $(D\pi_s)_\mathbb{C}$ since this is the complexification of the representation $D\pi_s \colon \mathfrak{su}(2) \to \mathfrak{u}(2s+1)$. However, this notation is cumbersome, so we will avoid it without any serious risk of confusion.

Higher Spin 105

Note that this is not an element of $\mathfrak{sl}_2(\mathbb{C})$, since it is built from the spin observables using multiplication, and cannot be obtained through the Lie bracket alone. Hence, when considering the squared total spin operator for a spin-s particle, we must actually compute

$$D\pi_s(S_x)^2 + D\pi_s(S_y)^2 + D\pi_s(S_z)^2.$$

Recall the notation from theorem 5.25:

$$X := D\pi_s(\sigma_1 + i\sigma_2), \qquad Y := D\pi_s(\sigma_1 - i\sigma_2), \qquad Z := D\pi_s(\sigma_3).$$

The element S_z acts as $\frac{\hbar}{2}Z$, so its square acts as $\frac{\hbar^2}{4}Z^2$. Similarly, S_x acts as $\frac{\hbar}{4}(X+Y)$ and S_y acts as $\frac{\hbar}{4i}(X-Y)$, so $S_x^2 + S_y^2$ acts as

$$
\begin{aligned}
\frac{\hbar^2}{16}\left((X+Y)^2 - (X-Y)^2\right) &= \frac{\hbar^2}{8}(XY + YX) \\
&= \frac{\hbar^2}{8}([X,Y] + 2YX) \\
&= \frac{\hbar^2}{4}(2Z + YX).
\end{aligned}
$$

It follows that $\mathbf{S}^2$ acts on $W_{2s} = \mathbb{C}^{2s+1}$ as

$$\frac{\hbar^2}{4}(Z^2 + 2Z + YX).$$

Applying this operator to a basis ket $|j\rangle$ and remembering that $m = 2s$, we find that (see exercise 5.32)

$$
\begin{aligned}
|j\rangle \;\mapsto\; & \frac{\hbar^2}{4}(Z^2 + 2Z + YX)|j\rangle \\
=\; & \frac{\hbar^2}{4}(j^2 + 2j + m(m+2) - j(j+2))|j\rangle \\
=\; & \frac{\hbar^2}{4}m(m+2)|j\rangle \\
=\; & \hbar^2 s(s+1)|j\rangle.
\end{aligned}
$$

Hence, the squared total spin operator $\mathbf{S}^2$ acts as the scalar $\hbar^2 s(s+1)$ on spinor space $W_{2s} = \mathbb{C}^{2s+1}$ for a spin-s particle.

The previous example has the following consequence: if W is a representation of $\mathfrak{sl}_2(\mathbb{C})$, then we can determine if the decomposition of W into irreducibles contains a copy of W_{2s} by finding the eigenvalues for the action of $\mathbf{S}^2$ on W. We will use this strategy in our analysis of the irreducible representations of $SO(3)$ in the next section.

5.6 Representations of $SO(3)$

In Section 5.2, we saw that $SU(2)$ acts in a natural way on the polynomial ring $\mathbb{C}[w_1, w_2]$, and the grading by degree provides a decomposition of this infinite-dimensional representation as a direct sum of irreducibles:

$$\mathbb{C}[w_1, w_2] = \bigoplus_{m=0}^{\infty} W_m.$$

Moreover, the results in Sections 5.3 and 5.4 imply that this decomposition contains one copy of each irreducible representation of $SU(2)$. In proposition 5.28, we saw that the irreducible representations of $SO(3)$ are exactly the odd-dimensional $SU(2)$-representations W_{2d}. This raises the following question: is there a way to produce the sequence of irreducible $SO(3)$-representations in a natural way, just as $\mathbb{C}[w_1, w_2]$ produced the sequence of irreducible $SU(2)$-representations?

As a first guess, we might try the $SO(3)$-action on the polynomial ring $\mathbb{C}[x, y, z]$ defined by composition:

$$A \star p := p \circ A^{-1} \qquad \text{for } A \in SO(3) \text{ and } p = p(x, y, z) \in \mathbb{C}[x, y, z].$$

As in the case of $SU(2)$, this action preserves degrees, so we get a direct sum decomposition

$$\mathbb{C}[x, y, z] = \bigoplus_{d=0}^{\infty} V_d,$$

where $V_d = \{\text{homogeneous polynomials of degree } d\}$. Once again, we have the trivial representation $V_0 = \mathbb{C}$, and the defining representation $V_1 = \mathbb{C}^3$, both of which are irreducible. But the representation V_2 is reducible, having a one-dimensional invariant subspace spanned by the quadratic polynomial $x^2 + y^2 + z^2$.

Exercise 5.36. *Show that for all $A \in SO(3)$, we have*

$$A \star (x^2 + y^2 + z^2) = x^2 + y^2 + z^2.$$

In fact, all of the remaining $SO(3)$-representations V_d are reducible. Indeed, for $d \geq 2$, consider the subspace of degree d homogeneous polynomials that are multiples of $x^2 + y^2 + z^2$:

$$(x^2 + y^2 + z^2)V_{d-2} := \{(x^2 + y^2 + z^2)q(x, y, z) \mid q \in V_{d-2}\} \subset V_d.$$

This subspace is invariant under the action of $SO(3)$. To see this, suppose that $p = (x^2 + y^2 + z^2)q$ for some $q \in V_{d-2}$. Then for any $A \in SO(3)$ we have (using exercise 5.36):

$$A \star p = A \star (x^2 + y^2 + z^2)q = (x^2 + y^2 + z^2)(A \star q) \in (x^2 + y^2 + z^2)V_{d-2},$$

Higher Spin

which establishes the $SO(3)$-invariance.

So, the polynomial ring $\mathbb{C}[x, y, z]$ doesn't produce the sequence of irreducible $SO(3)$-representations in the way that $\mathbb{C}[w_1, w_2]$ produces the irreducible $SU(2)$-representations—each graded piece V_d is too big. To fix this, we need to restrict our attention to *harmonic polynomials*, which are introduced in the next definition.

Definition 5.37. *The* Laplacian operator $\Delta\colon C^\infty(\mathbb{R}^3) \to C^\infty(\mathbb{R}^3)$ *is the linear operator on the space of smooth functions $f(x, y, z)$ defined by*

$$\Delta(f) = \frac{\partial^2 f}{\partial x^2} + \frac{\partial^2 f}{\partial y^2} + \frac{\partial^2 f}{\partial z^2}.$$

For each $d \geq 2$, the Laplacian restricts to a linear transformation $\Delta\colon V_d \to V_{d-2}$, taking homogeneous polynomials of degree d to homogeneous polynomials of degree $d - 2$; it acts as the zero operator on polynomials of degree at most 1. For each $d \geq 0$, define $H_d := \ker(\Delta) \subset V_d$. The elements of H_d are called harmonic polynomials *of degree d.*

The next exercise describes the product rule for the Laplacian.

Exercise 5.38. *Show that if f and g are smooth functions on $\mathbb{R}^3$, then*

$$\Delta(fg) = \Delta(f)g + 2\nabla f \cdot \nabla g + f\Delta(g).$$

Proposition 5.39. *For each $d \geq 2$, the Laplacian $\Delta\colon V_d \to V_{d-2}$ is a surjective linear transformation.*

Proof. Order the monomial basis of V_{d-2} in lexicographical order. Then the first monomial is x^{d-2} which is clearly in the image of Δ. Now consider $x^a y^b z^c \in V_{d-2}$, and assume that all previous monomials are in the image. In particular, all monomials containing more than a factors of x are in the image. Then consider the monomial $x^{a+2} y^b z^c \in V_d$. Using the previous exercise, compute

$$
\begin{aligned}
\Delta(x^{a+2} y^b z^c) &= \Delta(x^{a+2}) y^b z^c + x^{a+2} \Delta(y^b z^c) \\
&= (a+2)(a+1) x^a y^b z^c + x^{a+2} \Delta(y^b z^c).
\end{aligned}
$$

Hence,

$$x^a y^b z^c = \frac{1}{(a+2)(a+1)} \left(\Delta(x^{a+2} y^b z^c) - x^{a+2} \Delta(y^b z^c) \right),$$

and the right-hand side is in the image of Δ since the last term contains $a+2$ factors of x. $\qquad\square$

Proposition 5.40. *For each $d \geq 0$, the space of homogeneous polynomials of degree d in x, y, z has dimension $\frac{(d+2)(d+1)}{2}$. It follows that H_d, the space of harmonic polynomials of degree d, has dimension $2d + 1$.*

108   *Symmetry and Quantum Mechanics*

Proof. Note that V_d has a basis consisting of monomials in x, y, z of exact degree d:

$$\beta_d := \{x^i y^j z^k \mid i + j + k = d\}.$$

But this basis has the same cardinality as the set of monomials in x, y having degree *at most d*:

$$\gamma_{\leq d} := \{x^i y^j \mid i + j \leq d\}.$$

Indeed, the map $\beta_d \to \gamma_{\leq d}$ defined by setting $z = 1$ is a bijection. By exercise 5.41 below, $\gamma_{\leq d}$ has $\frac{(d+2)(d+1)}{2}$ elements.

By rank-nullity (see theorem A.8), we have

$$\dim(H_d) = \dim(V_d) - \dim(V_{d-2}) = \frac{(d+2)(d+1)}{2} - \frac{d(d-1)}{2} = 2d + 1.$$

$\square$

Exercise 5.41 (♣). *Show that the number of monomials of degree at most d in two variables x, y is $\frac{(d+2)(d+1)}{2}$.*

The next proposition shows that H_d is an $SO(3)$-invariant subspace of V_d.

Proposition 5.42. *For each $d \geq 2$, the Laplacian $\Delta \colon V_d \to V_{d-2}$ is a morphism of $SO(3)$-representations in the sense of definition 5.1:*

$$\Delta(A \star p) = A \star \Delta(p) \qquad \text{for all } A \in SO(3) \text{ and } p \in V_d.$$

In particular, the space of harmonic polynomials H_d is invariant under $SO(3)$.

Proof. In order to simplify the notation, we write $\partial_1 = \frac{\partial}{\partial x}, \partial_2 = \frac{\partial}{\partial y}, \partial_3 = \frac{\partial}{\partial z}$. Setting $q(\mathbf{x}) = p(A\mathbf{x})$, we wish to show that $\Delta q(\mathbf{x}) = (\Delta p)(A\mathbf{x})$ for all $A \in SO(3)$. We begin by computing the second partial derivatives of q:

$$
\begin{aligned}
\partial_j \partial_i q(\mathbf{x}) &= \partial_j \left(\sum_{k=1}^{3} (\partial_k p)(A\mathbf{x}) A_{ki} \right) \\
&= \sum_{k,l=1}^{3} (\partial_l \partial_k p)(A\mathbf{x}) A_{lj} A_{ki}.
\end{aligned}
$$

Thus, we have

$$\Delta q(\mathbf{x}) = \sum_{j=1}^{3} \partial_j \partial_j q(\mathbf{x})$$

$$= \sum_{j,k,l=1}^{3} (\partial_l \partial_k p)(A\mathbf{x}) A_{lj} A_{kj}$$

$$= \sum_{k,l=1}^{3} (\partial_l \partial_k p)(A\mathbf{x}) \sum_{j=1}^{3} (A_{lj} A_{kj})$$

$$= \sum_{k,l=1}^{3} (\partial_l \partial_k p)(A\mathbf{x}) \delta_{lk} \qquad (\text{since } AA^T = I)$$

$$= \sum_{k=1}^{3} (\partial_k \partial_k p)(A\mathbf{x})$$

$$= (\Delta p)(A\mathbf{x}).$$

The fact that Δ is a morphism of $SO(3)$-representations immediately implies the $SO(3)$-invariance of the kernel: if $p \in H_d = \ker(\Delta)$, then

$$\Delta(A \star p) = A \star \Delta(p) = A \star 0 = 0,$$

so that $A \star p \in H_d$ for all $A \in SO(3)$. $\qquad \square$

Hence, H_d is an $SO(3)$-invariant subspace of V_d and has the correct dimension to be isomorphic to the irreducible representation W_{2d} of $SO(3)$. Recall from example 5.35 that the squared total spin operator $\mathbf{S}^2$ acts as the scalar $\hbar^2 d(d+1)$ on W_{2d}. Since the $SO(3)$-representations H_d and W_{2d} have the same dimension, in order to show that they are isomorphic, we just need to show that $\mathbf{S}^2$ also acts as the scalar $\hbar^2 d(d+1)$ on the space H_d. In the next section, we prepare the ground for this computation by providing an explicit description of the action of the Lie algebra $\mathfrak{so}(3)$ on polynomials.

5.6.1 The $\mathfrak{so}(3)$-action

Recall the double cover $f \colon SU(2) \to SO(3)$ from theorem 2.39. In proposition 5.29 of Section 5.5, we saw that $Df \colon \mathfrak{su}(2) \to \mathfrak{so}(3)$ is an isomorphism of Lie algebras, and that the elements $L_j := Df(\frac{1}{2i}\sigma_j)$ provide a basis for $\mathfrak{so}(3)$. From Section 3.2, we know that these matrices are tangent vectors at

110   *Symmetry and Quantum Mechanics*

the identity to the following curves in $SO(3)$:

$$\beta_1(t) = f\left(\exp\left(\frac{t}{2i}\sigma_1\right)\right) = \begin{bmatrix} 1 & 0 & 0 \\ 0 & \cos(t) & -\sin(t) \\ 0 & \sin(t) & \cos(t) \end{bmatrix},$$

$$\beta_2(t) = f\left(\exp\left(\frac{t}{2i}\sigma_2\right)\right) = \begin{bmatrix} \cos(t) & 0 & \sin(t) \\ 0 & 1 & 0 \\ -\sin(t) & 0 & \cos(t) \end{bmatrix},$$

$$\beta_3(t) = f\left(\exp\left(\frac{t}{2i}\sigma_3\right)\right) = \begin{bmatrix} \cos(t) & -\sin(t) & 0 \\ \sin(t) & \cos(t) & 0 \\ 0 & 0 & 1 \end{bmatrix}.$$

Hence, we have

$$L_1 = \dot\beta_1(0) = \begin{bmatrix} 0 & 0 & 0 \\ 0 & 0 & -1 \\ 0 & 1 & 0 \end{bmatrix},$$

$$L_2 = \dot\beta_2(0) = \begin{bmatrix} 0 & 0 & 1 \\ 0 & 0 & 0 \\ -1 & 0 & 0 \end{bmatrix},$$

$$L_3 = \dot\beta_3(0) = \begin{bmatrix} 0 & -1 & 0 \\ 1 & 0 & 0 \\ 0 & 0 & 0 \end{bmatrix}.$$

In the next proposition, we provide explicit formulas for the action of $\mathfrak{so}(3)$ on polynomials.

Proposition 5.43. *The elements $L_j \in \mathfrak{so}(3)$ act on the polynomial ring $\mathbb{C}[x, y, z]$ as differential operators:*

$$L_1 \star p(x, y, z) = \left(z\frac{\partial}{\partial y} - y\frac{\partial}{\partial z}\right)p$$

$$L_2 \star p(x, y, z) = -\left(z\frac{\partial}{\partial x} - x\frac{\partial}{\partial z}\right)p$$

$$L_3 \star p(x, y, z) = \left(y\frac{\partial}{\partial x} - x\frac{\partial}{\partial y}\right)p.$$

Proof. To compute the action of L_j on a polynomial $p(x, y, z)$, we need to differentiate the action of the curve $\beta_j(t) \in SO(3)$, which acts by composition: $\beta_j(t) \star p := p \circ \beta_j(t)^{-1} = p \circ \beta_j(-t)$. By the chain rule we have:

$$\frac{d}{dt}(\beta_j(t) \star p)\,|_{t=0} = \frac{d}{dt}(p \circ \beta_j(-t)(x, y, z)^T)\,|_{t=0}$$
$$= \nabla p \cdot (-L_j(x, y, z)^T).$$

Higher Spin

But

$$-L_1(x,y,z)^T = \begin{bmatrix} 0 & 0 & 0 \\ 0 & 0 & 1 \\ 0 & -1 & 0 \end{bmatrix} \begin{bmatrix} x \\ y \\ z \end{bmatrix} = \begin{bmatrix} 0 \\ z \\ -y \end{bmatrix}$$

$$-L_2(x,y,z)^T = \begin{bmatrix} 0 & 0 & -1 \\ 0 & 0 & 0 \\ 1 & 0 & 0 \end{bmatrix} \begin{bmatrix} x \\ y \\ z \end{bmatrix} = \begin{bmatrix} -z \\ 0 \\ x \end{bmatrix}$$

$$-L_3(x,y,z)^T = \begin{bmatrix} 0 & 1 & 0 \\ -1 & 0 & 0 \\ 0 & 0 & 0 \end{bmatrix} \begin{bmatrix} x \\ y \\ z \end{bmatrix} = \begin{bmatrix} y \\ -x \\ 0 \end{bmatrix}.$$

Hence, we find that

$$L_1 \star p(x,y,z) = \nabla p \cdot \left(-L_1(x,y,z)^T\right) = \left(z\frac{\partial}{\partial y} - y\frac{\partial}{\partial z}\right)p$$

$$L_2 \star p(x,y,z) = \nabla p \cdot \left(-L_2(x,y,z)^T\right) = -\left(z\frac{\partial}{\partial x} - x\frac{\partial}{\partial z}\right)p$$

$$L_3 \star p(x,y,z) = \nabla p \cdot \left(-L_3(x,y,z)^T\right) = \left(y\frac{\partial}{\partial x} - x\frac{\partial}{\partial y}\right)p.$$

$\square$

Since $SO(3)$ is the rotation group, it will be convenient to introduce spherical coordinates on $\mathbb{R}^3$ via the function $g\colon \mathbb{R}_{\geq 0} \times [0,\pi] \times [0,2\pi] \to \mathbb{R}^3$ defined by (see Figure 5.3)

$$(x,y,z) = g(r,\theta,\phi) = (r\sin(\theta)\cos(\phi), r\sin(\theta)\sin(\phi), r\cos(\theta)).$$

Writing $p(r,\theta,\phi)$ for the composition $p \circ g$, the chain rule A.37 yields:

$$\begin{bmatrix} \partial_r p \\ \partial_\theta p \\ \partial_\phi p \end{bmatrix} = \begin{bmatrix} \sin(\theta)\cos(\phi) & \sin(\theta)\sin(\phi) & \cos(\theta) \\ r\cos(\theta)\cos(\phi) & r\cos(\theta)\sin(\phi) & -r\sin(\theta) \\ -r\sin(\theta)\sin(\phi) & r\sin(\theta)\cos(\phi) & 0 \end{bmatrix} \begin{bmatrix} \partial_x p \\ \partial_y p \\ \partial_z p \end{bmatrix}.$$

Using Cramer's Rule A.21 to invert the matrix, we find that

$$\begin{bmatrix} \partial_x p \\ \partial_y p \\ \partial_z p \end{bmatrix} = \begin{bmatrix} \sin(\theta)\cos(\phi) & \frac{1}{r}\cos(\theta)\cos(\phi) & -\frac{1}{r}\csc(\theta)\sin(\phi) \\ \sin(\theta)\sin(\phi) & \frac{1}{r}\cos(\theta)\sin(\phi) & \frac{1}{r}\csc(\theta)\cos(\phi) \\ \cos(\theta) & -\frac{1}{r}\sin(\theta) & 0 \end{bmatrix} \begin{bmatrix} \partial_r p \\ \partial_\theta p \\ \partial_\phi p \end{bmatrix}.$$

Exercise 5.44. *Verify that*

$$z\frac{\partial p}{\partial y} - y\frac{\partial p}{\partial z} = \sin(\phi)\frac{\partial p}{\partial \theta} + \cot(\theta)\cos(\phi)\frac{\partial p}{\partial \phi}$$

$$z\frac{\partial p}{\partial x} - x\frac{\partial p}{\partial z} = \cos(\phi)\frac{\partial p}{\partial \theta} - \cot(\theta)\sin(\phi)\frac{\partial p}{\partial \phi}$$

$$x\frac{\partial p}{\partial y} - y\frac{\partial p}{\partial x} = \frac{\partial p}{\partial \phi}.$$

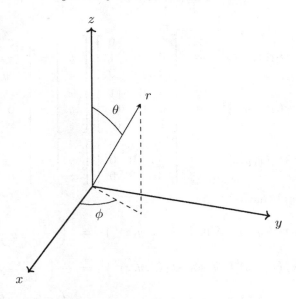

FIGURE 5.3: Spherical coordinates on $\mathbb{R}^3$.

Proposition 5.45. *In spherical coordinates, the $\mathfrak{so}(3)$-action on polynomials is given by*

$$L_1 \quad \text{acts as} \quad \left(\sin(\phi)\frac{\partial}{\partial \theta} + \cot(\theta)\cos(\phi)\frac{\partial}{\partial \phi}\right)$$

$$L_2 \quad \text{acts as} \quad -\left(\cos(\phi)\frac{\partial}{\partial \theta} - \cot(\theta)\sin(\phi)\frac{\partial}{\partial \phi}\right)$$

$$L_3 \quad \text{acts as} \quad -\frac{\partial}{\partial \phi}.$$

The corresponding action of $\mathfrak{sl}_2(\mathbb{C}) = \mathfrak{su}(2)_\mathbb{C} \simeq \mathfrak{so}(3)_\mathbb{C}$ obtained by complexification may then be described in terms of the generators $\frac{1}{2}\sigma_j$ by

$$\frac{1}{2}\sigma_1 \quad \text{acts as} \quad i\left(\sin(\phi)\frac{\partial}{\partial \theta} + \cot(\theta)\cos(\phi)\frac{\partial}{\partial \phi}\right)$$

$$\frac{1}{2}\sigma_2 \quad \text{acts as} \quad -i\left(\cos(\phi)\frac{\partial}{\partial \theta} - \cot(\theta)\sin(\phi)\frac{\partial}{\partial \phi}\right)$$

$$\frac{1}{2}\sigma_3 \quad \text{acts as} \quad -i\frac{\partial}{\partial \phi}.$$

Proof. The action of the elements L_j is the content of exercise 5.44. Multiplication by i then yields the action of $\frac{1}{2}\sigma_j$ due to the isomorphism $Df\colon \mathfrak{su}(2) \to \mathfrak{so}(3)$ given by $Df(\frac{1}{2i}\sigma_j) = L_j$. □

Higher Spin
113

Exercise 5.46. *Recall the spin observables* $S_x, S_y, S_z \in \mathfrak{sl}_2(\mathbb{C})$ *given by*

$$S_x = \frac{\hbar}{2}\sigma_1, \qquad S_y = \frac{\hbar}{2}\sigma_2, \qquad S_z = \frac{\hbar}{2}\sigma_3.$$

Show that the squared total spin operator, $\mathbf{S}^2 = S_x^2 + S_y^2 + S_z^2$, *acts on polynomials as the differential operator* $\hbar^2 C$, *where* C *is the dimensionless operator*

$$C := -\left(\frac{\partial^2}{\partial\theta^2} + \cot(\theta)\frac{\partial}{\partial\theta} + \csc^2(\theta)\frac{\partial^2}{\partial\phi^2}\right).$$

After this preliminary work, we now return to our goal of showing that $\mathbf{S}^2$ acts as the scalar $\hbar^2 d(d+1)$ on the space of harmonic polynomials H_d. We begin by expressing the Laplacian Δ in terms of the dimensionless version of the squared total spin operator, C, introduced in the previous exercise.

Exercise 5.47 (♣). *Show that the Laplacian in spherical coordinates is given by*

$$\Delta = \frac{1}{r^2}\frac{\partial}{\partial r}\left(r^2\frac{\partial}{\partial r}\right) - \frac{1}{r^2}C,$$

where C *is the dimensionless scaling of the squared total spin operator from exercise 5.46.*

Now note that if $p(x, y, z) \in V_d$ is homogeneous of degree d, scaling each variable by the same factor $\lambda \in \mathbb{R}$ simply scales the output by λ^d:

$$p(\lambda x, \lambda y, \lambda z) = \lambda^d p(x, y, z).$$

In terms of spherical coordinates, this just says that $p(r, \theta, \phi) = r^d F(\theta, \phi)$, where $F(\theta, \phi) := p(1, \theta, \phi)$ is the restriction of $p(r, \theta, \phi)$ to the unit sphere $S^2 \subset \mathbb{R}^3$. The restriction map $p \mapsto F$ defines an isomorphism of V_d with a subspace of smooth functions on the sphere S^2, which we denote by $\mathcal{F}_d$.

Proposition 5.48. *The squared total spin operator* $\mathbf{S}^2$ *acts as the scalar* $\hbar^2 d(d+1)$ *on the space* H_d *of harmonic polynomials of degree* d. *Thus,* $H_d \simeq W_{2d}$ *as representations of* $SO(3)$.

Proof. Since $\mathbf{S}^2$ acts as the differential operator $\hbar^2 C$, we just need to show that C acts as $d(d+1)$ on H_d. So suppose that $p \in H_d$, so that $\Delta(p) = 0$. Writing $p(r, \theta, \phi) = r^d F(\theta, \phi)$ as above and using exercise 5.47, we find that

$$
\begin{aligned}
0 &= \Delta(r^d F) \\
&= \frac{1}{r^2}\frac{\partial}{\partial r}\left(r^2\frac{\partial r^d}{\partial r}\right)F - r^{d-2}C(F) \\
&= \frac{d}{r^2}\frac{\partial r^{d+1}}{\partial r}F - r^{d-2}C(F) \\
&= d(d+1)r^{d-2}F - r^{d-2}C(F).
\end{aligned}
$$

Symmetry and Quantum Mechanics

Dividing by r^{d-2}, we see that $C(F) = d(d+1)F$. But then

$$C(p) = C(r^d F) = r^d C(F) = d(d+1)r^d F = d(d+1)p.$$

$\square$

We have now achieved our original goal of generating the sequence of irreducible $SO(3)$-representations in a natural way. Namely, denote by $\mathrm{HarmPoly}(\mathbb{R}^3) \subset \mathbb{C}[x, y, z]$ the space of all harmonic polynomials in the variables x, y, z. Then $SO(3)$ acts on this infinite-dimensional space via composition:

$$A \star p(\mathbf{x}) := p(A^{-1}\mathbf{x}).$$

The grading by degree yields the decomposition into irreducible representations of $SO(3)$:

$$\mathrm{HarmPoly}(\mathbb{R}^3) = \bigoplus_{d=0}^{\infty} H_d \simeq \bigoplus_{d=0}^{\infty} W_{2d}. \tag{5.2}$$

This is exactly analogous to the decomposition of the $SU(2)$-representation $\mathbb{C}[w_1, w_2]$ into irreducibles:

$$\mathbb{C}[w_1, w_2] = \bigoplus_{m=0}^{\infty} W_m. \tag{5.3}$$

5.6.2 Comments about analysis

The decompositions in (5.2) and (5.3) involve only polynomials, and hence are completely algebraic, involving no analytic notions such as convergence of infinite series, etc. This algebraic treatment has been sufficient for the study of spin, which is the *intrinsic* angular momentum of a particle, described by the finite-dimensional representations of $SU(2)$. But as we will see in Chapter 8, the representations of $SO(3)$ arise naturally in the study of *orbital* angular momentum, which is the angular momentum of a particle due to its motion through space. Moreover, as we extend our model to account for the motion of particles through space, we will find ourselves studying infinite-dimensional function spaces rather than the finite-dimensional spinor spaces of the previous chapters. In that context, we will want to have an explicit description of the irreducible $SO(3)$-representations as spaces of functions on the unit sphere S^2. We have already begun this process by establishing the isomorphism of $SO(3)$-representations $W_{2d} \simeq H_d$, where H_d is the space of degree-d harmonic polynomials on $\mathbb{R}^3$. But we would now like to go further and exploit the isomorphism of H_d with a space of functions on the unit sphere S^2 given by the restriction map $p(r, \theta, \phi) \mapsto F(\theta, \phi) := p(1, \theta, \phi)$. This will lead us into the world of analysis and our first encounter with some analytic subtleties that lie at the core of quantum mechanics. Our aim will be to expose some

Higher Spin   115

difficulties without getting bogged down in technical details, leaving a full-blown exposition of the analytic theory to another text, such as [10].

To begin, note that the $SO(3)$-action on the space of polynomials $\mathbb{C}[x, y, z]$ may be extended to the space of all smooth complex-valued functions $C^\infty(\mathbb{R}^3)$. Moreover, we obtain a corresponding action of the Lie algebra $\mathfrak{so}(3)$ on $C^\infty(\mathbb{R}^3)$, where the generators L_j act via the differential operators described in proposition 5.43. Similarly, working in spherical coordinates, we may consider the action of $SO(3)$ on the space of smooth functions on S^2, together with the corresponding action of $\mathfrak{so}(3)$ on $C^\infty(S^2)$ by the differential operators described in proposition 5.45.

In proposition 5.48, we saw[5] that for a degree l homogeneous polynomial $p \in V_l$, the harmonic condition $\Delta(p) = 0$ corresponds to the following eigenvector condition on the function $F = p(1, \theta, \phi)$:

$$C(F) = l(l+1)F.$$

In order to describe these eigenfunctions explicitly, we would like to solve the differential equation (see exercise 5.46)

$$-C(F) = \frac{\partial^2 F}{\partial \theta^2} + \cot(\theta)\frac{\partial F}{\partial \theta} + \csc^2(\theta)\frac{\partial^2 F}{\partial \phi^2} = -l(l+1)F. \qquad (5.4)$$

Moreover, since we are attempting to describe the irreducible representation $D\pi_l$ of $\mathfrak{so}(3)_\mathbb{C} \simeq \mathfrak{sl}_2(\mathbb{C})$, and since we know the structure of this representation from theorem 5.30, we look for solutions that are also eigenvectors for $\frac{1}{\hbar}S_z = \frac{1}{2}\sigma_3$, which (by proposition 5.45) acts as $-i\frac{\partial}{\partial \phi}$ with eigenvalues $m = -l, -l+1, \ldots, 0, \ldots, l-1, l$. Such solutions are separable of the form

$$F(\theta, \phi) = e^{im\phi}G(\theta).$$

Plugging into the differential equation (5.4) and rearranging, we find that $G(\theta)$ must satisfy (using primes to denote differentiation)

$$\sin^2(\theta)G'' + \sin(\theta)\cos(\theta)G' + \left[l(l+1)\sin^2(\theta) - m^2\right]G = 0.$$

Now make the change of variables $w = \cos(\theta)$ and write $g(w) = G(\theta)$, so that

$$G'(\theta) = \frac{dG}{d\theta} = \frac{dg}{dw}\frac{dw}{d\theta} = -\sin(\theta)g'(w)$$

and

$$G''(\theta) = \frac{d}{d\theta}(-\sin(\theta)g'(w)) = -\cos(\theta)g'(w) + \sin^2(\theta)g''(w).$$

Making the substitutions yields the *general Legendre equation*:

$$(1 - w^2)g'' - 2wg' + \left[l(l+1) - \frac{m^2}{(1 - w^2)}\right]g = 0.$$

---

[5]Here we make a slight notational adjustment to connect with the standard description: we replace the degree d with the index l, and then use $m = -l, -l+1, \ldots, l$ for the eigenvalues of $\frac{1}{\hbar}S_z$ in the irreducible representation $D\pi_l$.

Symmetry and Quantum Mechanics

This equation has two fundamental solutions, but only one of them is bounded on the sphere S^2, and this is the one we are interested in, since we hope that it is obtained by restriction from a harmonic polynomial on $\mathbb{R}^3$. In order to describe this bounded solution, we first introduce the *Legendre polynomials* via *Rodrigues' formula*:

$$P_l(w) = \frac{1}{2^l l!} \frac{d^l}{dw^l} (w^2 - 1)^l.$$

Exercise 5.49. *Check that, for each* $m = 0, 1, \ldots, l$, *the associated Legendre function provides a solution to the general Legendre equation:*

$$P_l^m(w) := (1 - w^2)^{\frac{m}{2}} \frac{d^m}{dw^m} P_l(w).$$

Multiplying by the complex exponentials $e^{\pm im\phi}$, we obtain $2l + 1$ linearly independent solutions to our original differential equation (5.4):

$$F_l^{\pm m}(\theta, \phi) := e^{\pm im\phi} P_l^m(\cos(\theta)), \qquad m = 0, 1, \ldots, l.$$

While it is not immediately obvious, the functions $F_l^{\pm m}(\theta, \phi)$ are actually restrictions of harmonic polynomials to the unit sphere. To show this, consider the function on $\mathbb{R}^3$ defined by $h(r, \theta, \phi) = r^l F_l^{\pm m}(\theta, \phi)$. Our claim is that $h(r, \theta, \phi) = p(x, y, z)$ for some harmonic polynomial $p \in H_l$. Note that the function h is certainly harmonic, so we just need to show that it is a polynomial.

First observe that the Legendre polynomial $P_l(w)$ is of degree l and contains only exponents of the same parity as l. Setting $Q_l^m(w) = \frac{d^m}{dw^m} P_l(w)$, it follows that $Q_l^m(w)$ is a polynomial of degree $l - m$ containing only exponents of the same parity as $l - m$. This simple fact will play a key role in proving that h is a polynomial. Recall the relationship between Cartesian and polar coordinates:

$$
\begin{aligned}
x &= r\sin(\theta)\cos(\phi) \\
y &= r\sin(\theta)\cos(\phi) \\
z &= r\cos(\theta).
\end{aligned}
$$

Making these replacements in $h(r, \theta, \phi)$, we find

$$
\begin{aligned}
h(r, \theta, \phi) &= r^l e^{\pm im\phi} (1 - \cos^2(\theta))^{\frac{m}{2}} Q_l^m(\cos(\theta)) \\
&= r^l (\cos(\phi) \pm i\sin(\phi))^m \sin^m(\theta) Q_l^m(\cos(\theta)) \\
&= (r\sin(\theta)\cos(\phi) \pm ir\sin(\theta)\sin(\phi))^m r^{l-m} Q_l^m(\cos(\theta)) \\
&= (x \pm iy)^m r^{l-m} Q_l^m(\cos(\theta)).
\end{aligned}
$$

A general term of $r^{l-m} Q_l^m(\cos(\theta))$ looks like a constant times

$$r^{l-m}(\cos(\theta))^n = r^{l-m-n}(r\cos(\theta))^n = (x^2 + y^2 + z^2)^{\frac{l-m-n}{2}} z^n.$$

Higher Spin

Here the exponent $n \leq l - m$ has the same parity as $l - m$. It follows that $\frac{l-m-n}{2} \geq 0$ is an integer, so that each term of Q_l^m yields a homogeneous polynomial of degree $l - m$ in the variables x, y, z. It follows that

$$h(r, \theta, \phi) = (x \pm iy)^m q_l^m(x, y, z) =: p_l^{\pm m}(x, y, z),$$

where q_l^m is a homogeneous polynomial of degree $l - m$, so that $p_l^{\pm m}$ is homogeneous of degree l. In fact, the polynomials $p_l^{\pm m}$ form an L_3-eigenbasis for the $\mathfrak{so}(3)$-representation H_l. The restriction mapping $p_l^{\pm m} \mapsto F_l^{\pm m}$ defines an isomorphism of H_l with a space of harmonic functions $\mathcal{Y}_l$ on the unit sphere S^2, thereby realizing the irreducible $SO(3)$-representation π_l explicitly as a space of smooth functions on the unit sphere.

From corollary 5.10 from Section 5.1, we know that there is an $SO(3)$-invariant inner product (unique up to a constant) on $\mathcal{Y}_l$, for which $SO(3)$ acts unitarily (equivalently, $\mathfrak{so}(3)$ acts via skew-Hermitian transformations). Explicitly, we can describe this inner product as

$$\langle f | g \rangle = \int_{S^2} f^* g \, d\Omega = \int_0^{2\pi} \int_0^{\pi} f(\theta, \phi)^* g(\theta, \phi) \sin(\theta) d\theta d\phi. \tag{5.5}$$

Here $d\Omega = \sin(\theta)d\theta d\phi$ is the solid angle measure on S^2, obtained by restriction from the usual Lebesgue measure on $\mathbb{R}^3$. With respect to this inner product, the functions $F_l^{\pm m}$ form an orthogonal basis for $\mathcal{Y}_l$ since they are eigenfunctions for the skew-Hermitian operator $D\pi_l(L_3)$. Normalizing these functions, we obtain the *spherical harmonics of degree l*, denoted $Y_l^{\pm m}$ (see [9, equation 4.32]).

Note that the integral formula for the inner product is independent of the degree l of the representation. Since all the representations $\mathcal{Y}_l$ are subspaces of the space of smooth functions on S^2, we may consider the infinite dimensional complex inner product space $(\mathcal{C}^\infty(S^2), \langle | \rangle)$, where the inner product is defined by the integral formula (5.5). The subspaces $\mathcal{Y}_l$ are mutually orthogonal with respect to this inner product, so the orthogonal direct sum embeds as a subspace:

$$\bigoplus_{l=0}^{\infty} \mathcal{Y}_l \subset \mathcal{C}^\infty(S^2).$$

Note that elements of the direct sum are *finite* linear combinations of spherical harmonics. In order to consider *infinite* series of spherical harmonics, we must pass to a larger inner product space, namely the space of square-integrable functions on S^2, where the inner product is again given by the integral formula (5.5):

$$L^2(S^2) = \left\{ f : S^2 \to \mathbb{C} \mid \int_{S^2} |f|^2 d\Omega < \infty \right\}.$$

The virtue of this space (compared with $\mathcal{C}^\infty(S^2)$) is that it is *complete*[6], and

---

[6]The requirement of completeness necessitates that we use the Lebesgue integral, rather than the Riemann integral. However, the reader unfamiliar with measure theory and Lebesgue integration will not lose much by thinking of the Riemann integral throughout.

Symmetry and Quantum Mechanics

hence a *Hilbert space* (see Appendix A.3.1). In fact, the orthogonal direct sum of the spaces $\mathcal{Y}_l$ is dense in $L^2(S^2)$, so that its completion is equal to the entire space:

$$L^2(S^2) = \widehat{\bigoplus}_{l=0}^{\infty} \mathcal{Y}_l. \tag{5.6}$$

Explicitly, this means that every square-integrable function on S^2 may be expressed uniquely as a convergent infinite series of spherical harmonics.

Note that the $SO(3)$-action extends to the space of square-integrable functions on $S^2 \subset \mathbb{R}^3$:

$$A \star f := f \circ A^{-1} \qquad \text{for all } f \in L^2(S^2).$$

The completed orthogonal direct sum in (5.6) expresses the decomposition of $L^2(S^2)$ into irreducible representations of $SO(3)$. This is an analytic version of the purely algebraic decomposition of the space of harmonic polynomials on $\mathbb{R}^3$ from (5.2):

$$\text{HarmPoly}(\mathbb{R}^3) = \bigoplus_{l=0}^{\infty} H_l.$$

One worrisome aspect of the move to square-integrable functions is that it is now unclear exactly how to understand the $\mathfrak{so}(3)$-action. In particular, if we attempt to obtain an $\mathfrak{so}(3)$-action by differentiating the $SO(3)$-action as usual, we run into a problem: not all square-integrable functions are differentiable! This is the first hint of substantial analytic difficulties that arise in the study of quantum mechanics. It turns out that a rich theory of self-adjoint operators on Hilbert spaces exists, and in that framework we may understand the sense in which $\mathfrak{so}(3)$ acts on $L^2(S^2)$. Roughly speaking, the idea is that a self-adjoint operator A is generally defined only on a dense subspace of the Hilbert space $\boldsymbol{H}$ under consideration. For instance, the subspace of smooth functions $\mathcal{C}^{\infty}(S^2)$ is dense in $L^2(S^2)$. Nonetheless, the Spectral Theorem for self-adjoint operators provides a *functional calculus* that allows us to assign a one-parameter group of unitary operators e^{itA} (on the full Hilbert space $\boldsymbol{H}$) to every self-adjoint operator A (defined only on a dense subspace of $\boldsymbol{H}$). By Stone's Theorem A.51, this assignment is actually a bijection between self-adjoint operators and strongly continuous one-parameter unitary groups. In this way, the Lie algebra $\mathfrak{so}(3)$ acts on $L^2(S^2)$ via (densely defined) skew-adjoint operators.

The issue of specifying the exact domain of a self-adjoint operator is a tricky and technical one. In this book, we will largely ignore the issue, being content to describe operators by their actions on "sufficiently nice functions." In this way we will be able to develop the general mathematical structure of the theory, focusing mainly on the algebraic aspects and notions of symmetry, giving only a brief nod to the analytic difficulties. Comparing the mathematical theory of quantum mechanics to a house, one might say that we are framing it and laying most of the bricks, but omitting almost all of the mortar.

My hope is that the somewhat unstable structure that results might provide some helpful motivation to the beginning student as she pursues her study of functional analysis.

Chapter 6

Multiple Particles

In which M and P learn about the tensor product.

6.1 Tensor products of representations

P wonders how to model a system of several particles. For concreteness (and looking ahead to the hydrogen atom), he starts with the case of one electron and one proton. The proton is also a spin-$\frac{1}{2}$ particle, and it has an electric charge of equal magnitude but opposite sign to the electron. However, the proton is much more massive than the electron, with $m_p/m_e \approx 1836$. The proton and the electron spins are each separately described by spin-$\frac{1}{2}$ representations of $SU(2)$, which we denote by $(W_p, \langle|\rangle_p)$ and $(W_e, \langle|\rangle_e)$ respectively. To describe the possible spin states of the composite system of particles, note that the specification of a spin-state for each individual particle should certainly determine a spin-state for the system. Thus, every pair of kets $(|\psi\rangle_p, |\phi\rangle_e) \in W_p \times W_e$ will determine a spin-state for the system. But distinct pairs of kets won't necessarily determine distinct states of the composite. Indeed, since $|\psi\rangle_p$ and $e^{i\theta} |\psi\rangle_p$ correspond to the same spin-state of the proton, the pairs $(|\psi\rangle_p, |\phi\rangle_e)$ and $(e^{i\theta} |\psi\rangle_p, |\phi\rangle_e)$ must correspond to the same composite spin-state. Of course, a similar comment applies to phases on the electron kets in the second component.

More generally, suppose that $|\psi\rangle_p = c_1 |\psi_1\rangle_p + c_2 |\psi_2\rangle_p$ is a superposition of proton spin-states. Then it seems reasonable to assume that the pair $(|\psi\rangle_p, |\phi\rangle_e)$ should describe a composite spin-state that is linearly related to those described by the pairs $(|\psi_1\rangle_p, |\phi\rangle_e)$ and $(|\psi_2\rangle_p, |\phi\rangle_e)$, and similarly if $|\phi\rangle_e$ is a superposition. M agrees with this assumption, and offers the following construction as a simple way of satisfying this requirement. Begin by considering the complex vector space $\mathbb{C}\{W_p \times W_e\}$ with basis given by the set $W_p \times W_e$. This is a huge vector space, of uncountably infinite dimension equal to the cardinality of $W_p \times W_e$. But we will construct a small quotient by imposing some equivalences coming from the previous remarks. So consider the subspace $R \subset \mathbb{C}\{W_p \times W_e\}$ defined as the span of all vectors of the form

$$(\mathbf{a}, \alpha\mathbf{b}_1 + \beta\mathbf{b}_2) - \alpha(\mathbf{a}, \mathbf{b}_1) - \beta(\mathbf{a}, \mathbf{b}_2),$$
$$(\alpha\mathbf{a}_1 + \beta\mathbf{a}_2, \mathbf{b}) - \alpha(\mathbf{a}_1, \mathbf{b}) - \beta(\mathbf{a}_2, \mathbf{b}).$$

122   *Symmetry and Quantum Mechanics*

When we mod out by the subspace R, we are setting all such vectors equal to zero, thereby ensuring that superpositions of electron kets yield superpositions of the composite system, and similarly for superpositions of proton kets.

Definition 6.1. *The* tensor product *of W_p and W_e is the quotient vector space*

$$W_p \otimes W_e := \mathbb{C}\{W_p \times W_e\}/R.$$

If $\mathbf{a} \in W_p$ and $\mathbf{b} \in W_e$, then $\mathbf{a} \otimes \mathbf{b}$ denotes the image of $(\mathbf{a}, \mathbf{b})$ in $W_p \otimes W_e$, and a general element of the tensor product is a finite linear combination of such decomposable elements:

$$\sum_{k=1}^{n} c_k(\mathbf{a}_k \otimes \mathbf{b}_k) \quad for \ \mathbf{a}_k \in W_p, \mathbf{b}_k \in W_e, c_k \in \mathbb{C}.$$

The relations in R imply that the tensor product is bilinear:

$$\mathbf{a} \otimes (\alpha\mathbf{b}_1 + \beta\mathbf{b}_2) = \alpha(\mathbf{a} \otimes \mathbf{b}_1) + \beta(\mathbf{a} \otimes \mathbf{b}_2),$$
$$(\alpha\mathbf{a}_1 + \beta\mathbf{a}_2) \otimes \mathbf{b} = \alpha(\mathbf{a}_1 \otimes \mathbf{b}) + \beta(\mathbf{a}_2 \otimes \mathbf{b}).$$

As a first result about the tensor product, we compute its dimension.

Proposition 6.2. *Let $\{\mathbf{p}_1, \mathbf{p}_2\}$ be a basis for W_p and $\{\mathbf{e}_1, \mathbf{e}_2\}$ a basis for W_e. Then $\{\mathbf{p}_1 \otimes \mathbf{e}_1, \mathbf{p}_1 \otimes \mathbf{e}_2, \mathbf{p}_2 \otimes \mathbf{e}_1, \mathbf{p}_2 \otimes \mathbf{e}_2\}$ is a basis for $W_p \otimes W_e$, so that the tensor product of two 2-dimensional vector spaces has dimension 4.*

Proof. We first show that the four putative basis elements span the tensor product. For this, note that since a general element of $W_p \otimes W_e$ is a linear combination of decomposable elements, it will suffice to show that every decomposable element $\mathbf{a} \otimes \mathbf{b}$ is in the span. We may write $\mathbf{a} = \alpha_1\mathbf{p}_1 + \alpha_2\mathbf{p}_2$ and $\mathbf{b} = \beta_1\mathbf{e}_1 + \beta_2\mathbf{e}_2$ for some complex scalars α_i, β_i. Using the bilinearity of the tensor product, we find that

$$\begin{aligned}
\mathbf{a} \otimes \mathbf{b} &= (\alpha_1\mathbf{p}_1 + \alpha_2\mathbf{p}_2) \otimes (\beta_1\mathbf{e}_1 + \beta_2\mathbf{e}_2) \\
&= \alpha_1\mathbf{p}_1 \otimes (\beta_1\mathbf{e}_1 + \beta_2\mathbf{e}_2) + \alpha_2\mathbf{p}_2 \otimes (\beta_1\mathbf{e}_1 + \beta_2\mathbf{e}_2) \\
&= \alpha_1\beta_1(\mathbf{p}_1 \otimes \mathbf{e}_1) + \alpha_1\beta_2(\mathbf{p}_1 \otimes \mathbf{e}_2) \\
&\quad + \alpha_2\beta_1(\mathbf{p}_2 \otimes \mathbf{e}_1) + \alpha_1\beta_2(\mathbf{p}_2 \otimes \mathbf{e}_2).
\end{aligned}$$

In order to show linear independence, suppose that we have a linear dependence:

$$c_1(\mathbf{p}_1 \otimes \mathbf{e}_1) + c_2(\mathbf{p}_1 \otimes \mathbf{e}_2) + c_3(\mathbf{p}_2 \otimes \mathbf{e}_1) + c_4(\mathbf{p}_2 \otimes \mathbf{e}_2) = 0. \tag{6.1}$$

We wish to show that each $c_i = 0$. For this, begin by defining a function $g \colon W_p \times W_e \to W_e$ via the following recipe:

$$g(\alpha_1\mathbf{p}_1 + \alpha_2\mathbf{p}_2, \mathbf{b}) := \alpha_1\mathbf{b}.$$

Multiple Particles 123

Note that the function g is well defined on the set $W_p \times W_e$, since $\{\mathbf{p}_1, \mathbf{p}_2\}$ is a basis for W_p. Thus, g extends to a linear mapping $g\colon \mathbb{C}\{W_p \times W_e\} \to W_e$, since the set $W_p \times W_e$ forms a basis of $\mathbb{C}\{W_p \times W_e\}$. It is straightforward to check that the subspace of relations R is in the kernel of g, so that we get a linear transformation $g\colon W_p \otimes W_e \to W_e$. Applying this mapping g to the linear dependence (6.1) yields a linear dependence in W_e:

$$c_1 \mathbf{e}_1 + c_2 \mathbf{e}_2 = 0.$$

Since $\{\mathbf{e}_1, \mathbf{e}_2\}$ is a basis for W_e, it follows that $c_1 = c_2 = 0$. A similar argument applied to the function $h\colon W_p \times W_e \to W_e$ defined by

$$h(\alpha_1 \mathbf{p}_1 + \alpha_2 \mathbf{p}_2, \mathbf{b}) := \alpha_2 \mathbf{b}$$

yields $c_3 = c_4 = 0$. $\square$

Exercise 6.3. *In the notation of proposition 6.2, show that the element $\mathbf{p}_1 \otimes \mathbf{e}_1 + \mathbf{p}_2 \otimes \mathbf{e}_2$ is not decomposable, i.e., cannot be written as $\mathbf{a} \otimes \mathbf{b}$ for any $\mathbf{a} \in W_p$ and $\mathbf{b} \in W_e$.*

Exercise 6.4. *Observe that definition 6.1 makes sense for vector spaces of arbitrary dimensions. Then adapt the proof of proposition 6.2 to show that if W_1 has dimension n and W_2 has dimension m, then the tensor product $W_1 \otimes W_2$ has dimension nm.*

Definition 6.5. *The inner products on the spaces W_p and W_e endow the tensor product $W_p \otimes W_e$ with an inner product $\langle | \rangle$ given by $\langle \mathbf{a}_1 \otimes \mathbf{b}_1 | \mathbf{a}_2 \otimes \mathbf{b}_2 \rangle :=$ $\langle \mathbf{a}_1 | \mathbf{a}_2 \rangle_p \langle \mathbf{b}_1 | \mathbf{b}_2 \rangle_e$ on decomposable elements and extended to be linear in the second component and conjugate linear in the first component.*

Exercise 6.6. *Show that this prescription is well defined, and does indeed yield an inner product on $W_p \otimes W_e$.*

Definition 6.7. *The set of spin-states for the proton-electron pair is the set of unit vectors in $(W_p \otimes W_e, \langle | \rangle)$ modulo the action of the phase group $U(1)$. That is, two unit vectors correspond to the same spin-state if and only if they differ by a phase $e^{i\theta}$.*

For ease of notation we often write $|\psi \otimes \phi\rangle := |\psi\rangle_p \otimes |\phi\rangle_e$ for the tensor product of two kets, which is a decomposable ket representing the composite spin-state in which the proton has spin-state ψ and the electron has spin-state ϕ. As shown in exercise 6.3, not all elements of $W_p \otimes W_e$ are decomposable. Hence, this model predicts the existence of *entangled* spin-states—these are the spin-states of the composite system that cannot be determined by specifying a spin-state for each particle separately.

To investigate further, P chooses a basis $\{\mathbf{u}_1, \mathbf{u}_2, \mathbf{u}_3\}$ for physical space and sets up his Stern-Gerlach device SGz, thus providing an identification of

124   *Symmetry and Quantum Mechanics*

each spinor space W_p and W_e with the defining representation of $SU(2)$ on
$\mathbb{C}^2$ by means of the z-bases:

$$W_p = \mathbb{C}|+z\rangle_p \oplus \mathbb{C}|-z\rangle_p \simeq \mathbb{C}^2$$
$$W_e = \mathbb{C}|+z\rangle_e \oplus \mathbb{C}|-z\rangle_e \simeq \mathbb{C}^2.$$

It follows that an orthonormal basis for the tensor product is given by

$$\{|+z\rangle_p \otimes |+z\rangle_e, |+z\rangle_p \otimes |-z\rangle_e, |-z\rangle_p \otimes |+z\rangle_e, |-z\rangle_p \otimes |-z\rangle_e\}. \qquad (6.2)$$

Hence, an arbitrary spin-state of the proton-electron system may be expressed
as a superposition of these particular z-states, and the identifications with $\mathbb{C}^2$
yield an identification of the tensor product with $\mathbb{C}^4$. Moreover, $SU(2)$ acts
on $W_p \otimes W_e \simeq \mathbb{C}^4$ via it's action on each factor of the tensor product:

$$B \star (|\psi\rangle_p \otimes |\phi\rangle_e) = (B|\psi\rangle_p) \otimes (B|\phi\rangle_e) \quad \text{for } B \in SU(2).$$

The interpretation of this $SU(2)$-action is the same as in the one-particle case:
if $A \in SO(3)$ describes the rotation from P's basis for physical space to M's,
and if $f(B) = A$ (where $f: SU(2) \to SO(3)$ is the double cover), then the
action of B on $W_p \otimes W_e$ sends P's z-basis for the tensor product to M's z'-
basis for the tensor product. This $SU(2)$-action preserves the inner product on
$W_p \otimes W_e$, so we see that the elements of $SU(2)$ act via unitary matrices on $\mathbb{C}^4$.
Thus, the proton-electron system is described by a 4-dimensional unitary rep-
resentation of $SU(2)$, which we denote by $\pi_{\frac{1}{2}} \otimes \pi_{\frac{1}{2}}: SU(2) \to U(4) \subset GL(4, \mathbb{C})$.
At this point, it is irresistible to ask how this representation decomposes into
irreducible representations of $SU(2)$. From the discussion in Chapter 5, it
will suffice to determine the decomposition of the associated representation of
$\mathfrak{su}(2)_\mathbb{C} = \mathfrak{sl}_2(\mathbb{C})$.

For this, we begin by computing the corresponding representation of the
Lie algebra $\mathfrak{su}(2)$. As usual, we will compute the action of an element $iX \in \mathfrak{su}(2)$ on $W_p \otimes W_e$ by differentiating the action of $\exp(itX) \in SU(2)$. We
will need a Leibniz rule for differentiation of the tensor product, which is
established in the next exercise.

Exercise 6.8 (♣). *Suppose that* $\mathbf{a}(t) \in W_p$ *and* $\mathbf{b}(t) \in W_e$ *are differentiable vector-valued functions. Then adapt the proof of the product rule from
1-variable calculus to show that*

$$\frac{d}{dt}(\mathbf{a}(t) \otimes \mathbf{b}(t)) = \dot{\mathbf{a}}(t) \otimes \mathbf{b}(t) + \mathbf{a}(t) \otimes \dot{\mathbf{b}}(t).$$

Using this result, we may compute the action of $iX \in \mathfrak{su}(2)$:

$$\begin{aligned}
iX \star (|\psi\rangle_p \otimes |\phi\rangle_e) &= \frac{d}{dt}\left(\exp(itX) \star (|\psi\rangle_p \otimes |\phi\rangle_e)\right)|_{t=0} \\
&= \frac{d}{dt}\left(\exp(itX)|\psi\rangle_p \otimes \exp(itX)|\phi\rangle_e\right)|_{t=0} \\
&= (iX|\psi\rangle_p) \otimes |\phi\rangle_e + |\psi\rangle_p \otimes (iX|\phi\rangle_e).
\end{aligned}$$

Multiple Particles

Denoting the representation $\pi_{\frac{1}{2}} \otimes \pi_{\frac{1}{2}}$ by $\pi \colon SU(2) \to U(4)$, we see that $D\pi \colon \mathfrak{su}(2) \to \mathfrak{u}(4) = iH(4)$ may be written as

$$D\pi(iX) = iX \otimes I + I \otimes iX,$$

where I denotes the 2×2 identity matrix. As in Chapter 5 (see exercise 5.19), we now complexify this representation to obtain the representation $D\pi \colon \mathfrak{sl}_2(\mathbb{C}) \to \mathfrak{gl}_4(\mathbb{C})$ given by the same formula

$$D\pi(K) = K \otimes I + I \otimes K \quad \text{for all } K \in \mathfrak{sl}_2(\mathbb{C}) = M_0(2, \mathbb{C}).$$

(As mentioned earlier, we write $D\pi$ rather than $D\pi_{\mathbb{C}}$ in order to simplify the notation.) We would like to determine how this four-dimensional representation of $\mathfrak{sl}_2(\mathbb{C})$ decomposes into a direct sum of the irreducible representations $D\pi_s$ described in Chapter 5.

Observer M sounds a note of caution: the Lie algebra $\mathfrak{gl}_4(\mathbb{C}) = M(4, \mathbb{C})$ is also a ring under matrix multiplication. Hence, if $K_1, K_2 \in \mathfrak{sl}_2(\mathbb{C})$, then we can consider the linear transformation $D\pi(K_1)D\pi(K_2) \colon \mathbb{C}^4 \to \mathbb{C}^4$. But $\mathfrak{sl}_2(\mathbb{C}) = M_0(2, \mathbb{C})$ is *not* closed under matrix multiplication, since the product of two traceless matrices isn't traceless in general. Moreover, there is no reason to think that $D\pi$ respects the operation of matrix multiplication—it is a homomorphism of Lie algebras, not of rings. This situation can lead to real confusion if we attempt to determine the action of $K_1 K_2$ on X by applying $D\pi$ to the product $K_1 K_2 \in M(2, \mathbb{C})$. Instead, we must first determine $D\pi(K_1)$ and $D\pi(K_2)$, and then compute their product in $M(4, \mathbb{C})$.

Having made this comment, M now turns to the problem of decomposing the 4-dimensional representation of $\mathfrak{sl}_2(\mathbb{C})$ on $W_p \otimes W_e$ into irreducibles. Using the remark after example 5.35, it will suffice to study the action of the squared total spin squared operator, $\mathbf{S}^2$, since this element acts as the scalar $\hbar^2 s(s+1)$ on the irreducible representation $W_{2s} \simeq \mathbb{C}^{2s+1}$.

Recall that $\mathbf{S} := (S_x, S_y, S_z)$ denotes the vector of spin observables, where (expressed in the z-basis as 2×2 matrices):

$$S_x = \frac{\hbar}{2}\sigma_1, \quad S_y = \frac{\hbar}{2}\sigma_2, \quad S_z = \frac{\hbar}{2}\sigma_3.$$

Then the squared total spin operator is

$$\mathbf{S}^2 := \mathbf{S} \cdot \mathbf{S} = S_x^2 + S_y^2 + S_z^2.$$

Note that as a 2×2 matrix, we have $\mathbf{S}^2 = \frac{3\hbar^2}{4}I$, since each $\sigma_i^2 = I$. But (heeding M's earlier warning) in order to determine the action of $\mathbf{S}^2$ on $W_p \otimes W_e$, we need to compute $D\pi(\mathbf{S}) \cdot D\pi(\mathbf{S}) \in \mathfrak{gl}_4(\mathbb{C})$. We have

$$D\pi(\mathbf{S}) = (S_x \otimes I, S_y \otimes I, S_z \otimes I) + (I \otimes S_x, I \otimes S_y, I \otimes S_z) =: \hat{\mathbf{S}}_p + \hat{\mathbf{S}}_e,$$

where we introduce the symbols $\hat{\mathbf{S}}_p$ and $\hat{\mathbf{S}}_e$ to denote the vectors of spin-operators acting only on the proton or electron factor of the tensor product

126   *Symmetry and Quantum Mechanics*

respectively. Hence,

$$D\pi(\mathbf{S}) \cdot D\pi(\mathbf{S}) = (\hat{\mathbf{S}}_p + \hat{\mathbf{S}}_e) \cdot (\hat{\mathbf{S}}_p + \hat{\mathbf{S}}_e) = \hat{\mathbf{S}}_p \cdot \hat{\mathbf{S}}_p + 2\hat{\mathbf{S}}_p \cdot \hat{\mathbf{S}}_e + \hat{\mathbf{S}}_e \cdot \hat{\mathbf{S}}_e,$$

where we have used the fact that the components of $\hat{\mathbf{S}}_p$ and $\hat{\mathbf{S}}_e$ commute, since they act on different factors of the tensor product. The operator $\hat{\mathbf{S}}_p \cdot \hat{\mathbf{S}}_p$ acts as the scalar $\frac{3\hbar^2}{4}$ on the tensor product, as does $\hat{\mathbf{S}}_e \cdot \hat{\mathbf{S}}_e$:

$$\hat{\mathbf{S}}_p \cdot \hat{\mathbf{S}}_p = S_x^2 \otimes I + S_y^2 \otimes I + S_z^2 \otimes I = (S_x^2 + S_y^2 + S_z^2) \otimes I = \frac{3\hbar^2}{4} I \otimes I.$$

On the other hand, the operator $\hat{\mathbf{S}}_p \cdot \hat{\mathbf{S}}_e$ acts in a more complicated fashion. Applied to the first basis element:

$$\begin{aligned}
\hat{\mathbf{S}}_p \cdot \hat{\mathbf{S}}_e(|+z\rangle_p \otimes |+z\rangle_e) &= S_x|+z\rangle_p \otimes S_x|+z\rangle_e + S_y|+z\rangle_p \otimes S_y|+z\rangle_e \\
&\quad + S_z|+z\rangle_p \otimes S_z|+z\rangle_e \\
&= \frac{\hbar^2}{4}(|-z\rangle_p \otimes |-z\rangle_e + i^2|-z\rangle_p \otimes |-z\rangle_e \\
&\quad + |+z\rangle_p \otimes |+z\rangle_e) \\
&= \frac{\hbar^2}{4}|+z\rangle_p \otimes |+z\rangle_e.
\end{aligned}$$

Putting these computations together establishes the first column of the following matrix expressing the action of $\mathbf{S}^2$ on $W_p \otimes W_e$ in the z-basis:

$$\hbar^2 \begin{bmatrix} 2 & 0 & 0 & 0 \\ 0 & 1 & 1 & 0 \\ 0 & 1 & 1 & 0 \\ 0 & 0 & 0 & 2 \end{bmatrix}.$$

Exercise 6.9. *Verify that the other columns of this matrix are correct.*

The eigenvalues of the matrix above are 0 and $2\hbar^2$. The eigenspace for 0 is one-dimensional with basis ket

$$\frac{1}{\sqrt{2}}\left(|+z\rangle_p \otimes |-z\rangle_e - |-z\rangle_p \otimes |+z\rangle_e\right),$$

while the eigenspace for $2\hbar^2$ has dimension 3 with basis

$$\left\{ |+z\rangle_p \otimes |+z\rangle_e, \frac{1}{\sqrt{2}}\left(|+z\rangle_p \otimes |-z\rangle_e + |-z\rangle_p \otimes |+z\rangle_e\right), |-z\rangle_p \otimes |-z\rangle_e \right\}.$$

From example 5.35, we know that the eigenspace for $2\hbar^2$ is isomorphic to the irreducible $SU(2)$-representation $W_2 \cong \mathbb{C}^3$, and the eigenspace for 0 is isomorphic to the trivial $SU(2)$-representation $W_0 \cong \mathbb{C}$. Thus, we obtain the decomposition into irreducibles

$$W_p \otimes W_e \cong \mathbb{C} \oplus \mathbb{C}^3.$$

Multiple Particles 127

In the notation introduced in Section 5.5, where π_s denotes the irreducible spin-s representation of dimension $2s + 1$, we see that

$$\pi_{\frac{1}{2}} \otimes \pi_{\frac{1}{2}} \simeq \pi_0 \oplus \pi_1.$$

Moreover, we have chosen the bases for π_0 and π_1 to be simultaneous eigenkets for the action of S_z and $\mathbf{S}^2$, so that we may write the isomorphism $\pi_0 \oplus \pi_1 \to \pi_{\frac{1}{2}} \otimes \pi_{\frac{1}{2}}$ explicitly as

$$|0\rangle_0 \;\mapsto\; \frac{1}{\sqrt{2}} \left(|+z\rangle_p \otimes |-z\rangle_e - |-z\rangle_p \otimes |+z\rangle_e \right)$$

$$|-1\rangle_1 \;\mapsto\; |-z\rangle_p \otimes |-z\rangle_e$$

$$|0\rangle_1 \;\mapsto\; \frac{1}{\sqrt{2}} \left(|+z\rangle_p \otimes |-z\rangle_e + |-z\rangle_p \otimes |+z\rangle_e \right)$$

$$|1\rangle_1 \;\mapsto\; |+z\rangle_p \otimes |+z\rangle_e.$$

The coefficients displayed above that explicitly define this isomorphism in terms of the tensor product basis $\{|\pm z\rangle_p \otimes |\pm z\rangle_e\}$ are called *Clebsch-Gordan coefficients*.

Observer P provides the physical interpretation of this representation theory: the spin-states of the composite proton-electron system may be described as the superpositions of a spin-0 and spin-1 particle. Thus, the representations π_0 and π_1 arise naturally in the study of *composite* systems of spin-$\frac{1}{2}$ particles.

Observer P points out that the interaction between the proton and electron has not been specified in any way, and thus the previous computations remain valid for any choice of Hamiltonian to describe the interaction. As a specific example, P decides to model only the spin-spin interaction of the proton-electron system.

Example 6.10. *The fact that the proton is a charged particle of nonzero spin means that it has a magnetic dipole moment generating a magnetic field which interacts with the spin of the electron (see [9, Section 6.5]). P introduces the following Hamiltonian operator on the tensor product $W_p \otimes W_e$ to model this spin-spin interaction:*

$$\mathcal{H} = \frac{b}{\hbar^2} \hat{\mathbf{S}}_p \cdot \hat{\mathbf{S}}_e = \frac{b}{\hbar^2} \left(S_x \otimes S_x + S_y \otimes S_y + S_z \otimes S_z \right).$$

Here b is a constant with the dimensions of energy. Using the identity

$$\hat{\mathbf{S}}_p \cdot \hat{\mathbf{S}}_e = \frac{1}{2} \left(D\pi(\mathbf{S}) \cdot D\pi(\mathbf{S}) - \hat{\mathbf{S}}_p \cdot \hat{\mathbf{S}}_p - \hat{\mathbf{S}}_e \cdot \hat{\mathbf{S}}_e \right)$$

it follows that the eigenvectors for $\mathcal{H}$ are the same as those found above for the action of $\mathbf{S}^2$, and P concludes that the spin-1 representation describes higher energy states than the spin-0 representation. Specifically, there is a single state with energy $-3b$ and a three-dimensional space of states with energy b. In general, the state of the system is given by a superposition of these energy states. This phenomena is called the hyperfine splitting *of the ground state of the hydrogen atom. We will return to this example in Section 8.7.1.*

128   *Symmetry and Quantum Mechanics*

6.2   The Clebsch-Gordan problem

We have just discovered one entry in a "multiplication table" for the irreducible representations of $SU(2)$:

$$\pi_{\frac{1}{2}} \otimes \pi_{\frac{1}{2}} \simeq \pi_0 \oplus \pi_1.$$

Moreover, the analysis in the previous section actually specified the isomorphism explicitly, by identifying S_z-eigenbases for the irreducible representations π_0 and π_1 inside the tensor product. The obvious next step is to investigate the *Clebsch-Gordan problem* for $SU(2)$: given two irreducible representations π_a and π_b, how does the tensor product $\pi_a \otimes \pi_b$ decompose into irreducible representations? The answer turns out to be quite simple:

$$\pi_a \otimes \pi_b \simeq \bigoplus_{s=|a-b|}^{a+b} \pi_s. \tag{6.3}$$

Observer P interprets this "multiplication table" as tabulating the rules for adding spins in quantum mechanics. For example, a composite system consisting of a spin-1 and a spin-2 particle decomposes as

$$\pi_1 \otimes \pi_2 \simeq \pi_1 \oplus \pi_2 \oplus \pi_3.$$

In this section, we will establish the existence of the isomorphism (6.3), but we will not describe it explicitly by finding the Clebsch-Gordan coefficients as in the case of $\pi_{\frac{1}{2}} \otimes \pi_{\frac{1}{2}}$ in the previous section.

Proposition 6.11. *For all $a, b \geq 0$, the tensor product of the irreducible $SU(2)$-representations π_a and π_b has a direct-sum decomposition as follows:*

$$\pi_a \otimes \pi_b \simeq \bigoplus_{s=|a-b|}^{a+b} \pi_s.$$

Proof. To begin, recall from Section 5.5 that π_a has a basis of σ_3-eigenkets (labeled by their eigenvalues)

$$|-2a\rangle, |-2a+2\rangle, \ldots, |2a\rangle.$$

Similarly, π_b has eigenbasis

$$|-2b\rangle, |-2b+2\rangle, \ldots, |2b\rangle.$$

The pairwise tensor products of these basis elements yield a basis of σ_3-eigenkets for $\pi_a \otimes \pi_b$, where $|\alpha\rangle \otimes |\beta\rangle$ is an eigenket for σ_3 with eigenvalue

Multiple Particles

$\alpha + \beta$:

$$
\begin{aligned}
\sigma_3 \star (|\alpha\rangle \otimes |\beta\rangle) &= (\sigma_3 \star |\alpha\rangle) \otimes |\beta\rangle + |\alpha\rangle \otimes (\sigma_3 \star |\beta\rangle) \\
&= \alpha|\alpha\rangle \otimes |\beta\rangle + |\alpha\rangle \otimes \beta|\beta\rangle \\
&= (\alpha + \beta)|\alpha\rangle \otimes |\beta\rangle.
\end{aligned}
$$

Suppose without loss of generality that $a \geq b$. Then there is a unique eigenket with largest eigenvalue $2(a+b)$, namely $|2a\rangle \otimes |2b\rangle$. The next largest eigenvalue is $2(a + b - 1)$, with 2-dimensional eigenspace spanned by

$$|2a\rangle \otimes |2b - 2\rangle \text{ and } |2a - 2\rangle \otimes |2b\rangle.$$

This pattern continues (the eigenvalue decreasing by 2 at each step and the number of eigenkets increasing by 1) until we reach the eigenvalue $2(a-b) \geq 0$, with eigenspace of dimension $2b + 1$ spanned by

$$|2a\rangle \otimes |-2b\rangle, \ |2a - 2\rangle \otimes |-2b + 2\rangle, \ \ldots, \ |2a - 4b\rangle \otimes |2b\rangle.$$

Now write

$$\pi_a \otimes \pi_b \simeq \pi_{s_1} \oplus \pi_{s_2} \oplus \cdots \oplus \pi_{s_k}, \tag{6.4}$$

for some as-yet-unknown numbers $s_j \in \frac{1}{2}\mathbb{Z}$. Assume that $s_1 \geq s_2 \geq \cdots \geq s_k$. From the direct sum decomposition, the largest eigenvalue for the action of σ_3 on the tensor product must be $2s_1$. But we know that the largest eigenvalue is $2(a + b)$, which implies that $s_1 = a + b$ as claimed. The representation π_{s_1} contributes one dimension to each of the following eigenspaces of the tensor product:

$$-2s_1, -2s_1 + 2, \ldots, 2s_1 - 2, 2s_1.$$

But the eigenspace for $2s_1 - 2 = 2(a+b-1)$ has dimension 2, so a 1-dimensional subspace must come from π_{s_2}. It follows that $s_2 = a + b - 1$. The pattern continues until we reach the eigenspace for $2(a - b)$: each time we decrease the eigenvalue by 2, we find that the previously identified representations do not quite account for the entire eigenspace, which allows us to identify the next representation. At this point we have established that

$$s_1 = a + b, s_2 = a + b - 1, \ldots, s_{2b+1} = a - b.$$

We now show that this list gives the full decomposition of the tensor-product by making a dimension count. Indeed, since $\dim(\pi_s) = 2s + 1$, taking dimensions in (6.4) yields

$$(2a + 1)(2b + 1) = (2s_1 + 1) + (2s_2 + 1) + \cdots + (2s_k + 1).$$

130 *Symmetry and Quantum Mechanics*

Noting that $s_j + j = a + b + 1$ for $j = 1, \ldots, 2b + 1$, we compute

$$
\begin{aligned}
\sum_{j=1}^{2b+1} (2s_j + 1) &= \sum_{j=1}^{2b+1} (2(a + b + 1 - j) + 1) \\
&= (2(a + b) + 3)(2b + 1) - 2 \sum_{j=1}^{2b+1} j \\
&= (2(a + b) + 3)(2b + 1) - (2b + 2)(2b + 1) \\
&= (2a + 1)(2b + 1).
\end{aligned}
$$

This equality of dimensions establishes the direct sum decomposition. $\qquad\square$

6.3 Identical particles—spin only

M is quite impressed with the foregoing analysis of two-particle systems, but he wonders about the case where the particles are identical, therefore theoretically indistinguishable. Take two identical spin-s particles for instance, each with spinor-space given by a copy of the $SU(2)$-representation π_s. Then it would seem that the spinor-space for the composite system should be modeled by $\pi_s \otimes \pi_s$. But this tensor product has the obvious symmetry defined by $Q(|\psi_1\rangle \otimes |\psi_2\rangle) = |\psi_2\rangle \otimes |\psi_1\rangle$. That is, Q simply switches the labels on the two states. Since the particles are indistinguishable (unlike the proton and electron in our earlier discussion), there is no way to *actually label* the two particles. That is, if observer P and observer M each independently assign mental labels to the two particles for the purpose of their modeling, then they may well disagree, and switching the labels shouldn't change the underlying physical state. Hence, the physical spin-states in $\pi_s \otimes \pi_s$ must be eigenvectors of Q, so that switching the labels only results in a change of phase. But the operator Q satisfies $Q^2 = I$, hence has eigenvalues ± 1. It follows that physical states are either symmetric or antisymmetric under label-switching.

The tensor product decomposes as a direct sum of antisymmetric and symmetric tensors:

$$
\pi_s \otimes \pi_s = \wedge^2 \pi_s \oplus \mathrm{Sym}^2 \pi_s.
$$

Here, the first summand consists of the antisymmetric tensors, which we may describe explicitly in terms of the basis kets $|j\rangle$ for π_s:

$$
\wedge^2 \pi_s = \mathrm{span}_{\mathbb{C}}\{|j\rangle \otimes |j'\rangle - |j'\rangle \otimes |j\rangle \mid j, j' = -2s, -2s + 2, \ldots, 2s\}.
$$

Similarly, the second summand consists of the symmetric tensors:

$$
\mathrm{Sym}^2 \pi_s = \mathrm{span}_{\mathbb{C}}\{|j\rangle \otimes |j'\rangle + |j'\rangle \otimes |j\rangle \mid j, j' = -2s, -2s + 2, \ldots, 2s\}.
$$

Multiple Particles

Exercise 6.12. *Show that* $\dim(\wedge^2 \pi_s) = s(2s+1)$ *and also that* $\dim(\text{Sym}^2 \pi_s) = (s+1)(2s+1)$. *Verify that the sum of these dimensions is* $(2s+1)^2 = \dim(\pi_s \otimes \pi_s)$.

The upshot of the discussion above is that the physical spin-states for the composite system of two identical spin-s particles must lie in one of these summands or the other. Observer P mentions something called the *Spin-Statistics Theorem*: particles with integral spin (*bosons*) only form symmetric tensors, while particles with half-integral spin (*fermions*) only form antisymmetric tensors. Actually, it is a bit premature to make this demand of our spin-states: the Spin-Statistics Theorem applies only to the full quantum states of identical particles, describing not only their spin, but also the more familiar observables position and momentum. In Part II we develop the mathematical framework for this description, and we will return to the question of identical particles at the end of Chapter 8.

Part II

Position & Momentum

Chapter 7

A One-Dimensional World

In which M and P discover the Heisenberg group, H_1.

M and P now want to incorporate the more familiar dynamical variables into their model: position and momentum. Before building a three-dimensional model, they decide to get their bearings by imagining a one-dimensional world. M starts things off with the following comment: while it might seem reasonable to employ the vector space $\mathbb{R}$ as a model of a one-dimensional world, there is a small problem: the vector space has a special point 0, while the physical world presumably has no such distinguished point. For this reason, she proposes the following definition, which amounts to "forgetting the origin" in the vector space $\mathbb{R}$.

Definition 7.1. One-dimensional physical space *is a real affine space*, $\mathbb{A}$, *of dimension* 1. *More precisely,* $\mathbb{A}$ *is a set endowed with a simply transitive action by the Lie group* $(\mathbb{R}, +)$:

$$\mathbb{R} \times \mathbb{A} \to \mathbb{A}, \qquad (r, a) \mapsto r + a.$$

Simple transitivity means that for all $a, a' \in \mathbb{A}$, there exists a unique $r \in \mathbb{R}$ such that $r + a = a'$. This formalism serves to capture the fact that although M and P are looking at the same physical space $\mathbb{A}$, they may be at different locations, and hence think of different points in space as the origin. Indeed, suppose that P is located at $a_0 \in \mathbb{A}$ and M is located at $a_0' \in \mathbb{A}$. By simple transitivity, there exists a unique w in $(\mathbb{R}, +)$ satisfying $a_0' = w + a_0$. Thus, M's location is obtained from P's through translating by w (see Figure 7.1).

The point a_0 yields a bijection $\mathbb{R} \to \mathbb{A}$ given by $r \mapsto r + a_0$. Hence, the choice of location a_0 allows P to think of physical space as the vector space $\mathbb{R}$. Let $\varphi \colon \mathbb{A} \to \mathbb{R}$ denote the inverse of this identification, so $\varphi(a_0) = 0$. Similarly, the choice of a_0' yields a bijection $\mathbb{R} \to \mathbb{A}$ given by $r \mapsto r + a_0'$. Let $\varphi' \colon \mathbb{A} \to \mathbb{R}$ denote the inverse of this map, so $\varphi'(a_0') = 0$. Then the composition $\varphi' \circ \varphi^{-1} \colon \mathbb{R} \to \mathbb{R}$ is given by $r \mapsto r - w$. This means that M's position values are obtained from P's by subtracting w, the translation necessary to move P's

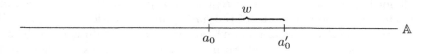

FIGURE 7.1: Observer M's location is translated from P's through a distance $w \in \mathbb{R}$.

136   Symmetry and Quantum Mechanics

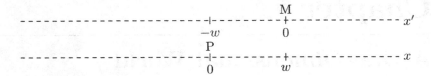

FIGURE 7.2: Observer M's coordinate system is translated from P's through a distance $w \in \mathbb{R}$, so $x' = x - w$.

location to M's (see Figure 7.2). Thus, P thinks of one-dimensional physical space as $\mathbb{R}$ endowed with the translation action of the Lie group $(\mathbb{R}, +)$, serving to connect his choice of location/origin to any other choice. This is completely analogous to the way that, in Chapter 1, observer P thought of the inner product space $(V, \langle, \rangle)$ as $(\mathbb{R}^3, \cdot)$ endowed with the rotation action of the Lie group $SO(3)$.

7.1 Position

Now P imagines that there is a particle in space. Thinking naively, if he were to measure the particle's position, he might obtain any value $x \in \mathbb{R}$. That is, the particle could be anywhere. Following the formalism that emerged from the investigation of spin, he denotes by the symbol δ_x the state of the particle with definite position x. Then a general position state should be modeled by a superposition of the states δ_x. As a first attempt at a mathematical model, P considers the complex vector space $\mathcal{X}$ having the set $\{\delta_x \mid x \in \mathbb{R}\}$ as a basis. Moreover, he defines an inner product on $\mathcal{X}$ by taking the basis vectors δ_x to be orthonormal. In this model, an arbitrary position state would be represented by a unit vector $\psi \in \mathcal{X}$, considered up to multiplication by phases $e^{i\theta} \in U(1)$. If ψ is such a position state, then there exist $n \in \mathbb{N}$, $x_1, \ldots, x_n \in \mathbb{R}$ and $c_1, \ldots, c_n \in \mathbb{C}$ such that

$$\psi = c_1 \delta_{x_1} + \cdots + c_n \delta_{x_n}.$$

Moreover, $|c_1|^2 + \cdots + |c_n|^2 = 1$, with $|c_i|^2$ giving the probability that a position measurement on ψ will return the value x_i. This model is unsatisfactory for the following reason: the state ψ represents a particle that when measured will definitely be found at one of the *finitely many* locations $x_1, \ldots, x_n$. But there are *uncountably many* different possible position values x, so the state ψ strikes P as quite unphysical. So he is left with the question: how to find a model for position states that allows for the superposition of uncountably many states?

M listens carefully to P and then proposes the following solution: *If physi-*

A One-Dimensional World

cal space consisted of only finitely many points $x_1, \ldots, x_n$, then the preceding model would be fine, and we could think of the state ψ as being a function $\psi \colon \{x_1, \ldots, x_n\} \to \mathbb{C}$ defined by $\psi(x_i) = c_i$. That is, instead of thinking of ψ as being a linear combination of the vectors δ_{x_i}, we could think of ψ as being a function defined on space, with $\psi(x_i) = $ "amplitude for being found at x_i." So, since P has modeled physical space as $\mathbb{R}$, he should define a position state to be a function $\psi \colon \mathbb{R} \to \mathbb{C}$, and interpret $\psi(x)$ as the "amplitude density" for the position of the particle. The requirement that the total probability of being found *somewhere* in space is 1 becomes the integral condition that $\int_{\mathbb{R}} |\psi|^2 dx = 1$. This discussion leads M to propose the definition:

Definition 7.2. One-dimensional position space *is the complex inner product space of square-integrable functions* $(L^2(\mathbb{R}), \langle | \rangle)$, *where*

$$L^2(\mathbb{R}) = \{\psi \colon \mathbb{R} \to \mathbb{C} \mid \int_{\mathbb{R}} |\psi|^2 dx < \infty\},$$

and $\langle \psi | \phi \rangle = \int_{\mathbb{R}} \psi^* \phi \, dx$. Position states[1] *are unit vectors in* $L^2(\mathbb{R})$ *considered up to multiplication by a phase in* $U(1)$. *If* ψ *is a position state and* $E \subset \mathbb{R}$ *is a measurable subset, then* $\int_E |\psi|^2 dx$ *is the probability that the particle, if measured, will be found in the subset* E. *We refer to the functions* ψ *as* wavefunctions.

Note that this definition of position space depends on P's choice of location $a_0 \in A$, which identifies one-dimensional physical space A with $\mathbb{R}$ endowed with the translation action of $(\mathbb{R}, +)$ given by $x \mapsto x + w$ for $x \in \mathbb{R}$ and $w \in (\mathbb{R}, +)$. To account for this, the position space $L^2(\mathbb{R})$ also carries an action of $(\mathbb{R}, +)$.

Definition 7.3. *The* translation action *of* $(\mathbb{R}, +)$ *on position space* $L^2(\mathbb{R})$ *is given by*[2]
$$\mathbb{R} \times L^2(\mathbb{R}) \to L^2(\mathbb{R}), \qquad (w, \psi(x)) \mapsto \psi(x - w).$$
We denote the translation action by $T_w \psi(x) := \psi(x - w)$.

Note that the translation action preserves the inner product:

$$\langle T_w \psi | T_w \phi \rangle = \int_{\mathbb{R}} \psi(x - w)^* \phi(x - w) dx = \int_{\mathbb{R}} \psi(x)^* \phi(x) dx = \langle \psi | \phi \rangle.$$

In fact (see Appendix A.3.2), the map $w \mapsto T_w$ is a *strongly continuous one-parameter unitary group* acting on the Hilbert space $L^2(\mathbb{R})$. Thus, Stone's

---

[1] In our treatment of spin, we were careful to distinguish notationally between a spin-state ψ and a ket $|\psi\rangle$ representing it. In order to ease the notation, we now discontinue this practice, and use ψ to denote both a position state and a function representing it, always remembering that the representing function can be multiplied by a global phase $e^{i\theta} \in U(1)$ without changing the state.

[2] Recall from Section 5.2 that when a group G acts on a space X, it also acts in a natural way on the space of functions with domain X by pre-composition with the inverse: $g \star F(x) := F(g^{-1}x)$. See Section 7.3.1 for further thoughts about this phenomenon.

138   *Symmetry and Quantum Mechanics*

Theorem A.51 applies, as described briefly at the end of Section 5.6. As we will see in due course, the corresponding self-adjoint operator on position space will provide the *linear momentum observable*.

From our experience with spin, we want to model the observable "position" by a self-adjoint operator $\hat{x}$ on position space. Thinking informally, the position state ψ is a superposition of the states δ_λ having definite position values $\lambda \in \mathbb{R}$. These should be eigenstates for the position operator $\hat{x}$, satisfying $\hat{x}\delta_\lambda = \lambda\delta_\lambda$. This leads us to define the position operator to be multiplication by the function x:

$$\hat{x}\psi(x) := x\psi(x).$$

Of course, this operator is not defined on all of $L^2(\mathbb{R})$, since $x\psi$ may not be square-integrable even though ψ is. Nevertheless, the operator $\hat{x}$ is defined and self-adjoint on the dense subspace

$$\left\{\psi \in L^2(\mathbb{R}) \mid x\psi \in L^2(\mathbb{R})\right\},$$

and that is all we need in order to have a good analytic theory. Note, however, that the operator $\hat{x}$ has no actual eigenfunctions in $L^2(\mathbb{R})$. Indeed, if $\delta_\lambda(x)$ were such an eigenfunction with eigenvalue $\lambda \in \mathbb{R}$, then it would satisfy $(x - \lambda)\delta_\lambda(x) = 0$ for all x, which would imply that $\delta_\lambda(x) = 0$ unless $x = \lambda$. But the only such function in $L^2(\mathbb{R})$ is the zero function (we identify square-integrable functions that differ on a set of measure zero). To get around this fact, physicists often speak of the *Dirac delta function* $\delta_\lambda(x)$ which has the seemingly contradictory properties of being zero except at $x = \lambda$ but having an integral equal to 1:

$$\int_{\mathbb{R}} \delta_\lambda(x)dx = 1.$$

In fact, such an object does exist, but it is not an honest function. Rather, it is a *distribution*: a continuous linear functional on the dense subspace of compactly supported smooth functions, $\mathcal{C}_c^\infty(\mathbb{R}) \subset L^2(\mathbb{R})$. Specifically, the Dirac delta function δ_λ is the linear mapping $\mathcal{C}_c^\infty(\mathbb{R}) \to \mathbb{C}$ defined by evaluation at $x = \lambda$:

$$\delta_\lambda(f) := f(\lambda).$$

As a bit of suggestive notation, we generally write the preceding formula as an integral:

$$\int_{\mathbb{R}} \delta_\lambda(x)f(x)dx := f(\lambda).$$

We will not develop the theory of distributions in this book, although we will mention them occasionally in a motivational way as above. For future reference, we record the definition of the position operator below:

Definition 7.4. *The* position operator $\hat{x}$ *is the self-adjoint operator on position space* $L^2(\mathbb{R})$ *defined as multiplication by* x:

$$\hat{x}\psi(x) := x\psi(x).$$

The domain of $\hat{x}$ *is the dense subspace* $\left\{\psi \in L^2(\mathbb{R}) \mid x\psi \in L^2(\mathbb{R})\right\}$.

A One-Dimensional World

Before going on with the development of one-dimensional position space, we stop to highlight some parallels with the model of electron spin developed in the first half of the book. We began in Chapter 2 with a discussion of the Stern-Gerlach experiment, which led to the notion of the observable "spin in the z-direction." To model this observable, we introduced a complex inner product space $\mathbb{C}^2 = \mathbb{C}|+z\rangle \oplus \mathbb{C}|-z\rangle$ spanned by an orthonormal basis corresponding to the states of definite spin in the z-direction. We defined the observable itself as the Hermitian operator $S_z = \frac{\hbar}{2}\sigma_3$, obtained as the sum of the orthogonal projections onto the states of definite spin, weighted by the observed values $\pm\frac{\hbar}{2}$. Of course, this description depends on the choice of z-direction, and different choices are accounted for by the $SU(2)$-action on $\mathbb{C}^2$ corresponding to the rotation action of $SO(3)$ on $\mathbb{R}^3$. Moreover, the operator $-\frac{i}{\hbar}S_z$ is an element of the Lie algebra $\mathfrak{su}(2)$, and it generates a one-parameter unitary group via exponentiation:

$$\exp\left(-\frac{i}{\hbar}\theta S_z\right) = \exp\left(\frac{-i\theta}{2}\sigma_3\right) = \begin{bmatrix} e^{-\frac{i\theta}{2}} & 0 \\ 0 & e^{\frac{i\theta}{2}} \end{bmatrix} \in SU(2).$$

The original operator $-\frac{i}{\hbar}S_z$ may be recovered as the generator of this unitary action, obtained as the θ-derivative (evaluated at $\theta = 0$). Moreover, this one-parameter group maps to the group of rotations around the z-axis under the double cover $f\colon SU(2) \to SO(3)$. Hence, in the case of spin, the observable S_z ultimately arises from a symmetry of spinor space, which in turn arises from a symmetry of the physical space $\mathbb{R}^3$ of Chapter 1.

Analogously, the observable "position" is given by the self-adjoint operator $\hat{x}$ on position space $L^2(\mathbb{R})$. If we allow ourselves to think about distributions, then we may regard position space (informally) as being the orthogonal span of the Dirac delta functions δ_λ, thought of as states of definite position. Then the operator $\hat{x}$ is just the sum of the orthogonal projections onto these states, weighted by the observed values λ. Again, this description depends on the choice of origin in physical space, and different choices are accounted for by the translation action on $L^2(\mathbb{R})$ coming from the action of $(\mathbb{R}, +)$ on $\mathbb{R}$. Pushing the analogy further, we see that the operator $-\frac{i}{\hbar}\hat{x}$ arises via differentiation (with respect to the parameter v) from the unitary $(\mathbb{R}, +)$-action on $L^2(\mathbb{R})$ defined by

$$\mathbb{R} \times L^2(\mathbb{R}) \to L^2(\mathbb{R}), \qquad (v, \psi(x)) \mapsto e^{-\frac{ivx}{\hbar}}\psi(x).$$

(This is Stone's Theorem A.51 for the self-adjoint position operator $\hat{x}$.) Hence, the operator $\hat{x}$ does arise from a symmetry of position space. But note that the $(\mathbb{R}, +)$-action generated by $\hat{x}$ is *not* the translation action from definition 7.3. Nevertheless, we will see in Section 7.3 that this additional symmetry of position space will combine with translation to yield a larger symmetry group (the Heisenberg group) that has fundamental importance for the theory of quantum mechanics. Moreover, we will provide a physical interpretation of the Heisenberg group action in Section 7.3.1.

140   *Symmetry and Quantum Mechanics*

7.2   Momentum

Just as M and P obtained the spin operators S_x, S_y, S_z by differentiating the action of $SU(2)$ on spinor space, they now apply the same procedure to the translation action of $(\mathbb{R}, +)$ on $L^2(\mathbb{R})$. Clearly, the tangent space to $(\mathbb{R}, +)$ at the identity is $\mathbb{R}$. Moreover, since $(\mathbb{R}, +)$ is an abelian group, it follows that the Lie algebra of $(\mathbb{R}, +)$ is $\mathfrak{r} = \mathbb{R}$ with trivial Lie bracket (i.e., all brackets are zero). In order to determine the exponential map $\exp \colon \mathfrak{r} \to \mathbb{R}$, we use the characterization from Section 5.3 (see the discussion before proposition 5.20): if $p \in \mathfrak{r} = \mathbb{R}$ is a tangent vector at $0 \in (\mathbb{R}, +)$, then $\exp(p) := c_p(1)$, where $c_p \colon (\mathbb{R}, +) \to (\mathbb{R}, +)$ is the unique homomorphism satisfying $\dot{c}_p(0) = p$. In this case we clearly have $c_p(t) = pt$, so we find that $\exp(p) := c_p(1) = p$.

It may seem strange that the exponential mapping for the Lie group $(\mathbb{R}, +)$ is simply the identity function rather than the usual infinite series. But this is because we have expressed the group additively, rather than as a multiplicative matrix group. The next exercise reveals the hidden exponential series by expressing $(\mathbb{R}, +)$ as a matrix group.

Exercise 7.5. *Consider the set $\mathcal{R}$ of 2×2 upper-triangular real matrices with 1's on the diagonal:*

$$\mathcal{R} := \left\{ \begin{bmatrix} 1 & a \\ 0 & 1 \end{bmatrix} \mid a \in \mathbb{R} \right\}.$$

Show that $\mathcal{R}$ is a group under matrix multiplication, isomorphic to the additive group of real numbers $(\mathbb{R}, +)$. Show that the Lie algebra of $\mathcal{R}$ is the space of strictly upper triangular real matrices. Finally, use the power series for the exponential mapping to show that

$$\exp \begin{bmatrix} 0 & p \\ 0 & 0 \end{bmatrix} = \begin{bmatrix} 1 & p \\ 0 & 1 \end{bmatrix}.$$

To determine the action of $\mathfrak{r}$ on $L^2(\mathbb{R})$, we take the derivative of the translation action of $(\mathbb{R}, +)$. For $p \in \mathfrak{r} = \mathbb{R}$ and a sufficiently nice function $\psi(x) \in L^2(\mathbb{R})$ we have

$$p \star \psi(x) = \frac{d}{dt} \left(T_{c_p(t)} \psi(x) \right) \Big|_{t=0} = \frac{d}{dt} \psi(x - pt) \Big|_{t=0} = -p \frac{d\psi}{dx}.$$

In particular, the element $1 \in \mathfrak{r} = \mathbb{R}$ acts on $L^2(\mathbb{R})$ as $-\frac{d}{dx}$. Thinking of 1 as the generator of translation and remembering the correspondence $S_{\mathbf{u}} \leftrightarrow -\frac{i}{\hbar} S_{\mathbf{u}}$ between spin-observables and generators of rotations, M makes the following definition.

Definition 7.6. *The linear momentum operator $\hat{p}$ is the self-adjoint operator*

A One-Dimensional World

on^3 *position space $L^2(\mathbb{R})$ given by the differential operator*

$$\hat{p} := -\frac{\hbar}{i}\left(-\frac{d}{dx}\right) = \frac{\hbar}{i}\frac{d}{dx}.$$

Note that this operator does have the dimensions of angular momentum divided by length, or linear momentum. The interpretation of $\hat{p}$ as the momentum operator will be justified more fully below.

To understand the structure of the operator $\hat{p} = \frac{\hbar}{i}\frac{d}{dx}$, let's look for it's eigenfunctions ψ_p, which are solutions to the differential equation

$$\frac{d\psi_p}{dx} = \frac{i}{\hbar}p\psi_p \qquad p \in \mathbb{R}.$$

Clearly, for each eigenvalue $p \in \mathbb{R}$, the corresponding solution is

$$\psi_p(x) = N_p e^{\frac{i}{\hbar}px},$$

where $N_p \in \mathbb{C}$ is an arbitrary constant. In order to represent a position-state, ψ_p must have unit norm, which should determine the choice of N_p. But

$$\int_{\mathbb{R}} \psi_p^* \psi_p dx = \int_{\mathbb{R}} N_p^* e^{-\frac{i}{\hbar}px} N_p e^{\frac{i}{\hbar}px} dx = \int_{\mathbb{R}} |N_p|^2 dx = \infty,$$

unless $N_p = 0$. Thus, the eigenfunctions ψ_p are not elements of $L^2(\mathbb{R})$, and the model predicts that states with definite momentum values are unphysical. Nevertheless, the momentum eigenfunctions *may be thought of* as a basis for $L^2(\mathbb{R})$, via the theory of the Fourier transform.

To explain this, note that for any $\psi(x) \in L^1(\mathbb{R}) \cap L^2(\mathbb{R})$, the integral $\langle \psi_p | \psi \rangle = \int_{\mathbb{R}} \psi_p^*(x)\psi(x)dx = \int_{\mathbb{R}} N_p^* e^{-\frac{i}{\hbar}px}\psi(x)dx$ exists[4]. We thus obtain a function $\widetilde{\psi} \colon \mathbb{R} \to \mathbb{C}$ defined by $\widetilde{\psi}(p) := \langle \psi_p | \psi \rangle$, called the *Fourier transform* of ψ. Choosing the normalization constants to be $N_p = \frac{1}{\sqrt{2\pi\hbar}}$, the *Fourier Inversion Theorem* says that (assuming $\widetilde{\psi}(p) \in L^1(\mathbb{R})$):

$$\psi(x) = \int_{\mathbb{R}} \widetilde{\psi}(p)\psi_p(x)dp = \frac{1}{\sqrt{2\pi\hbar}} \int_{\mathbb{R}} \widetilde{\psi}(p)e^{\frac{i}{\hbar}px}dp.$$

In this way, we can think of sufficiently nice position states $\psi(x)$ as superpositions of the momentum eigenstates ψ_p, with the amplitude density for ψ to be found with momentum p given by the Fourier transform $\widetilde{\psi}(p)$. In fact, the Fourier transform may be extended to a unitary automorphism of the entire Hilbert space $L^2(\mathbb{R})$; see [4, Section 8.3].

To investigate the relationship between position and momentum, we begin by computing the action of the commutator $[\hat{x}, \hat{p}]$ on position space:

---

[3] Just as for the position operator $\hat{x}$, the domain of $\hat{p}$ is a dense subspace of $L^2(\mathbb{R})$.

[4] Here, $L^1(\mathbb{R}) = \{\psi \colon \mathbb{R} \to \mathbb{C} \mid \int_{\mathbb{R}} |\psi|dx < \infty\}$ denotes the space of *absolutely integrable* functions on $\mathbb{R}$.

142 *Symmetry and Quantum Mechanics*

Proposition 7.7. *The commutator $[\hat{x}, \hat{p}]$ acts as the scalar $i\hbar$ on sufficiently nice functions in position space $L^2(\mathbb{R})$.*

Proof. This is essentially the product rule:

$$
\begin{aligned}
[\hat{x}, \hat{p}]\psi(x) &= (\hat{x}\hat{p} - \hat{p}\hat{x})\psi(x) \\
&= \frac{\hbar}{i}\left(x\frac{d\psi}{dx} - \frac{d}{dx}(x\psi(x)) \right) \\
&= \frac{\hbar}{i}\left(x\frac{d\psi}{dx} - \psi(x) - x\frac{d\psi}{dx} \right) \\
&= i\hbar\psi(x).
\end{aligned}
$$

$\square$

Just as in definitions 3.14 and 3.16, we define the expectation value and uncertainty of the position observable in a state ψ as follows:

$$
\begin{aligned}
\langle\hat{x}\rangle_\psi &:= \langle\psi|\hat{x}\psi\rangle = \int_{\mathbb{R}} x|\psi(x)|^2 dx \\
\Delta_\psi\hat{x} &:= \langle\psi|(\hat{x} - \langle\hat{x}\rangle_\psi)^2\psi\rangle.
\end{aligned}
$$

Similarly, for the momentum observable we have

$$
\begin{aligned}
\langle\hat{p}\rangle_\psi &:= \langle\psi|\hat{p}\psi\rangle = -i\hbar\int_{\mathbb{R}} \psi(x)^*\frac{d\psi}{dx} dx \\
\Delta_\psi\hat{p} &:= \langle\psi|(\hat{p} - \langle\hat{p}\rangle_\psi)^2\psi\rangle.
\end{aligned}
$$

For a given wavefunction ψ, the expectation values should be interpreted as the average values we would obtain if we made many measurements on identical copies of the state ψ. The uncertainties then indicate the "spread" of the measured values around the expectation value.

Although one must be careful to restrict attention to position states ψ such that $\hat{x}\psi, \hat{p}\psi, \hat{x}\hat{p}\psi$ and $\hat{p}\hat{x}\psi$ are in $L^2(\mathbb{R})$, the argument leading to theorem 3.20 extends to establish the *Heisenberg Uncertainty Principle* for position and momentum (for further details see [10, Chapter 12]):

Theorem 7.8 (Heisenberg Uncertainty Principle). *For sufficiently nice position states $\psi \in L^2(\mathbb{R})$, we have the inequality:*

$$
\Delta_\psi\hat{x}\Delta_\psi\hat{p} \geq \frac{\hbar}{2}. \tag{7.1}
$$

The Heisenberg Uncertainty Principle places a fundamental limit on the simultaneous determination of the position and momentum of a particle: the less uncertainty there is in the particle's position, the more there must be in the particle's momentum, and vice versa. Moreover, unlike in the case of spin, there are no physical states of definite position or definite momentum. However, it is possible to construct *minimum uncertainty states*

A *One-Dimensional World* 143

which achieve equality in (7.1); see [10, Section 12.4] for details. Moreover, there exist such minimum uncertainty states with arbitrarily small (but nonzero!) position-uncertainty and therefore arbitrarily large momentum-uncertainty. Similarly, there are minimum uncertainty states of arbitrarily small momentum-uncertainty and corresponding large position-uncertainty.

7.3 The Heisenberg Lie algebra and Lie group

From proposition 7.7, it follows that the three skew-Hermitian operators

$$-\frac{i}{\hbar}\hat{x} \qquad -\frac{i}{\hbar}\hat{p} \qquad -\frac{i}{\hbar}I$$

generate a real Lie algebra with all brackets between basis elements equal to zero, except for the commutation relation

$$\left[-\frac{i}{\hbar}\hat{x}, -\frac{i}{\hbar}\hat{p}\right] = -\frac{i}{\hbar}I.$$

As the next exercise shows, this Lie algebra is isomorphic to a matrix Lie algebra called the *Heisenberg algebra*.

Exercise 7.9. *Let* $\mathfrak{h}_1$ *denote the real vector space of* 3×3 *strictly upper triangular real matrices:*

$$\mathfrak{h}_1 = \left\{ \begin{bmatrix} 0 & v & a \\ 0 & 0 & w \\ 0 & 0 & 0 \end{bmatrix} \mid v, w, a \in \mathbb{R} \right\}.$$

Check that $\mathfrak{h}_1$ *is closed under commutation, so that it forms a 3-dimensional real Lie algebra. Denote by* E_{ij} *the* 3×3 *matrix with 1 in the ith row and jth column, and zeros elsewhere. Show that the following map defines an isomorphism of Lie algebras:*

$$E_{12} \mapsto -\frac{i}{\hbar}\hat{x} \qquad E_{23} \mapsto -\frac{i}{\hbar}\hat{p} \qquad E_{13} \mapsto -\frac{i}{\hbar}I.$$

Exercise 7.10. *Show that the exponential of a matrix in* $\mathfrak{h}_1$ *is given as follows:*

$$\exp \begin{bmatrix} 0 & v & a \\ 0 & 0 & w \\ 0 & 0 & 0 \end{bmatrix} = \begin{bmatrix} 1 & v & a + \frac{1}{2}vw \\ 0 & 1 & w \\ 0 & 0 & 1 \end{bmatrix}.$$

Verify that $\mathfrak{h}_1$ *is the Lie algebra of the Heisenberg group, the 3-dimensional Lie group of* 3×3 *upper triangular real matrices with 1's on the diagonal:*

$$H_1 := \left\{ \begin{bmatrix} 1 & v & \alpha \\ 0 & 1 & w \\ 0 & 0 & 1 \end{bmatrix} \mid a, b, c \in \mathbb{R} \right\}.$$

144   *Symmetry and Quantum Mechanics*

Show that H_1 is isomorphic to the group $(\mathbb{R}^3, \bullet)$ with operation

$$(v, w, \alpha) \bullet (v', w', \alpha') := (v + v', w + w', \alpha + \alpha' + vw'),$$

and that the inverse of an element (v, w, α) is given by

$$(v, w, \alpha)^{-1} = (-v, -w, -\alpha + vw).$$

We have seen that the Lie algebra $\mathfrak{h}_1$ acts on position space $L^2(\mathbb{R})$ via skew-Hermitian operators, and it is natural to wonder whether this action arises from a unitary action of H_1. To show that this is indeed the case, note that H_1 has three obvious subgroups isomorphic to $(\mathbb{R}, +)$, and we know how each of these acts on position space to yield the operators $-\frac{i}{\hbar}\hat{x}$, $-\frac{i}{\hbar}\hat{p}$, and $-\frac{i}{\hbar}I$ via differentiation. Identifying H_1 with $(\mathbb{R}^3, \bullet)$ as in the previous exercise, we have:

$$(v, 0, 0) \quad \text{acts as} \quad \psi(x) \mapsto e^{-\frac{ivx}{\hbar}}\psi(x)$$
$$(0, w, 0) \quad \text{acts as} \quad \psi(x) \mapsto \psi(x - w)$$
$$(0, 0, \alpha) \quad \text{acts as} \quad \psi(x) \mapsto e^{-\frac{i\alpha}{\hbar}}\psi(x).$$

Observe that the element $(0, 0, \alpha)$ acts by the *constant* phase $e^{-\frac{i\alpha}{\hbar}}$ and therefore does not actually change the position state, but only the representing function. In contrast (and despite superficial appearances), the element $(v, 0, 0)$ acts by the *function* $e^{-\frac{ivx}{\hbar}}$, thereby changing the position state (see Section 7.3.1 for an interpretation of this action). To emphasize this sort of distinction, we say that $(0, 0, \alpha)$ acts via a *global phase*, i.e., a phase that does not depend on the position variable x.

Now note that an arbitrary element of H_1 may be factored as a product:

$$(v, w, \alpha) = (0, 0, \alpha) \bullet (0, w, 0) \bullet (v, 0, 0).$$

Hence, applying each factor in turn, we see that a general element (v, w, α) must act as follows:

$$\psi(x) \mapsto e^{-\frac{ivx}{\hbar}}\psi(x) \mapsto e^{-\frac{iv}{\hbar}(x-w)}\psi(x - w) \quad \mapsto \quad e^{-\frac{i\alpha}{\hbar}}e^{-\frac{iv}{\hbar}(x-w)}\psi(x - w)$$
$$= \quad e^{-\frac{i}{\hbar}(\alpha + v(x-w))}\psi(x - w).$$

Exercise 7.11. *Check that the preceding formula defines a unitary action of H_1 on $L^2(\mathbb{R})$. This representation is called the* Schrödinger representation *of H_1.*

The *Stone-von Neumann Theorem* A.54 states that the Schrödinger representation is the *unique* irreducible strongly continuous unitary representation of H_1 in which the central element $(0, 0, 1)$ acts by the scalar $e^{-\frac{i}{\hbar}}$. Observer M is very pleased with the analysis: the representation theory of the Heisenberg group H_1 provides a canonical status for position space.

A One-Dimensional World 145

7.3.1 The meaning of the Heisenberg group action

Suppose that $\psi \in L^2(\mathbb{R})$ is a wavefunction, describing for P the position state of a particle. Hence, for a particular position-value $x \in \mathbb{R}$, P interprets the complex number $\psi(x)$ as the amplitude density for the particle to be found at position x. Note that if observer M is translated from P through a distance w, then she would describe the same position as $x' = x - w$. Moreover, she would also describe the particle's position state via a wavefunction ψ' with a *different* functional form than ψ. Indeed, since they are describing the same particle, we expect[5] that $\psi'(x') = \psi(x)$. To determine M's wavefunction ψ', we temporarily introduce an indeterminate ξ in order to emphasize the distinction between the functional form $\psi'(\xi)$ and the numerical values $\psi'(x')$. We require that

$$\psi'(x') = \psi(x) = \psi(x' + w),$$

which implies that $\psi'(\xi) = \psi(\xi + w)$. We summarize as follows: if M's actual position is obtained from P's through translation by w, then M's position-values x' are obtained from P's position-values x via the inverse translation: $x' = x - w$. The same reversal goes for the wavefunctions: M's wavefunction ψ' is obtained from P's wavefunction ψ via the inverse translation T_{-w}:

$$\psi'(x) = T_{-w}\psi(x) = \psi(x - (-w)) = \psi(x + w).$$

Note that we have abandoned the symbol ξ in favor of x, now thought of as an indeterminant rather than as P's numerical position values. This can lead to some confusion, but it allows us to interpret the translation action on wavefunctions as providing the connection between P's position states and the position states of translated observers such as M. Explicitly, if P describes a particular location with the position-value x_0, then M will describe it with $x_0' = x_0 - w$, and the corresponding amplitudes are given by

$$\psi'(x_0') := \psi'(x)|_{x=x_0'} = \psi(x + w)|_{x=x_0-w} = \psi(x_0 - w + w) = \psi(x_0).$$

Thus, our two observers agree on the amplitude densities and hence on the probability densities. Since the translations form a subgroup of the Heisenberg group, it is natural to ask for a similar physical interpretation of the H_1-action on wavefunctions. So suppose that $(v, w, \alpha) \in H_1$ and $\psi(x)$ is P's description of a position state. Then acting via the inverse element yields a new wavefunction:

$$
\begin{aligned}
\psi'(x) &= (v, w, \alpha)^{-1} \star \psi(x) \\
&= (-v, -w, -\alpha + vw) \star \psi(x) \\
&= e^{-\frac{i}{\hbar}(-\alpha + vw - v(x+w))}\psi(x + w) \\
&= e^{\frac{i}{\hbar}(\alpha + vx)}\psi(x + w).
\end{aligned}
$$

---

[5]Note that, strictly speaking, it is only the probability densities $|\psi'(x')|^2$ and $|\psi(x)|^2$ that must be equal, which leaves room to adjust by a phase. We will return to this issue in just a moment.

146   *Symmetry and Quantum Mechanics*

Again, using $x' = x - w$, we find that $\psi'(x') = e^{\frac{i}{\hbar}(\alpha + v(x-w))}\psi(x)$, so the two wavefunctions again yield the same position-probability densities, although they now differ by a position-dependent phase. To understand the physical meaning of this phase, consider the Fourier transform of P's wavefunction, expressing the position state $\psi(x)$ as a superposition of momentum eigenstates $\psi_p(x)$:

$$\psi(x) = \frac{1}{\sqrt{2\pi\hbar}} \int_{\mathbb{R}} \widetilde{\psi}(p) e^{\frac{i}{\hbar}px} dp.$$

It follows that

$$\begin{aligned}
\psi'(x) &= \frac{1}{\sqrt{2\pi\hbar}} e^{\frac{i}{\hbar}(\alpha + vx)} \int_{\mathbb{R}} \widetilde{\psi}(p) e^{\frac{i}{\hbar}p(x+w)} dp \\
&= \frac{1}{\sqrt{2\pi\hbar}} e^{\frac{i}{\hbar}\alpha} \int_{\mathbb{R}} \widetilde{\psi}(p) e^{\frac{i}{\hbar}(p+v)x} e^{\frac{i}{\hbar}pw} dp \\
&= \frac{1}{\sqrt{2\pi\hbar}} e^{\frac{i}{\hbar}\alpha} \int_{\mathbb{R}} \widetilde{\psi}(p-v) e^{\frac{i}{\hbar}px} e^{\frac{i}{\hbar}(p-v)w} dp \\
&= e^{\frac{i}{\hbar}\alpha} \int_{\mathbb{R}} e^{\frac{i}{\hbar}(p-v)w} \widetilde{\psi}(p-v) \psi_p(x) dp.
\end{aligned}$$

Thus, the Fourier transform of ψ' is related to the Fourier transform of ψ as follows:

$$\widetilde{\psi'}(p) = e^{\frac{i}{\hbar}(\alpha + (p-v)w)} \widetilde{\psi}(p-v).$$

Taking the squared modulus to compare the momentum-probability densities, we find that

$$|\widetilde{\psi'}(p)|^2 = |\widetilde{\psi}(p-v)|^2,$$

which means that the probability that ψ' will be measured with momentum p is the same as the probability that ψ will be measured with momentum $p - v$. Thus, ψ' represents a particle whose momentum has been translated by v with respect to the particle represented by ψ. In particular, $\psi'(x)$ is the wavefunction that the translated observer M would use to describe a particle that only differs from the particle described by $\psi(x)$ in that its momentum has been shifted by v. Note that this is exactly what we would expect if M were not only translated from P, but also moving toward P. To be precise (and replacing v by mv where m is the particle's mass): suppose that M is moving toward P with speed v, and that at time $t = 0$, M is located at $x = w$ (which M would describe as $x' = 0$). Then if P describes a particle at time $t = 0$ via the wavefunction $\psi(x)$, then (at the same instant $t = 0$), M would describe the particle via the wavefunction $\psi'(x) = (mv, w, \alpha)^{-1} \star \psi(x)$. We will return to the question of the relation between the time evolutions $\psi(x,t)$ and $\psi'(x,t)$ for observers in relative motion in Chapter 9.

7.4 Time-evolution

While all this is quite pretty, P would like a bit more justification for the definition of the momentum operator $\hat{p}$. To make contact with some actual physics, P decides to introduce a Hamiltonian operator on position space $L^2(\mathbb{R})$ corresponding to the energy of a particle in one-dimensional space. Classically, the energy E of such a particle is given by the sum of the kinetic and potential energies:

$$E = \frac{p^2}{2m} + V(x).$$

Here, p is the momentum of the particle, m is the mass, and $V(x)$ is the potential energy function. Quantizing this classical expression leads to the time-independent Hamiltonian

$$\mathcal{H} = \frac{\hat{p}^2}{2m} + V(\hat{x}),$$

where we write $V(\hat{x})$ for the self-adjoint operator on $L^2(\mathbb{R})$ defined by multiplication by $V(x)$. The Schrödinger equation $i\hbar\dot{\psi} = \mathcal{H}\psi$ giving the time-evolution of an initial position-state $\psi(x) = \psi(x, 0)$ is then the partial differential equation

$$i\hbar\frac{\partial\psi(x, t)}{\partial t} = \left(\frac{\hat{p}^2}{2m} + V(\hat{x})\right)\psi(x, t) = -\frac{\hbar^2}{2m}\frac{\partial^2\psi(x, t)}{\partial x^2} + V(x)\psi(x, t).$$

Before going on, we take a moment to check that solutions to the Schrödinger equation transform correctly under the translation action on $L^2(\mathbb{R})$. In particular, we wish to show that if observer M is translated by w from P, and if P describes a solution to the Schrödinger equation as $\psi(x)$, then M's corresponding wavefunction $\psi'(x) = \psi(x + w)$ is also a solution to the Schrödinger equation. The key observation (see Figure 7.3) is that M would describe the potential by the translated function $V'(x) = V(x + w)$, so that $V'(x') = V(x)$ for $x' = x - w$. (Note that primes here denote M's version of the functions, *not* derivatives.) But then ψ' is manifestly a solution to the Schrödinger equation with potential V':

$$
\begin{aligned}
i\hbar\frac{\partial\psi'(x, t)}{\partial t} &= i\hbar\frac{\partial\psi(x + w, t)}{\partial t} \\
&= -\frac{\hbar^2}{2m}\frac{\partial^2\psi(x + w, t)}{\partial x^2} + V(x + w)\psi(x + w, t) \\
&= -\frac{\hbar^2}{2m}\frac{\partial^2\psi'(x, t)}{\partial x^2} + V'(x)\psi'(x, t).
\end{aligned}
$$

We now ask about the time-dependence of the position and momentum expectation values. Using the results from Section 4.3 and the notation

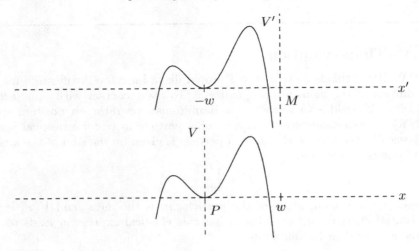

FIGURE 7.3: The same potential viewed by P (bottom) and by the translated observer M (top). The potential has a local minimum at the origin for P, but for M the local minimum is at the position-value $x' = -w$.

$\psi(x,t) = \psi_t(x)$, we have

$$\begin{aligned}
\frac{d}{dt}\langle\psi_t|\hat{x}\psi_t\rangle &= \frac{i}{\hbar}\langle\psi_t|[\mathcal{H},\hat{x}]\psi_t\rangle \\
&= \frac{i}{\hbar}\left\langle\psi_t\left|\left[\frac{\hat{p}^2}{2m}+V(\hat{x}),\hat{x}\right]\psi_t\right.\right\rangle \\
&= \frac{i}{2m\hbar}\langle\psi_t|(\hat{p}^2\hat{x}-\hat{x}\hat{p}^2)\psi_t\rangle \\
&= \frac{i}{2m\hbar}\langle\psi_t|(\hat{p}[\hat{p},\hat{x}]+[\hat{p},\hat{x}]\hat{p})\psi_t\rangle \\
&= \frac{\langle\psi_t|\hat{p}\psi_t\rangle}{m},
\end{aligned}$$

where we have used the commutation relation $[\hat{p},\hat{x}] = -i\hbar I$ and the fact that $V(\hat{x})$ commutes with $\hat{x}$. This computation shows that the time derivative of the expectation value of position is equal to the expectation value of momentum divided by the mass. Thus, the model includes the classical relationship between position and momentum, $\dot{x} = \frac{p}{m}$, at the level of expectation values.

A One-Dimensional World

Similarly, the time-dependence of the momentum expectation value is

$$
\begin{aligned}
\frac{d}{dt} \langle \psi_t | \hat{p} \psi_t \rangle &= \frac{i}{\hbar} \langle \psi_t | [\mathcal{H}, \hat{p}] \psi_t \rangle \\
&= \frac{i}{\hbar} \left\langle \psi_t \left| \left[\frac{\hat{p}^2}{2m} + V(\hat{x}), \hat{p} \right] \psi_t \right. \right\rangle \\
&= \frac{i}{\hbar} \langle \psi_t | [V(\hat{x}), \hat{p}] \psi_t \rangle \\
&= \left\langle \psi_t \left| \left(V(x) \frac{d\psi_t(x)}{dx} - \frac{d}{dx} (V(x)\psi_t(x)) \right) \right. \right\rangle \\
&= \left\langle \psi_t \left| -\frac{dV}{dx}(\hat{x})\psi_t \right. \right\rangle,
\end{aligned}
$$

where we have used the product rule in the last step. This shows that the time derivative of the expectation value of momentum is equal to the expectation value of minus the spatial derivative of the potential energy. Hence, the model includes Newton's Second Law for conservative forces, $\dot{p} = -\frac{dV}{dx}$, at the level of expectation values[6].

7.4.1 The free particle

To investigate the time-evolution of actual position states (as opposed to expectation values), we start by considering the free particle, with Hamiltonian $\mathcal{H} = \frac{\hat{p}^2}{2m}$. Then every momentum eigenfunction $\psi_p(x) = \frac{1}{\sqrt{2\pi\hbar}} e^{\frac{i}{\hbar}px}$ is also an energy eigenfunction:

$$
\mathcal{H}\psi_p = \frac{\hat{p}^2}{2m}\psi_p = \frac{p^2}{2m}\psi_p.
$$

But note that the energy eigenspaces are two dimensional, since $\psi_{\pm p}$ both have eigenvalue $\frac{p^2}{2m}$. Also observe that these energy eigenfunctions are unphysical, since they are not elements of $L^2(\mathbb{R})$. But if we use the Fourier transform to express an initial position state as $\psi(x) = \psi(x, 0) = \int_{\mathbb{R}} \tilde{\psi}(p)\psi_p dp$, then its time-evolution is given by

$$
\begin{aligned}
\psi(x, t) &= e^{-\frac{i}{\hbar}\mathcal{H}t}\psi(x, 0) \\
&= e^{-\frac{i}{\hbar}\mathcal{H}t} \int_{\mathbb{R}} \tilde{\psi}(p)\psi_p dp \\
&= \int_{\mathbb{R}} \tilde{\psi}(p)e^{-\frac{i}{\hbar}\mathcal{H}t}\psi_p dp \\
&= \int_{\mathbb{R}} \tilde{\psi}(p)e^{-\frac{ip^2 t}{2m\hbar}}\psi_p dp.
\end{aligned}
$$

---

[6]These two results are collectively called *Ehrenfest's Theorem*.

150   *Symmetry and Quantum Mechanics*

7.4.2   The infinite square well

Now suppose that the particle is constrained to reside in a region of length L. Informally, we could model this with a "potential function"

$$V(x) = \begin{cases} 0 & \text{if } 0 < x < L \\ \infty & \text{if } x \leq 0 \text{ or } x \geq L. \end{cases}$$

Alternatively (and more rigorously) we can restrict attention to the subspace $L^2([0, L])$ of position space consisting of functions that are zero outside of the open interval $(0, L)$. The particle is free inside this interval, so the Hamiltonian is again $\mathcal{H} = \frac{\hat{p}^2}{2m} = -\frac{\hbar^2}{2m}\frac{d^2}{dx^2}$, now considered as a differential operator on (a dense subspace of) $L^2([0, L])$. The restricted domain has the effect of "killing off" some of the eigenvalues of $\mathcal{H}$, while at the same time making the surviving eigenfunctions physical. Indeed, for $p > 0$, the restriction of the eigenfunction $\phi(x) = c_+\psi_p(x) + c_-\psi_{-p}(x)$ to $[0, L]$ is an element of $L^2([0, L])$ if and only it takes the value 0 at the endpoints $x = 0$ and $x = L$. Vanishing at $x = 0$ requires that $c_+ = -c_-$, so that $\phi(x) = c(\psi_p(x) - \psi_{-p}(x))$. Vanishing at $x = L$ then requires $\psi_p(L) = \psi_{-p}(L)$ or

$$1 = \psi_p(L)\psi_{-p}(L)^{-1} = e^{\frac{2ipL}{\hbar}},$$

which implies that $\frac{2pL}{\hbar} = 2\pi n$, or $p = \frac{\hbar\pi n}{L}$. Thus, the eigenvalues for $\mathcal{H}$ on $L^2([0, L])$ are given by

$$E_n = \frac{\hbar^2\pi^2 n^2}{2mL^2}, \qquad n = 1, 2, 3, \ldots$$

Choosing the normalization constant $c = -i\sqrt{\frac{\pi\hbar}{L}}$ to obtain a real function of unit norm, the stationary state with energy $E_n = \frac{p_n^2}{2m}$ is given by

$$\begin{aligned} \psi_n(x) &= -i\sqrt{\frac{\pi\hbar}{L}}\left(\psi_{p_n}(x) - \psi_{-p_n}(x)\right) \\ &= \sqrt{\frac{2}{L}}\sin\left(\frac{p_n x}{\hbar}\right) \\ &= \sqrt{\frac{2}{L}}\sin\left(\frac{\pi n x}{L}\right). \end{aligned}$$

The plots of the first four stationary states together with the corresponding probability densities are shown in Figure 7.4.

Something very interesting has happened: the *geometry* of physical space (in this case the length L) has led to a quantization of the observable energy levels. The set of stationary states forms an orthonormal Hilbert space basis of the space $L^2([0, L])$, meaning that an arbitrary initial position state $\psi(x) = \psi(x, 0)$ can be written uniquely as an infinite sum

$$\psi(x) = \sum_{n=1}^{\infty} c_n \psi_n(x),$$

A One-Dimensional World

151

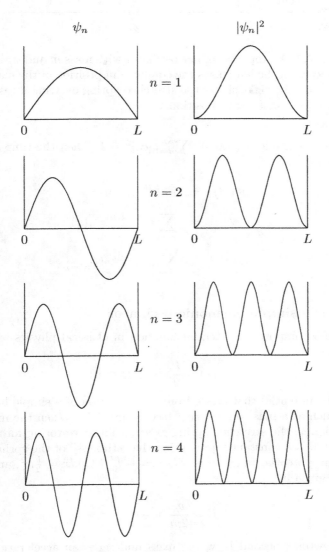

FIGURE 7.4: The first four energy eigenfunctions for the infinite spherical well. The plots on the left are the wavefunctions ψ_n; the plots on the right are the corresponding probability densities $|\psi_n|^2$.

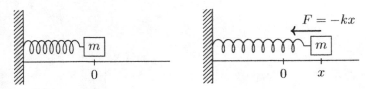

FIGURE 7.5: A simple harmonic oscillator with mass m and spring-constant k. The picture on the left shows the equilibrium position of the mass at $x = 0$; the picture on the right shows the stretched spring exerting a restoring force $F = -kx$ on the mass m at position x.

where the amplitudes c_n satisfy $\sum_{n=1}^{\infty} |c_n|^2 = 1$. Then the time-evolution of $\psi(x)$ is given by

$$\begin{aligned}\psi(x,t) &= e^{-\frac{i}{\hbar}\mathcal{H}t}\psi(x,0) \\ &= \sum_{n=1}^{\infty} c_n e^{-\frac{i}{\hbar}\mathcal{H}t}\psi_n(x) \\ &= \sum_{n=1}^{\infty} c_n e^{-\frac{i}{\hbar}E_n t}\psi_n(x).\end{aligned}$$

7.4.3 The simple harmonic oscillator

The most important potential function in classical physics is given by a pure quadratic:

$$V(x) = \frac{k}{2}x^2 \quad (k > 0).$$

This is the potential that arises from consideration of a simple harmonic oscillator such as a mass on a spring (see Figure 7.5). When the mass m is at the position x, the stretched spring exerts a linear restoring force $F = -kx$ where $k > 0$ is a constant representing the "stiffness" of the spring. The corresponding potential function is $V(x) := -\int_0^x F(x)dx = \frac{k}{2}x^2$, and the total energy is given by

$$E = \frac{p^2}{2m} + \frac{k}{2}x^2. \tag{7.2}$$

By Newton's Second Law, the mass undergoes an acceleration given by $\ddot{x} = \frac{F}{m} = \frac{-k}{m}x$. The general solution of this second-order differential equation is given by

$$x(t) = A\cos(\omega t + \phi),$$

describing (as expected) an oscillating motion of frequency $\omega := \sqrt{\frac{k}{m}}$. Here, the amplitude A and the phase ϕ are determined by the initial position and momentum. The choice of such initial conditions determines the total energy

A One-Dimensional World

of the system, which remains constant throughout the time-evolution. In particular, for a given energy E, the endpoints of the oscillator's motion occur where $p = 0$, namely $x = \pm\sqrt{\frac{2E}{k}}$. Expressing these endpoints in terms of the frequency ω, we see that an oscillator with energy E moves within the *classical region*

$$-\sqrt{\frac{2E}{m\omega^2}} \le x \le \sqrt{\frac{2E}{m\omega^2}}. \tag{7.3}$$

The importance of the simple harmonic potential stems from its ubiquity. Indeed, suppose that $V(x)$ is an arbitrary (analytic) potential function, and suppose that $x = 0$ is a stable equilibrium for the corresponding system. Using a prime to denote differentiation, the force $F(x) = -V'(x)$ is zero at $x = 0$, so we have $V'(0) = 0$. Since we are considering a *stable* equilibrium, F must act as a restoring force near $x = 0$, which implies that $F(x) = -V'(x)$ is positive to the left of 0 and negative to the right. Hence, $V(x)$ has a local minimum at $x = 0$, so $V''(0) \ge 0$. After adjusting by an additive constant (which doesn't affect the dynamics), we may also assume that $V(0) = 0$. Hence, the Taylor expansion of $V(x)$ around $x = 0$ looks like

$$V(x) = \frac{1}{2}V''(0)x^2 + \text{higher-order terms}.$$

As argued above, the stability of the equilibrium implies that $V''(0) \ge 0$. Thus, throwing away all higher-order terms (and assuming that $V''(0) > 0$), we see that we may approximate $V(x)$ near $x = 0$ by a simple harmonic potential $V(x) \approx \frac{1}{2}V''(0)x^2$.

After this brief discussion of the classical situation, we now turn to the quantum simple harmonic oscillator. Quantization of the classical expression (7.2) for the energy of a particle of mass m in a pure quadratic potential yields the Hamiltonian

$$\mathcal{H} = \frac{\hat{p}^2}{2m} + \frac{m\omega^2}{2}\hat{x}^2.$$

Note that we have expressed the Hamiltonian in terms of the frequency $\omega := \sqrt{\frac{k}{m}}$ from the classical solution. We can factor the classical energy function over the complex numbers as follows:

$$E = \frac{p^2}{2m} + \frac{m\omega^2}{2}x^2 = \frac{m\omega^2}{2}\left(x + \frac{i}{m\omega}p\right)\left(x - \frac{i}{m\omega}p\right).$$

This motivates the introduction of the following dimensionless operators:

$$\hat{a} := \sqrt{\frac{m\omega}{2\hbar}}\left(\hat{x} + \frac{i}{m\omega}\hat{p}\right), \qquad \hat{a}^\dagger := \sqrt{\frac{m\omega}{2\hbar}}\left(\hat{x} - \frac{i}{m\omega}\hat{p}\right).$$

These operators are not self-adjoint, and thus do not represent observables on

154 *Symmetry and Quantum Mechanics*

position space. Nevertheless, they will be extremely useful, just as the raising and lowering operators $S_x \pm iS_y \in \mathfrak{su}(2)_\mathbb{C}$ were helpful in our study of spin. In fact, we will see that $\hat{a}^\dagger$ and $\hat{a}$ also act as raising and lowering operators on energy eigenfunctions of the quantum harmonic oscillator.

First note that the product $\hat{N} := \hat{a}^\dagger \hat{a}$ is self-adjoint, and thus yields an observable on position space. Moreover, the commutation relation for $\hat{x}$ and $\hat{p}$ yields the relation

$$[\hat{a}, \hat{a}^\dagger] = \hat{a}\hat{a}^\dagger - \hat{a}^\dagger\hat{a} = \frac{m\omega}{2\hbar}\frac{i}{m\omega}(2\hat{p}\hat{x} - 2\hat{x}\hat{p}) = \frac{i}{\hbar}[\hat{p}, \hat{x}] = I.$$

From this we can derive the further commutation relations:

$$[\hat{N}, \hat{a}] = (\hat{a}^\dagger\hat{a}\hat{a} - \hat{a}\hat{a}^\dagger\hat{a}) = [\hat{a}^\dagger, \hat{a}]\hat{a} = -\hat{a}$$

and

$$[\hat{N}, \hat{a}^\dagger] = (\hat{a}^\dagger\hat{a}\hat{a}^\dagger - \hat{a}^\dagger\hat{a}^\dagger\hat{a}) = \hat{a}^\dagger[\hat{a}, \hat{a}^\dagger] = \hat{a}^\dagger.$$

Solving for the position and momentum operators in terms of the a-operators yields

$$\hat{x} = \sqrt{\frac{\hbar}{2m\omega}}\left(\hat{a} + \hat{a}^\dagger\right), \qquad \hat{p} = -i\sqrt{\frac{m\omega\hbar}{2}}\left(\hat{a} - \hat{a}^\dagger\right).$$

Using these relations, we may rewrite the Hamiltonian for the simple harmonic oscillator as

$$
\begin{aligned}
\mathcal{H} &= -\frac{\hbar\omega}{4}(\hat{a} - \hat{a}^\dagger)^2 + \frac{\hbar\omega}{4}(\hat{a} + \hat{a}^\dagger)^2 \\
&= \frac{\hbar\omega}{2}(\hat{a}\hat{a}^\dagger + \hat{a}^\dagger\hat{a}) \\
&= \frac{\hbar\omega}{2}([\hat{a}, \hat{a}^\dagger] + 2\hat{a}^\dagger\hat{a}) \\
&= \frac{\hbar\omega}{2}(I + 2\hat{N}) \\
&= \hbar\omega(\hat{N} + \frac{1}{2}I).
\end{aligned}
$$

Thus, to find the energy eigenstates for the quantum harmonic oscillator, we just need to find the eigenstates for the operator $\hat{N} = \hat{a}^\dagger\hat{a}$.

So suppose that ψ is a unit-norm eigenfunction for $\hat{N}$ with eigenvalue λ. The first thing to note is that λ must be nonnegative:

$$\lambda = \lambda\langle\psi|\psi\rangle = \langle\psi|\lambda\psi\rangle = \langle\psi|\hat{N}\psi\rangle = \langle\psi|\hat{a}^\dagger\hat{a}\psi\rangle = \langle\hat{a}\psi|\hat{a}\psi\rangle \geq 0.$$

Moreover, the commutation relations derived above imply that $\hat{a}\psi$, if nonzero, is also an eigenfunction for $\hat{N}$, with eigenvalue $\lambda - 1$:

$$\hat{N}\hat{a}\psi = (\hat{a}\hat{N} + [\hat{N}, \hat{a}])\psi = \hat{a}\hat{N}\psi - \hat{a}\psi = \lambda\hat{a}\psi - \hat{a}\psi = (\lambda - 1)\hat{a}\psi.$$

By the non-negativity of the eigenvalues, it follows that there is a minimal

A One-Dimensional World

integer $k \geq 0$ such that $\hat{a}^{k+1}\psi = 0$. Then $\hat{a}^k\psi$ will be a nonzero element of the kernel of $\hat{a}$, and thus an eigenfunction for $\hat{N}$ with eigenvalue 0. It follows that $\lambda = k$, so that all eigenvalues for $\hat{N}$ are integers.

Since the previous paragraph began with the *assumption* of the existence of an eigenfunction for $\hat{N}$, we now pause to explicitly determine a nonzero function in the kernel of $\hat{a}$. When written out explicitly, the condition $0 = \hat{a}\psi$ becomes

$$0 = \left(\hat{x} + \frac{i}{m\omega}\hat{p} \right)\psi(x) = x\psi(x) + \frac{\hbar}{m\omega}\frac{d\psi(x)}{dx}.$$

Hence, we must solve the linear differential equation

$$\frac{d\psi(x)}{dx} = -\frac{m\omega}{\hbar}x\psi(x).$$

By separation of variables, we see that the general solution is given by

$$\psi(x) = C \exp\left(-\frac{m\omega}{2\hbar}x^2 \right),$$

where C is an arbitrary complex constant. Choosing $C = \sqrt{\frac{\pi m\omega}{\hbar}}$, we obtain the unique real solution of unit norm:

$$\psi_0(x) := \sqrt{\frac{\pi m\omega}{\hbar}} \exp\left(-\frac{m\omega}{2\hbar}x^2 \right).$$

We now claim that for each $n \geq 0$, the function $(\hat{a}^\dagger)^n\psi_0$ is an eigenfunction for $\hat{N}$ with eigenvalue n. The claim is certainly true for $n = 0$, and we establish the general case by induction:

$$
\begin{aligned}
\hat{N}(\hat{a}^\dagger)^n\psi_0 &= (\hat{a}^\dagger\hat{N} + [\hat{N}, \hat{a}^\dagger])(\hat{a}^\dagger)^{n-1}\psi_0 \\
&= (\hat{a}^\dagger\hat{N} + \hat{a}^\dagger)(\hat{a}^\dagger)^{n-1}\psi_0 \\
&= (n-1)(\hat{a}^\dagger)^n\psi_0 + (\hat{a}^\dagger)^n\psi_0 \\
&= n(\hat{a}^\dagger)^n\psi_0.
\end{aligned}
$$

Thus, the claim will be established once we show that all of the functions $(\hat{a}^\dagger)^n\psi_0$ are nonzero. This will follow from a computation of the norms:

$$
\begin{aligned}
\langle(\hat{a}^\dagger)^n\psi_0|(\hat{a}^\dagger)^n\psi_0\rangle &= \langle(\hat{a}\hat{a}^\dagger)(\hat{a}^\dagger)^{n-1}\psi_0|(\hat{a}^\dagger)^{n-1}\psi_0\rangle \\
&= \langle(\hat{a}^\dagger\hat{a} + [\hat{a}, \hat{a}^\dagger])(\hat{a}^\dagger)^{n-1}\psi_0|(\hat{a}^\dagger)^{n-1}\psi_0\rangle \\
&= \langle(\hat{N} + I)(\hat{a}^\dagger)^{n-1}\psi_0|(\hat{a}^\dagger)^{n-1}\psi_0\rangle \\
&= n\langle(\hat{a}^\dagger)^{n-1}\psi_0|(\hat{a}^\dagger)^{n-1}\psi_0\rangle.
\end{aligned}
$$

Since ψ_0 has norm 1, it follows by induction that

$$\|(\hat{a}^\dagger)^n\psi_0\| = \sqrt{n!}.$$

Thus, we have found an infinite sequence of eigenfunctions for the operator

156 *Symmetry and Quantum Mechanics*

$\hat{N}$, one for each integer eigenvalue $n \geq 0$. Moreover, since $\hat{N}$ is self-adjoint, we know that these eigenfunctions are mutually orthogonal. Normalizing, we obtain an orthonormal sequence of eigenfunctions:

$$\psi_n := \frac{1}{\sqrt{n!}}(\hat{a}^\dagger)^n \psi_0.$$

In fact (see [10, Section 11.4]), $\{\psi_n \mid n = 0, 1, 2, \dots\}$ forms an orthonormal Hilbert space basis for $L^2(\mathbb{R})$; the first four are plotted in Figure 7.6. The operator $\hat{N}$ is called the *number operator* since its eigenvalues are the non-negative integers: $\hat{N}\psi_n = n\psi_n$. The operators $\hat{a}^\dagger$ and $\hat{a}$ act as *raising* and *lowering* operators on these number eigenstates:

$$\hat{a}^\dagger \psi_n = \frac{1}{\sqrt{n!}}(\hat{a}^\dagger)^{n+1}\psi_0 = \sqrt{\frac{(n+1)!}{n!}}\psi_{n+1} = \sqrt{n+1}\psi_{n+1},$$

and

$$
\begin{aligned}
\hat{a}\psi_n &= \frac{1}{\sqrt{n!}}(\hat{a}\hat{a}^\dagger)(\hat{a}^\dagger)^{n-1}\psi_0 \\
&= \frac{1}{\sqrt{n!}}(\hat{N}+I)(\hat{a}^\dagger)^{n-1}\psi_0 \\
&= \frac{n}{\sqrt{n!}}(\hat{a}^\dagger)^{n-1}\psi_0 \\
&= \frac{\sqrt{n}}{\sqrt{(n-1)!}}(\hat{a}^\dagger)^{n-1}\psi_0 \\
&= \sqrt{n}\psi_{n-1}.
\end{aligned}
$$

Returning to the Hamiltonian $\mathcal{H} = \hbar\omega(\hat{N} + \frac{1}{2}I)$ for the quantum harmonic oscillator, we see that $\{\psi_n\}$ forms an orthonormal Hilbert space basis of energy eigenstates for $L^2(\mathbb{R})$, with eigenvalues given by

$$\mathcal{H}\psi_n = \hbar\omega(\hat{N} + \frac{1}{2}I)\psi_n = \hbar\omega(n + \frac{1}{2})\psi_n.$$

Thus, the observable energies form the equally spaced set

$$E_n = \hbar\omega(n + \frac{1}{2}).$$

Since the operator $\hat{a}^\dagger$ transforms the state ψ_n into $\sqrt{n+1}\psi_{n+1}$, thus increasing the energy by one quantum $\hbar\omega$, it is called a *creation* operator. Similarly, the operator $\hat{a}$ destroys one quantum of energy in moving ψ_n to $\sqrt{n}\psi_{n-1}$, so it is called an *annihilation* operator.

Now suppose that $\psi(x) = \psi(x, 0) \in L^2(\mathbb{R})$ is an initial position state of the quantum harmonic oscillator. Then we may expand ψ as an infinite series in the eigenstates ψ_n:

$$\psi(x) = \sum_{n=0}^{\infty} c_n \psi_n(x).$$

A One-Dimensional World

The time-evolution of this initial state is given by

$$
\begin{aligned}
\psi(x,t) &= e^{-\frac{i}{\hbar}\mathcal{H}t}\psi(x,0) \\
&= \sum_{n=0}^{\infty} c_n e^{-\frac{i}{\hbar}\mathcal{H}t}\psi_n(x) \\
&= \sum_{n=0}^{\infty} c_n e^{-i\omega(n+\frac{1}{2})t}\psi_n(x) \\
&= e^{-\frac{i\omega t}{2}} \sum_{n=0}^{\infty} c_n e^{-i\omega n t}\psi_n(x).
\end{aligned}
$$

To end this section, we will find the actual functional form of the eigenstates $\psi_n(x)$. It turns out that each $\psi_n(x)$ is a polynomial of degree n in x times the function $\psi_0(x)$. Indeed, $\psi_0(x)$ certainly has this form for the constant polynomial $h_0(x) = 1$. Assume by induction that $\psi_n(x) = \frac{1}{\sqrt{n!}} h_n(x)\psi_0(x)$ for some degree-n polynomial $h_n(x)$. Then we compute

$$
\begin{aligned}
\sqrt{n+1}\psi_{n+1}(x) &= \hat{a}^{\dagger}\psi_n(x) \\
&= \sqrt{\frac{m\omega}{2\hbar}}\left(\hat{x} - \frac{i}{m\omega}\hat{p}\right)\psi_n(x) \\
&= \sqrt{\frac{m\omega}{2\hbar n!}}\left(x h_n(x)\psi_0(x) - \frac{\hbar}{m\omega}\frac{d}{dx}\left(h_n(x)\psi_0(x)\right)\right) \\
&= \sqrt{\frac{m\omega}{2\hbar n!}}\left(x h_n(x)\psi_0(x) - \frac{\hbar}{m\omega}h_n'(x)\psi_0(x)\right. \\
&\qquad\qquad \left. - \frac{\hbar}{m\omega}h_n(x)\psi_0'(x)\right) \\
&= \sqrt{\frac{m\omega}{2\hbar n!}}\left(x h_n(x)\psi_0(x) - \frac{\hbar}{m\omega}h_n'(x)\psi_0(x)\right. \\
&\qquad\qquad \left. + h_n(x)x\psi_0(x)\right) \\
&= \sqrt{\frac{m\omega}{2\hbar n!}}\left(2x h_n(x) - \frac{\hbar}{m\omega}h_n'(x)\right)\psi_0(x).
\end{aligned}
$$

It follows that $\psi_{n+1}(x) = \frac{1}{\sqrt{(n+1)!}} h_{n+1}(x)\psi_0(x)$, where h_{n+1} is the recursively defined polynomial of degree $n+1$:

$$
h_{n+1}(x) = \sqrt{\frac{m\omega}{2\hbar}}\left(2x h_n(x) - \frac{\hbar}{m\omega}h_n'(x)\right), \qquad h_0(x) = 1. \qquad (7.4)
$$

Exercise 7.12. *This exercise shows that (after a suitable change of variable and some normalization), the polynomials $h_n(x)$ are essentially the Hermite polynomials. To begin, we introduce the dimensionless scaling of the position variable $\tilde{x} := \sqrt{\frac{m\omega}{\hbar}}x$, and consider the polynomials $h_n(\tilde{x})$ as functions of*

158   *Symmetry and Quantum Mechanics*

$\widetilde{x}$. *Show that the recurrence (7.4) for the $h_n(x)$ translates to the following recurrence for the $h_n(\widetilde{x})$:*

$$h_{n+1}(\widetilde{x}) = \frac{1}{\sqrt{2}} \left(2\widetilde{x} h_n(\widetilde{x}) - h'_n(\widetilde{x}) \right), \qquad h_0(\widetilde{x}) = 1.$$

Now show that the leading coefficient of the polynomial $h_n(\widetilde{x})$ is $2^{\frac{n}{2}}$. By convention, we would instead like to normalize the polynomials so that the leading coefficients are 2^n. Hence, we define $H_n(\widetilde{x}) := 2^{\frac{n}{2}} h_n(\widetilde{x})$. Show that this has the effect of removing the factor of $\frac{1}{\sqrt{2}}$ from the recurrence:

$$H_{n+1}(\widetilde{x}) = 2\widetilde{x} H_n(\widetilde{x}) - H'_n(\widetilde{x}), \qquad H_0(\widetilde{x}) = 1.$$

This is the recurrence defining the Hermite polynomials. Work out the first four Hermite polynomials explicitly, and conclude that the first four energy eigenstates for the simple harmonic oscillator have the following form when expressed in the dimensionless variable $\widetilde{x}$:

$$
\begin{aligned}
\psi_0(\widetilde{x}) &= \sqrt{\pi} \exp\left(-\frac{1}{2}\widetilde{x}^2\right) \\
\psi_1(\widetilde{x}) &= \sqrt{2\pi}\,\widetilde{x} \exp\left(-\frac{1}{2}\widetilde{x}^2\right) \\
\psi_2(\widetilde{x}) &= \sqrt{\frac{\pi}{2}}(2\widetilde{x}^2 - 1) \exp\left(-\frac{1}{2}\widetilde{x}^2\right) \\
\psi_3(\widetilde{x}) &= \sqrt{\frac{\pi}{3}}(2\widetilde{x}^3 - 3\widetilde{x}) \exp\left(-\frac{1}{2}\widetilde{x}^2\right).
\end{aligned}
$$

Figure 7.6 displays plots of the first four energy eigenfunctions for the simple harmonic oscillator. On each plot, we have shaded the classical region (7.3) that would be accessible to a classical oscillator with energy $E_n = \hbar\omega(n + \frac{1}{2})$. Note that in each case, the wavefunction extends *outside* the classical region, so there is a non-zero probability of finding the quantum oscillator at a classically forbidden position.

A One-Dimensional World 159

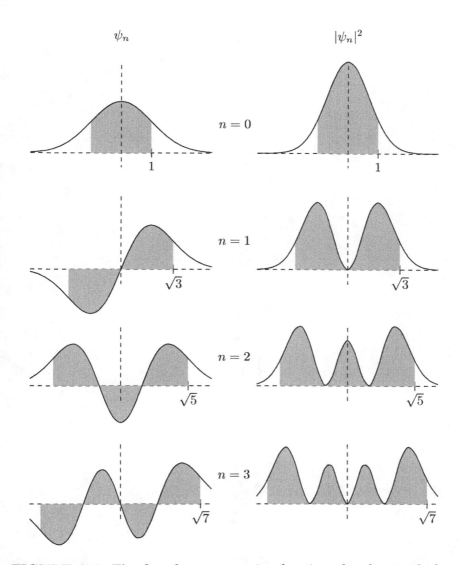

FIGURE 7.6: The first four energy eigenfunctions for the simple harmonic oscillator, expressed in terms of the dimensionless position variable $\tilde{x} = \sqrt{\frac{m\omega}{\hbar}}x$. The plots on the left are the wavefunctions ψ_n; the plots on the right are the corresponding probability densities $|\psi_n|^2$. The shaded portions indicate the classical region for an oscillator with energy $E_n = \hbar\omega(n+\frac{1}{2})$. Since the wavefunction extends outside the classical region, there is a non-zero probability of finding the oscillator at a classically forbidden position.

Chapter 8

A Three-Dimensional World

In which M and P combine their studies of the Heisenberg group H_3 and the rotation group $SO(3)$.

M and P are ready to construct a model for position and momentum in three dimensions. Their strategy is to generalize their one-dimensional model step-by-step. M begins with a definition that amounts to "forgetting the origin" of the oriented inner product space $(V, \langle, \rangle)$ of Chapter 1, thereby allowing for the possibility that M and P may be at different locations in physical space.

Definition 8.1. Physical space *is a real affine space, $\mathbb{A}^3$, of dimension 3. More precisely, the set $\mathbb{A}^3$ is endowed with a simply transitive action of the underlying Lie group $(V, +)$ of a three-dimensional oriented real inner product space $(V, \langle, \rangle)$:*

$$V \times \mathbb{A}^3 \to \mathbb{A}^3 \qquad (\mathbf{r}, \mathbf{a}) \mapsto \mathbf{r} + \mathbf{a}.$$

As in the case of one dimension, P's choice of a location $\mathbf{a}_0 \in \mathbb{A}^3$ yields an identification $\varphi \colon \mathbb{A}^3 \to V$ with $\varphi(\mathbf{a}_0) = \mathbf{0}$. Also, M's choice of a location $\mathbf{a}_0' \in \mathbb{A}^3$ yields a different identification $\varphi' \colon \mathbb{A}^3 \to V$ with $\varphi'(\mathbf{a}_0') = \mathbf{0}$. By simple transitivity, there is a unique $\mathbf{w} \in V$ such that $\mathbf{a}_0' = \mathbf{w} + \mathbf{a}_0$. Then the difference between the two descriptions is represented by the automorphism $\varphi' \circ \varphi^{-1} \colon V \to V$ given by $\mathbf{r} \mapsto \mathbf{r} - \mathbf{w}$. Thus, the choice of location $\mathbf{a}_0$ allows P to think of physical space as V endowed with the translation action of $(V, +)$.

Note that this really is a direct generalization of definition 7.1 in the one-dimensional case. Indeed, suppose that $(V, \langle, \rangle)$ is a *one-dimensional* oriented real inner product space. Then there is exactly one positively oriented orthonormal basis $\{\mathbf{u}\}$ for V and thus $(V, \langle, \rangle)$ is canonically isomorphic to $(\mathbb{R}, \cdot)$ via the map $r\mathbf{u} \mapsto r$. In this case the underlying Lie group $(V, +)$ is canonically isomorphic to $(\mathbb{R}, +)$, which leads to the definition of one-dimensional physical space given in the previous chapter.

8.1   Position

The motivation for the introduction of position-space in one dimension goes over word-for-word to the three-dimensional case, so M simply gives the following definition:

161

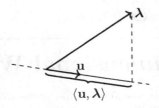

FIGURE 8.1: Projection of the position vector λ onto the line spanned by the unit vector **u**.

Definition 8.2. Position space *is the complex inner product space of square-integrable functions* $(L^2(V), \langle | \rangle)$, *where*

$$L^2(V) = \left\{ \psi : V \to \mathbb{C} \;\Big|\; \int_V |\psi|^2 < \infty \right\},$$

and $\langle \psi | \phi \rangle = \int_V \psi^* \phi$. *Position states are unit vectors in* $L^2(V)$ *up to the action of the phase group* $U(1)$. *That is, two unit vectors represent the same state if and only if they differ by a global phase* $e^{i\theta} \in U(1)$. *If* ψ *is a position state and* $E \subset V$ *is a measurable subset, then* $\int_E |\psi|^2$ *is the probability that the particle, if measured, will be found in the subset* E.

As before, this description depends on P's location $\mathbf{a}_0 \in \mathbb{A}^3$, and as a result, there is a corresponding unitary action of $(V, +)$ on position space $L^2(V)$.

Definition 8.3. *The* translation action *of* $(V, +)$ *on position space* $L^2(V)$ *is given by*
$$V \times L^2(V) \to L^2(V), \qquad (\mathbf{w}, \psi(\mathbf{r})) \mapsto \psi(\mathbf{r} - \mathbf{w}).$$
We denote the translation action by $T_\mathbf{w} \psi(\mathbf{r}) := \psi(\mathbf{r} - \mathbf{w})$.

As in the one-dimensional case, we will see that this translation action gives rise to the linear momentum operators.

But before talking about momentum, we introduce position operators as follows:

Definition 8.4. *If* **u** *is a unit vector in* $(V, \langle, \rangle)$, *the observable "position along the* **u***-axis" is given by the self-adjoint* position operator $\hat{\mathbf{u}}$ *on position space* $L^2(V)$ *defined by*
$$\hat{\mathbf{u}} \psi(\mathbf{r}) := \langle \mathbf{u}, \mathbf{r} \rangle \psi(\mathbf{r}).$$

The motivation for this definition goes as follows: if δ_λ represents the state of a particle with definite position $\lambda \in V$, then the projection of the particle's position unto the line spanned by **u** is given by $\langle \mathbf{u}, \lambda \rangle \mathbf{u}$ (see Figure 8.1). Hence, measuring the particle's position along the **u**-axis should return a measurement of $\langle \mathbf{u}, \lambda \rangle$. Since δ_λ should be an eigenvector for the operator $\hat{\mathbf{u}}$

A Three-Dimensional World

with eigenvalue the observed position measurement, we should have

$$\hat{\mathbf{u}}\delta_{\boldsymbol{\lambda}} = \langle \mathbf{u}, \boldsymbol{\lambda} \rangle \delta_{\boldsymbol{\lambda}}.$$

As described in the one-dimensional case, $\delta_{\boldsymbol{\lambda}}$ isn't really a function, but rather the linear functional on (a dense subspace of) $L^2(V)$ defined by evaluation at $\boldsymbol{\lambda}$. But informally, $\delta_{\boldsymbol{\lambda}}$ may be regarded as a unit-norm function that is zero everywhere except at $\mathbf{r} = \boldsymbol{\lambda}$. Thus, we may write the previous eigenvector condition as

$$\hat{\mathbf{u}}\delta_{\boldsymbol{\lambda}}(\mathbf{r}) = \langle \mathbf{u}, \mathbf{r} \rangle \delta_{\boldsymbol{\lambda}}(\mathbf{r}),$$

which suggests the definition of $\hat{\mathbf{u}}$ for general position states $\psi(\mathbf{r})$ given above. As in the one-dimensional case, these self-adjoint position operators are defined only on a dense subspace of $L^2(V)$ and they have no honest eigenfunctions. Nevertheless, we have seen that the three-dimensional Dirac delta function $\delta_{\boldsymbol{\lambda}}$ may be regarded as an eigenstate with definite position $\boldsymbol{\lambda} \in V$, and a general position state $\psi(\mathbf{r})$ may be thought of (informally) as a superposition of delta functions.

Note that for any unit vector $\mathbf{u} \in V$, the operator $-\frac{i}{\hbar}\hat{\mathbf{u}}$ arises via differentiation (with respect to the parameter v) from the unitary $(\mathbb{R}, +)$-action on $L^2(V)$ defined by

$$\mathbb{R} \times L^2(V) \to L^2(V), \qquad (v, \psi(\mathbf{r})) \mapsto e^{-\frac{iv\langle \mathbf{u}, \mathbf{r} \rangle}{\hbar}}\psi(\mathbf{r}).$$

In fact, these one-parameter groups fit together to define a unitary action of $(V, +)$ on $L^2(V)$:

$$(\mathbf{v}, \psi(\mathbf{r})) \mapsto e^{-\frac{i\langle \mathbf{v}, \mathbf{r} \rangle}{\hbar}}\psi(\mathbf{r}).$$

As in the one-dimensional case, this symmetry of position space will combine with the translation action to yield an irreducible representation of the three-dimensional Heisenberg group, H_3, discussed in Section 8.2.1.

In order to connect with the position operator $\hat{x}$ from the one-dimensional case, P lays out a right-handed coordinate system at the location $\mathbf{a}_0$, which corresponds to the choice of a positively oriented orthonormal basis $\{\mathbf{u}_1, \mathbf{u}_2, \mathbf{u}_3\}$ for $(V, \langle, \rangle)$ as in Chapter 1. Such a choice was unnecessary in the one-dimensional world because, as mentioned above, there is a unique orthonormal positively oriented basis in that case. Having made his choice of basis, P thinks of physical space as $\mathbb{R}^3$ endowed with the translation action of $(\mathbb{R}^3, +)$ together with the rotation action of the group $SO(3)$ as described in Chapter 1.

In terms of this identification, position-space becomes $L^2(\mathbb{R}^3)$, and we may write $\psi(\mathbf{r}) = \psi(x, y, z)$, where the vector $(x, y, z) \in \mathbb{R}^3$ is the coordinate representation of $\mathbf{r} \in V$ with respect to P's chosen basis:

$$\mathbf{r} = x\mathbf{u}_1 + y\mathbf{u}_2 + z\mathbf{u}_3.$$

We also write $\hat{\mathbf{u}}_1 = \hat{x}, \hat{\mathbf{u}}_2 = \hat{y}, \hat{\mathbf{u}}_3 = \hat{z}$ and note that

$$\begin{aligned}
\hat{x}\psi(x, y, z) &= x\psi(x, y, z) \\
\hat{y}\psi(x, y, z) &= y\psi(x, y, z) \\
\hat{z}\psi(x, y, z) &= z\psi(x, y, z).
\end{aligned}$$

8.2 Linear momentum

To obtain linear momentum operators, we consider the Lie algebra of the Lie group $(V, +)$, which acts via translation on position space $L^2(V)$. Clearly, the tangent space to $(V, +)$ is three-dimensional and can be identified with the vector space V itself. Moreover, since $(V, +)$ is an abelian group, the Lie bracket on V is trivial (i.e., all brackets are zero). Just as in the one-dimensional case, the exponential map is the identity function: if $\mathbf{p} \in V$ is a tangent vector at $\mathbf{0} \in V$, then the corresponding one-parameter subgroup of $(V, +)$ is given by $c_{\mathbf{p}}(t) = t\mathbf{p}$, and $\exp(\mathbf{p}) := c_{\mathbf{p}}(1) = \mathbf{p}$.

As usual, to determine the action of the Lie algebra V on $L^2(V)$, we take the derivative of the translation action by $(V, +)$: if $\mathbf{p} \in V$ is a unit tangent vector, then

$$\mathbf{p} \star \psi(\mathbf{r}) = \frac{d}{dt}\left(T_{c_{\mathbf{p}}(t)}\psi(\mathbf{r})\right)\Big|_{t=0} = \frac{d}{dt}\psi(\mathbf{r} - t\mathbf{p})\Big|_{t=0} = -D_{\mathbf{p}}\psi(\mathbf{r}),$$

where $D_{\mathbf{p}}$ denotes the directional derivative in the direction $\mathbf{p}$. Since $\mathbf{p}$ is the infinitesimal generator of translation in the $\mathbf{p}$-direction, we make the following definition for linear momentum.

Definition 8.5. *If* $\mathbf{p}$ *is a unit vector in* $(V, \langle , \rangle)$, *the observable "linear momentum in the* $\mathbf{p}$*-direction" is given by the self-adjoint* linear momentum operator $\hat{\mathbf{p}}$ *on* $L^2(V)$ *defined as the directional derivative*

$$\hat{\mathbf{p}} := -\frac{\hbar}{i}(-D_{\mathbf{p}}) = \frac{\hbar}{i}D_{\mathbf{p}}.$$

The eigenfunctions of $\hat{\mathbf{p}}$ are the solutions to the differential equation

$$D_{\mathbf{p}}\psi_{\mathbf{p},p} = \frac{i}{\hbar}p\psi_{\mathbf{p},p} \qquad p \in \mathbb{R}.$$

Similarly to the one-dimensional case, the solution is

$$\psi_{\mathbf{p},p}(\mathbf{r}) = N_p e^{\frac{i}{\hbar}p\langle\mathbf{p},\mathbf{r}\rangle}.$$

Just as before, there is no way of choosing the normalization constant N_p to make $\psi_{\mathbf{p},p}$ square-integrable. Nevertheless, setting $N_p = (2\pi\hbar)^{-\frac{3}{2}}$, the Fourier transform is obtained as follows: if $\psi(\mathbf{r}) \in L^1(V) \cap L^2(V)$, then define $\widetilde{\psi}: V \to \mathbb{C}$ by

$$\widetilde{\psi}(\mathbf{p}) := \left\langle \psi_{\frac{\mathbf{p}}{|\mathbf{p}|},|\mathbf{p}|} \Big| \psi \right\rangle = \int_V \psi^*_{\frac{\mathbf{p}}{|\mathbf{p}|},|\mathbf{p}|}\psi d\mathbf{r} = \frac{1}{(2\pi\hbar)^{\frac{3}{2}}} \int_V e^{-\frac{i}{\hbar}\langle\mathbf{p},\mathbf{r}\rangle}\psi(\mathbf{r})d\mathbf{r}.$$

Here, $\mathbf{p}$ is an arbitrary vector in V, not necessarily of unit length, and $\psi_{\frac{\mathbf{p}}{|\mathbf{p}|},|\mathbf{p}|}$

A Three-Dimensional World

denotes the state with linear momentum $\mathbf{p} = |\mathbf{p}|\frac{\mathbf{p}}{|\mathbf{p}|}$. The Fourier Inversion Theorem then asserts that (assuming $\widetilde{\psi} \in L^1(V)$):

$$\psi(\mathbf{r}) = \int_V \widetilde{\psi}(\mathbf{p})\psi_{\frac{\mathbf{p}}{|\mathbf{p}|},|\mathbf{p}|}\,d\mathbf{p} = \frac{1}{(2\pi\hbar)^{\frac{3}{2}}} \int_V \widetilde{\psi}(\mathbf{p})e^{\frac{i}{\hbar}\langle\mathbf{p},\mathbf{r}\rangle}\,d\mathbf{p}.$$

Thus, even though the linear momentum eigenfunctions are not square integrable, we can still think of sufficiently nice position states as being superpositions of linear momentum eigenstates via the Fourier transform. As mentioned in the one-dimensional case, the Fourier transform may be extended to a unitary automorphism of the entire Hilbert space $L^2(V)$; (see [4, Section 8.3]).

Proposition 8.6. *Let* $\mathbf{u}$ *and* $\mathbf{p}$ *be a unit vectors in* $(V, \langle,\rangle)$. *The commutator* $[\hat{\mathbf{u}}, \hat{\mathbf{p}}]$ *acts as the scalar* $i\hbar\langle\mathbf{u}, \mathbf{p}\rangle$ *on sufficiently nice functions in position space* $L^2(V)$.

Proof. We use the product rule for the directional derivative:

$$\begin{aligned}
[\hat{\mathbf{u}}, \hat{\mathbf{p}}]\psi(\mathbf{r}) &= (\hat{\mathbf{u}}\hat{\mathbf{p}} - \hat{\mathbf{p}}\hat{\mathbf{u}})\psi(\mathbf{r}) \\
&= \frac{\hbar}{i}\left(\langle\mathbf{u},\mathbf{r}\rangle D_{\mathbf{p}}\psi(\mathbf{r}) - D_{\mathbf{p}}\left(\langle\mathbf{u},\mathbf{r}\rangle\psi(\mathbf{r})\right)\right) \\
&= i\hbar D_{\mathbf{p}}\left(\langle\mathbf{u},\mathbf{r}\rangle\right)\psi(\mathbf{r}).
\end{aligned}$$

But

$$D_{\mathbf{p}}\langle\mathbf{u},\mathbf{r}\rangle = \frac{d}{dt}\langle\mathbf{u},\mathbf{r}+t\mathbf{p}\rangle\bigg|_{t=0} = \langle\mathbf{u},\mathbf{p}\rangle.$$

Thus, we conclude that

$$[\hat{\mathbf{u}}, \hat{\mathbf{p}}]\psi(\mathbf{r}) = i\hbar\langle\mathbf{u},\mathbf{p}\rangle\psi(\mathbf{r}).$$

$\square$

In particular, if $\mathbf{u} = \mathbf{p}$, then $[\hat{\mathbf{u}}, \hat{\mathbf{p}}] = i\hbar I$. If on the other hand $\mathbf{u}$ and $\mathbf{p}$ are orthogonal, we have $[\hat{\mathbf{u}}, \hat{\mathbf{p}}] = 0$. In any case, the commutation relation in proposition 8.6 implies a general Heisenberg Uncertainty Principle (see [24, Theorem 10.4]):

$$\Delta_\psi\hat{\mathbf{u}}\Delta_\psi\hat{\mathbf{p}} \geq \frac{\hbar}{2}|\langle\mathbf{u},\mathbf{p}\rangle|.$$

Here, we have defined the expectation values and uncertainties in the usual way:

$$\begin{aligned}
\langle\hat{\mathbf{u}}\rangle_\psi &:= \langle\psi|\hat{\mathbf{u}}\psi\rangle = \int_V \langle\mathbf{u},\mathbf{r}\rangle|\psi(\mathbf{r})|^2\,d\mathbf{r} \\
\Delta_\psi\hat{\mathbf{u}} &:= \langle\psi|(\hat{\mathbf{u}} - \langle\hat{\mathbf{u}}\rangle_\psi)^2\psi\rangle,
\end{aligned}$$

and similarly for linear momentum.

166   *Symmetry and Quantum Mechanics*

As described at the end of the previous section, when P chooses the basis $\{\mathbf{u}_1, \mathbf{u}_2, \mathbf{u}_3\}$ for $(V, \langle, \rangle)$, position space becomes $L^2(\mathbb{R}^3)$ endowed with the translation action of $(\mathbb{R}^3, +)$ and the rotation action of $SO(3)$. Now let $\mathbf{P}_1, \mathbf{P}_2, \mathbf{P}_3 \in V$ denote the unit vectors $\mathbf{u}_1, \mathbf{u}_2, \mathbf{u}_3$ considered as elements of the Lie algebra of $(V, +)$. Then the linear-momentum operators become partial derivatives, which we denote by

$$\hat{p}_x := \hat{\mathbf{P}}_1 = \frac{\hbar}{i} \frac{\partial}{\partial x}, \qquad \hat{p}_y := \hat{\mathbf{P}}_2 = \frac{\hbar}{i} \frac{\partial}{\partial y}, \qquad \hat{p}_z := \hat{\mathbf{P}}_3 = \frac{\hbar}{i} \frac{\partial}{\partial z}.$$

Proposition 8.6 then implies that these position and linear momentum operators satisfy the following commutation relations:

$$[\hat{x}, \hat{p}_x] = [\hat{y}, \hat{p}_y] = [\hat{z}, \hat{p}_z] = i\hbar I \tag{8.1}$$

with all other brackets being zero.

8.2.1 The Heisenberg group H_3 and its algebra $\mathfrak{h}_3$

M recognizes the commutation relations (8.1) as close to those defining the Heisenberg algebra $\mathfrak{h}_3$, which is the 7-dimensional real Lie algebra with basis

$$\{x, y, z, p_x, p_y, p_z, c\}$$

and bracket given by

$$[x, p_x] = [y, p_y] = [z, p_z] = c,$$

with all other brackets of basis elements being zero. As in the one-dimensional case, the map defined by

$$c \mapsto -\frac{i}{\hbar} I \qquad x \mapsto -\frac{i}{\hbar} \hat{x} \qquad p_x \mapsto -\frac{i}{\hbar} \hat{p}_x, \quad \text{etc.},$$

defines a skew-Hermitian representation of $\mathfrak{h}_3$ on $L^2(\mathbb{R}^3)$. The next exercise realizes $\mathfrak{h}_3$ as a matrix Lie algebra, computes the matrix exponential, and reveals the corresponding Lie group.

Exercise 8.7. *Consider the real vector space of 5×5 strictly upper triangular real matrices with nonzero entries only in the first row and last column:*

$$\left\{ \begin{bmatrix} 0 & \mathbf{v}^T & a \\ 0 & 0_{3\times 3} & \mathbf{w} \\ 0 & 0 & 0 \end{bmatrix} \mid \mathbf{v}, \mathbf{w} \in \mathbb{R}^3, a \in \mathbb{R} \right\}.$$

Check that this vector space is closed under commutation, so that it forms a 7-dimensional real Lie algebra. Denote by E_{ij} the 5×5 matrix with 1 in the

A Three-Dimensional World

ith row and jth column, and zeros elsewhere. Show that the following map defines an isomorphism with the Lie algebra $\mathfrak{h}_3$:

$$E_{12} \mapsto x \qquad E_{13} \mapsto y \qquad E_{14} \mapsto z$$
$$E_{25} \mapsto p_x \qquad E_{35} \mapsto p_y \qquad E_{45} \mapsto p_z$$
$$E_{15} \mapsto c.$$

By explicit computation, show that

$$\exp \begin{bmatrix} 0 & \mathbf{v}^T & a \\ 0 & 0_{3\times 3} & \mathbf{w} \\ 0 & 0 & 0 \end{bmatrix} = \begin{bmatrix} 1 & \mathbf{v}^T & a + \frac{1}{2}\mathbf{v}\cdot\mathbf{w} \\ 0 & 1 & 0 & 0 & \\ 0 & 0 & 1 & 0 & \mathbf{w} \\ 0 & 0 & 0 & 1 & \\ 0 & 0 & 0 & 0 & 1 \end{bmatrix}.$$

Conclude that $\mathfrak{h}_3$ is the Lie algebra of the Heisenberg group H_3, which consists of upper-triangular 5×5 real matrices of the following form:

$$\begin{bmatrix} 1 & \mathbf{v}^T & \alpha \\ 0 & 1 & 0 & 0 & \\ 0 & 0 & 1 & 0 & \mathbf{w} \\ 0 & 0 & 0 & 1 & \\ 0 & 0 & 0 & 0 & 1 \end{bmatrix}.$$

Finally, show that this matrix group is isomorphic to the set of 3-tuples $(\mathbf{v}, \mathbf{w}, \alpha) \in \mathbb{R}^3 \oplus \mathbb{R}^3 \oplus \mathbb{R}$ with operation

$$(\mathbf{v}, \mathbf{w}, \alpha) \bullet (\mathbf{v}', \mathbf{w}', \alpha') = (\mathbf{v} + \mathbf{v}', \mathbf{w} + \mathbf{w}', \alpha + \alpha' + \mathbf{v} \cdot \mathbf{w}'),$$

and that the inverse of an element $(\mathbf{v}, \mathbf{w}, \alpha)$ is given by

$$(\mathbf{v}, \mathbf{w}, \alpha)^{-1} = (-\mathbf{v}, -\mathbf{w}, -\alpha + \mathbf{v} \cdot \mathbf{w}).$$

The skew-Hermitian representation of $\mathfrak{h}_3$ described above comes from a unitary representation of H_3 on $L^2(\mathbb{R}^3)$. To discover this representation, we follow the pattern of our discussion in the one-dimensional case. The group H_3 has two obvious subgroups isomorphic to $(\mathbb{R}^3, +)$ and one isomorphic to $(\mathbb{R}, +)$, and we know how each of these acts on position space to yield (via differentiation) the position operators $-\frac{i}{\hbar}\hat{\mathbf{u}}$, the momentum operators $-\frac{i}{\hbar}\hat{\mathbf{p}}$, as well as the scalar operator $-\frac{i}{\hbar}I$. Indeed, identifying H_3 with $(\mathbb{R}^3 \oplus \mathbb{R}^3 \oplus \mathbb{R}, \bullet)$ as in the previous exercise, we have:

$$(\mathbf{v}, 0, 0) \quad \text{acts as} \quad \psi(\mathbf{r}) \mapsto e^{-\frac{i\mathbf{v}\cdot\mathbf{r}}{\hbar}}\psi(\mathbf{r})$$
$$(0, \mathbf{w}, 0) \quad \text{acts as} \quad \psi(\mathbf{r}) \mapsto \psi(\mathbf{r} - \mathbf{w})$$
$$(0, 0, \alpha) \quad \text{acts as} \quad \psi(\mathbf{r}) \mapsto e^{-\frac{i\alpha}{\hbar}}\psi(\mathbf{r}).$$

Now note that an arbitrary element of H_3 may be factored as a product:

$$(\mathbf{v}, \mathbf{w}, \alpha) = (0, 0, \alpha) \bullet (0, \mathbf{w}, 0) \bullet (\mathbf{v}, 0, 0).$$

168 *Symmetry and Quantum Mechanics*

Hence, applying each factor in turn, we see that a general element $(\mathbf{v}, \mathbf{w}, \alpha)$ must act as follows:

$$
\begin{aligned}
\psi(\mathbf{r}) \mapsto e^{-\frac{i\mathbf{v}\cdot\mathbf{r}}{\hbar}}\psi(\mathbf{r}) \quad &\mapsto \quad e^{-\frac{i}{\hbar}\mathbf{v}\cdot(\mathbf{r}-\mathbf{w})}\psi(\mathbf{r}-\mathbf{w}) \\
&\mapsto \quad e^{-\frac{i\alpha}{\hbar}}e^{-\frac{i}{\hbar}\mathbf{v}\cdot(\mathbf{r}-\mathbf{w})}\psi(\mathbf{r}-\mathbf{w}) \\
&= \quad e^{-\frac{i}{\hbar}(\alpha+\mathbf{v}\cdot(\mathbf{r}-\mathbf{w}))}\psi(\mathbf{r}-\mathbf{w}).
\end{aligned}
$$

Exercise 8.8. *Check that the preceding formula defines a unitary action of H_3 on $L^2(\mathbb{R}^3)$. This representation is called the* Schrödinger *representation of H_3.*

Just as in the one-dimensional case, the Stone-von Neumann Theorem A.54 states that the Schrödinger representation is the *unique* irreducible strongly continuous unitary representation of H_3 in which the central element $(\mathbf{0}, \mathbf{0}, 1)$ acts by the scalar $e^{-\frac{i}{\hbar}}$.

Exercise 8.9. *Copy the discussion in Section 7.3.1 to justify the following interpretation of the H_3-action on wavefunctions: suppose that observer M is moving through P's reference frame with velocity $-\mathbf{v}$ so that M is located at position $\mathbf{r} = \mathbf{w}$ at time $t = 0$. If at time $t = 0$ observer P describes a particle with mass m via the wavefunction $\psi(\mathbf{r})$, then M would describe the same particle (at time $t = 0$) via the wavefunction*

$$
\begin{aligned}
\psi'(\mathbf{r}) &= (m\mathbf{v}, \mathbf{w}, \alpha)^{-1} \star \psi(\mathbf{r}) \\
&= e^{\frac{i}{\hbar}(\alpha+m\mathbf{v}\cdot\mathbf{r})}\psi(\mathbf{r}+\mathbf{w}).
\end{aligned}
$$

8.3 Angular momentum

So far we have not made any use of the rotation action on $L^2(\mathbb{R}^3)$ defined by $A \star \psi = \psi \circ A^{-1}$ for $A \in SO(3)$. First note that rotations of $\mathbb{R}^3$ do not generally commute with translations:

$$
A^{-1}(T_{\mathbf{w}}(A\mathbf{r})) = A^{-1}(A\mathbf{r}+\mathbf{w}) = \mathbf{r} + A^{-1}\mathbf{w} = T_{A^{-1}\mathbf{w}}(\mathbf{r}).
$$

This shows that as automorphisms of $\mathbb{R}^3$, we have $A^{-1}T_{\mathbf{w}}A = T_{A^{-1}\mathbf{w}}$ for all $A \in SO(3)$ and $\mathbf{w} \in (\mathbb{R}^3, +)$. Thus, we can package these two group actions together into a single action by the *semi-direct product* $G_0 := \mathbb{R}^3 \rtimes SO(3)$. This Lie group consists of ordered pairs $(\mathbf{w}, A)$ with operation

$$
(\mathbf{w}, A)(\mathbf{w}', A') = (\mathbf{w} + A\mathbf{w}', AA'),
$$

and the action on $\mathbb{R}^3$ is given by $(\mathbf{w}, A) \star \mathbf{r} = A\mathbf{r} + \mathbf{w}$.

A Three-Dimensional World   169

Exercise 8.10. *Show that in the group G_0, the inverse of the element $(\mathbf{w}, A)$ is given by $(-A^{-1}\mathbf{w}, A^{-1})$.*

Thus, position-space $L^2(\mathbb{R}^3)$ carries a unitary action of G_0 defined by

$$(\mathbf{w}, A) \star \psi(\mathbf{r}) = \psi((\mathbf{w}, A)^{-1} \star \mathbf{r}) = \psi(A^{-1}\mathbf{r} - A^{-1}\mathbf{w}) = \psi(A^{-1}(\mathbf{r} - \mathbf{w})).$$

In terms of the G_0-action on $L^2(\mathbb{R}^3)$, the linear momentum operators arise from the Lie subalgebra $\mathbb{R}p_x \oplus \mathbb{R}p_y \oplus \mathbb{R}p_z$ corresponding to the translation subgroup $\mathbb{R}^3 \times \{I\} = \mathbb{R}^3$. To obtain the *angular momentum operators*, we investigate the action of the Lie subalgebra corresponding to the rotation subgroup $\{0\} \times SO(3) = SO(3)$.

First of all, recall from the investigation of spin in Chapter 2 that $SU(2)$ is the universal cover of $SO(3)$ via the double cover $f \colon SU(2) \to SO(3)$ from theorem 2.39. As shown in proposition 5.29 in Chapter 5, the mapping f induces an isomorphism of Lie algebras $Df \colon \mathfrak{su}(2) \to \mathfrak{so}(3)$, and we use this isomorphism to identify the two Lie algebras, writing

$$\mathfrak{so}(3) = \mathbb{R}L_1 \oplus \mathbb{R}L_2 \oplus \mathbb{R}L_3,$$

where $L_1 = Df\left(\frac{1}{2i}\sigma_1\right), L_2 = Df\left(\frac{1}{2i}\sigma_2\right)$, and $L_3 = Df\left(\frac{1}{2i}\sigma_3\right)$. In proposition 5.43 we also determined the action of L_j on functions by differentiating the $SO(3)$-action:

$$L_1 \qquad \text{acts via} \qquad z\frac{\partial}{\partial y} - y\frac{\partial}{\partial z}$$

$$L_2 \qquad \text{acts via} \qquad -z\frac{\partial}{\partial x} + x\frac{\partial}{\partial z}$$

$$L_3 \qquad \text{acts via} \qquad y\frac{\partial}{\partial x} - x\frac{\partial}{\partial y}.$$

Multiplying by $-\frac{\hbar}{i}$ as usual leads to the components of the angular momentum operator.

Definition 8.11. *The x-component of angular momentum observable is given by the operator $\hat{L}_x$ on position space $L^2(\mathbb{R}^3)$ defined by*

$$\hat{L}_x := -\frac{\hbar}{i}\left(z\frac{\partial}{\partial y} - y\frac{\partial}{\partial z}\right) = \hat{y}\hat{p}_z - \hat{z}\hat{p}_y.$$

Similarly, we have

$$\hat{L}_y := \hat{z}\hat{p}_x - \hat{x}\hat{p}_z, \qquad \hat{L}_z := \hat{x}\hat{p}_y - \hat{y}\hat{p}_x$$

for the y- and z-components of angular momentum. Note that these operators are simply the quantization of the definition of angular momentum in classical mechanics (see Section 2.1): $\mathbf{L} = \mathbf{r} \times \mathbf{p}$.

170   *Symmetry and Quantum Mechanics*

A natural question now arises: how does position space $L^2(\mathbb{R}^3)$ decompose into irreducible representations of the rotation subgroup $SO(3) \subset G_0$? Recall that in Section 5.6.2 we answered this question for the closely related space $L^2(S^2)$ of square-integrable functions on the unit sphere. There we explained that the Hilbert space $L^2(S^2)$ decomposes as the completed orthogonal direct sum of one copy of each irreducible representation of $SO(3)$:

$$L^2(S^2) = \widehat{\bigoplus_{l=0}^{\infty}} \mathcal{Y}_l \simeq \widehat{\bigoplus_{l=0}^{\infty}} \pi_l.$$

Here, the $2l+1$-dimensional subspace $\mathcal{Y}_l \subset C^{\infty}(S^2)$ is spanned by the spherical harmonics of degree l, denoted $Y_l^{\pm m}$ for $m = 0, 1, \ldots, l$. The total orbital angular momentum squared operator $\hat{\mathbf{L}}^2 = \hat{L}_x^2 + \hat{L}_y^2 + \hat{L}_z^2$ acts as the scalar $l(l+1)\hbar^2$ on $\mathcal{Y}_l$, and each basis function $Y_l^{\pm m}$ is an eigenvector for $\hat{L}_z$:

$$\hat{L}_z Y_l^{\pm m} = \pm m\hbar Y_l^{\pm m}.$$

We would now like to make a similar analysis of the $SO(3)$-action on $L^2(\mathbb{R}^3)$. For this, we once again (see Figure 5.3) use spherical coordinates on $\mathbb{R}^3$ given by the mapping $g \colon \mathbb{R}_{\geq 0} \times [0, \pi] \times [0, 2\pi] \to \mathbb{R}^3$ defined by

$$g(r, \theta, \phi) = (r\sin(\theta)\cos(\phi), r\sin(\theta)\sin(\phi), r\cos(\theta)).$$

By change of variables (see theorem A.41), we have

$$\int_{\mathbb{R}^3} F(\mathbf{r})d\mathbf{r} = \int_0^{\infty} \int_0^{2\pi} \int_0^{\pi} F(g(r, \theta, \phi))r^2 \sin(\theta)d\theta d\phi dr.$$

Note that if $F(g(r, \theta, \phi)) = f(r)Y(\theta, \phi)$ is a separable function, then the integral factors into a product of integrals:

$$\int_0^{\infty} f(r)r^2 dr \int_0^{2\pi} \int_0^{\pi} Y(\theta, \phi)\sin(\theta)d\theta d\phi.$$

This leads ultimately to the identification of position space $L^2(\mathbb{R}^3)$ with the completed tensor product

$$L^2(\mathbb{R}_{\geq 0}, r^2 dr) \hat{\otimes} L^2(S^2, d\Omega),$$

where $d\Omega = \sin(\theta)d\theta d\phi$ is the solid angle measure on the unit sphere S^2. Explicitly, this means that every square integrable function on $\mathbb{R}^3$ may be expressed as a convergent infinite series of separable functions of the form $f(r)Y(\theta, \phi)$, where $f \in L^2(\mathbb{R}_{\geq 0}, r^2 dr)$ and $Y \in L^2(S^2)$.

By proposition 5.45, the orbital angular momentum operators have the following expressions in spherical coordinates:

$$\hat{L}_x = \frac{\hbar}{i}\left(-\sin(\phi)\frac{\partial}{\partial\theta} - \cot(\theta)\cos(\phi)\frac{\partial}{\partial\phi}\right)$$

$$\hat{L}_y = \frac{\hbar}{i}\left(\cos(\phi)\frac{\partial}{\partial\theta} - \cot(\theta)\sin(\phi)\frac{\partial}{\partial\phi}\right)$$

$$\hat{L}_z = \frac{\hbar}{i}\frac{\partial}{\partial\phi}.$$

A Three-Dimensional World

The key fact to notice is that these operators act trivially on the first factor $L^2(\mathbb{R}_{\geq 0}, r^2 dr)$ of the tensor product decomposition given above. This means that $SO(3)$ acts as the identity on this first factor, and the decomposition of the second factor $L^2(S^2)$ determines the decomposition of position space $L^2(\mathbb{R}^3)$ as a representation of $SO(3)$. In particular, fix any element $f(r) \in L^2(\mathbb{R}_{\geq 0}, r^2 dr)$ and consider the $2l+1$-dimensional subspace of functions $f \mathcal{Y}_l \subset L^2(\mathbb{R}^3)$ obtained by multiplying the degree-l spherical harmonics by $f(r)$. Since $SO(3)$ acts trivially on the function f, it follows that this space is isomorphic to the irreducible representation π_l of $SO(3)$. Conversely, one can show that every finite-dimensional irreducible subrepresentation of $SO(3)$ contained in $L^2(\mathbb{R}^3)$ is of this form (see [10, Proposition 17.19]).

8.4 The Lie group $G = H_3 \rtimes SO(3)$ and its Lie algebra $\mathfrak{g}$

We have now determined the actions of the Lie subalgebras $\mathbb{R}^3$ and $\mathfrak{so}(3)$ of the Lie algebra $\mathfrak{g}_0$ of $G_0 = \mathbb{R}^3 \rtimes SO(3)$. To describe the full action of $\mathfrak{g}_0$ on $L^2(\mathbb{R}^3)$, all that remains is to find the commutation relations between the angular momentum and linear momentum operators. Straightforward computations reveal that

$$[\hat{L}_x, \hat{p}_x] = 0 \qquad [\hat{L}_x, \hat{p}_y] = i\hbar \hat{p}_z \qquad [\hat{L}_x, \hat{p}_z] = -i\hbar \hat{p}_y$$

$$[\hat{L}_y, \hat{p}_x] = -i\hbar \hat{p}_z \qquad [\hat{L}_y, \hat{p}_y] = 0 \qquad [\hat{L}_y, \hat{p}_z] = i\hbar \hat{p}_x$$

$$[\hat{L}_z, \hat{p}_x] = i\hbar \hat{p}_y \qquad [\hat{L}_z, \hat{p}_y] = -i\hbar \hat{p}_x \qquad [\hat{L}_x, \hat{p}_z] = 0.$$

Exercise 8.12. *Verify these commutation relations by acting on suitably nice functions ψ in $L^2(\mathbb{R}^3)$.*

We may summarize as follows: $\mathfrak{g}_0$ is the real six-dimensional Lie algebra with basis $\{p_x, p_y, p_z, L_x, L_y, L_z\}$ with bracket defined by

$$[L_x, L_y] = L_z \qquad [L_y, L_z] = L_x \qquad [L_z, L_x] = L_y$$

$$[L_x, p_x] = 0 \qquad [L_x, p_y] = p_z \qquad [L_x, p_z] = -p_y$$

$$[L_y, p_x] = -p_z \qquad [L_y, p_y] = 0 \qquad [L_y, p_z] = p_x$$

$$[L_z, p_x] = p_y \qquad [L_z, p_y] = -p_x \qquad [L_z, p_z] = 0,$$

all other brackets of basis elements being zero. The mapping defined by

$$p_x \mapsto -\frac{i}{\hbar}\hat{p}_x \qquad p_y \mapsto -\frac{i}{\hbar}\hat{p}_y \qquad p_z \mapsto -\frac{i}{\hbar}\hat{p}_z$$

$$L_x \mapsto -\frac{i}{\hbar}\hat{L}_x \qquad L_y \mapsto -\frac{i}{\hbar}\hat{L}_y \qquad L_z \mapsto -\frac{i}{\hbar}\hat{L}_z$$

172   *Symmetry and Quantum Mechanics*

defines a skew-Hermitian representation of $\mathfrak{g}_0$ on $L^2(\mathbb{R}^3)$. Moreover, this representation of $\mathfrak{g}_0$ is induced by the unitary representation of G_0 on $L^2(\mathbb{R}^3)$ defined by

$$(\mathbf{w}, A) \star \psi(\mathbf{r}) = \psi(A^{-1}(\mathbf{r} - \mathbf{w})).$$

This representation of G_0 may be combined with the Schrödinger representation of the Heisenberg group H_3, which also contains the translation group $\mathbb{R}^3$ as a subgroup. For this, consider the 10-dimensional real Lie algebra $\mathfrak{g}$ with basis $\{x, y, z, p_x, p_y, p_z, c, L_x, L_y, L_z\}$, containing the Lie subalgebras $\mathfrak{h}_3$ and $\mathfrak{g}_0$ with intersection $\mathfrak{h}_3 \cap \mathfrak{g}_0 = \mathbb{R}p_x \oplus \mathbb{R}p_y \oplus \mathbb{R}p_z$, the translation subalgebra. The bracket is defined by the brackets for the subalgebras given above together with

$$[L_x, x] = 0 \qquad [L_x, y] = z \qquad [L_x, z] = -y$$

$$[L_y, x] = -z \qquad [L_y, y] = 0 \qquad [L_y, z] = x$$

$$[L_z, x] = y \qquad [L_z, y] = -x \qquad [L_z, z] = 0,$$

and taking all brackets with c to be zero. With this definition, we obtain a skew-Hermitian representation of $\mathfrak{g}$ on $L^2(\mathbb{R}^3)$ extending the representations of $\mathfrak{h}_3$ and $\mathfrak{g}_0$ already defined. Note that this must be an irreducible representation of $\mathfrak{g}$, since it is already irreducible for the Heisenberg subalgebra $\mathfrak{h}_3$.

Exercise 8.13. *Check that the additional commutation relations given above for the Lie algebra $\mathfrak{g}$ are indeed satisfied by the corresponding angular momentum and position operators on $L^2(\mathbb{R}^3)$.*

Now $\mathfrak{g}$ is the Lie algebra of the Lie group $G = H_3 \rtimes SO(3)$ consisting of elements $(\mathbf{v}, \mathbf{w}, \alpha, A)$ with operation

$$(\mathbf{v}, \mathbf{w}, \alpha, A)(\mathbf{v}', \mathbf{w}', \alpha', A') = (\mathbf{v} + A\mathbf{v}', \mathbf{w} + A\mathbf{w}', \alpha + \alpha' + \mathbf{v} \cdot A\mathbf{w}', AA').$$

Exercise 8.14. *Show that the inverse of an element $(\mathbf{v}, \mathbf{w}, \alpha, A) \in G$ is given by*

$$(\mathbf{v}, \mathbf{w}, \alpha, A)^{-1} = (-A^{-1}\mathbf{v}, -A^{-1}\mathbf{w}, -\alpha + \mathbf{v} \cdot \mathbf{w}, A^{-1}).$$

The representation of $\mathfrak{g}$ constructed above comes from the irreducible unitary representation of G on $L^2(\mathbb{R}^3)$ defined by

$$(\mathbf{v}, \mathbf{w}, \alpha, A) \star \psi(\mathbf{r}) = e^{-\frac{i}{\hbar}(\alpha + \mathbf{v} \cdot (\mathbf{r} - \mathbf{w}))} \psi(A^{-1}(\mathbf{r} - \mathbf{w})).$$

The restriction of this representation to the Heisenberg group H_3 is the Schrödinger representation discussed in Section 8.2.1.

Thus, P's choice of location $\mathbf{a}_0$ in physical space $\mathbb{A}^3$, together with his choice of a right-handed orthonormal basis $\{\mathbf{u}_1, \mathbf{u}_2, \mathbf{u}_3\}$ for $(V, \langle , \rangle)$ has led to a description of position space $L^2(\mathbb{R}^3)$ as an irreducible unitary representation of $G = H_3 \rtimes SO(3)$ extending the Schrödinger representation of H_3.

A Three-Dimensional World

Exercise 8.15 (♣). *Extend the interpretation of the H_3-action developed in exercise 8.9 to the larger group G: suppose that observer M is rotated with respect to P according to $A \in SO(3)$. Moreover, suppose that M is also moving through P's reference frame with velocity $-\mathbf{v}$ so that M is located at position $\mathbf{r} = \mathbf{w}$ at time $t = 0$. If at time $t = 0$ observer P describes a particle with mass m via the wavefunction $\psi(\mathbf{r})$, then M would describe the same particle (at time $t = 0$) via the wavefunction*

$$\begin{aligned}
\psi'(\mathbf{r}) &= (m\mathbf{v}, \mathbf{w}, \alpha, A)^{-1} \star \psi(\mathbf{r}) \\
&= e^{\frac{i}{\hbar}(\alpha + m\mathbf{v} \cdot A\mathbf{r})} \psi(A\mathbf{r} + \mathbf{w}).
\end{aligned}$$

8.5 Time-evolution

Finally, we are ready to study quantum dynamics in three dimensions. To begin, we consider the classical expression for the energy of a particle of mass[1] M in a potential $V \colon \mathbb{R}^3 \to \mathbb{R}$:

$$E = \frac{\mathbf{p} \cdot \mathbf{p}}{2M} + V(\mathbf{r}).$$

Here $\mathbf{r} = (x, y, z)$ is the position and $\mathbf{p} = (p_x, p_y, p_z)$ is the momentum of the particle. Quantizing this expression, we obtain the Hamiltonian

$$\mathcal{H} = \frac{1}{2M} \hat{\mathbf{p}} \cdot \hat{\mathbf{p}} + V(\hat{\mathbf{r}}),$$

where $\hat{\mathbf{r}} = (\hat{x}, \hat{y}, \hat{z})$ and $\hat{\mathbf{p}} = (\hat{p}_x, \hat{p}_y, \hat{p}_z)$ are vector-observables. The Schrödinger equation governing the time-evolution of an initial state $\psi(\mathbf{r}) = \psi(\mathbf{r}, 0)$ is then

$$\begin{aligned}
i\hbar \frac{\partial \psi(\mathbf{r}, t)}{\partial t} &= \left(\frac{1}{2M} \left(\hat{p}_x^2 + \hat{p}_y^2 + \hat{p}_z^2 \right) + V(\hat{\mathbf{r}}) \right) \psi(\mathbf{r}, t) \\
&= -\frac{\hbar^2}{2M} \Delta \psi(\mathbf{r}, t) + V(\mathbf{r}) \psi(\mathbf{r}, t),
\end{aligned}$$

where $\Delta = \frac{\partial^2}{\partial x^2} + \frac{\partial^2}{\partial y^2} + \frac{\partial^2}{\partial z^2}$ is the Laplacian operator discussed in Section 5.6.

Exercise 8.16. *Suppose that observer M is rotated with respect to P according to $A \in SO(3)$, and also translated via $\mathbf{w} \in \mathbb{R}^3$. Hence, M's coordinates $\mathbf{r}'$ are obtained from P's according to $\mathbf{r}' = A^{-1}(\mathbf{r} - \mathbf{w})$. Suppose that $\psi(\mathbf{r}, t)$ is*

---

[1] Here we switch from lower-case to upper-case for the mass, in order to avoid conflict with the traditional symbol, m, for the *magnetic quantum number* labeling the eigenvalues of the operator $\hat{L}_z$. Context should prevent any confusion between the mass M and the observer M.

174   *Symmetry and Quantum Mechanics*

a solution (for observer P) to the Schrödinger equation with potential $V(\mathbf{r})$. Show that M's corresponding wavefunction $\psi'(\mathbf{r}, t) = \psi(A\mathbf{r}+\mathbf{w}, t)$ is a solution to the Schrödinger equation with transformed potential $V'(\mathbf{r}) = V(A\mathbf{r} + \mathbf{w})$. (Hint: see the discussion in Section 7.4.)

8.5.1   The free particle

Taking $V(\mathbf{r}) = 0$ yields the Hamiltonian for a free quantum particle:

$$\mathcal{H} = -\frac{\hbar^2}{2M}\Delta.$$

As in the one-dimensional case, every momentum eigenstate $\psi_{\frac{\mathbf{p}}{|\mathbf{p}|}, |\mathbf{p}|} = (2\pi\hbar)^{-\frac{3}{2}} e^{\frac{i}{\hbar}\langle \mathbf{p}, \mathbf{r}\rangle}$ is also an energy eigenstate:

$$\mathcal{H}\psi_{\frac{\mathbf{p}}{|\mathbf{p}|}, |\mathbf{p}|} = -\frac{\hbar^2}{2M}\Delta\psi_{\frac{\mathbf{p}}{|\mathbf{p}|}, |\mathbf{p}|} = \frac{|\mathbf{p}|^2}{2M}\psi_{\frac{\mathbf{p}}{|\mathbf{p}|}, |\mathbf{p}|}.$$

But now the energy eigenspaces are infinite-dimensional, since for any rotation $A \in SO(3)$, the function $\psi_{\frac{A\mathbf{p}}{|A\mathbf{p}|}, |A\mathbf{p}|}$ also has eigenvalue $\frac{|\mathbf{p}|^2}{2M}$. But if we use the Fourier transform to express an initial position state as

$$\psi(\mathbf{r}) = \psi(\mathbf{r}, 0) = \int_{\mathbb{R}^3} \tilde{\psi}(\mathbf{p})\psi_{\frac{\mathbf{p}}{|\mathbf{p}|}, |\mathbf{p}|} d\mathbf{p},$$

then its time-evolution is given by

$$
\begin{aligned}
\psi(\mathbf{r}, t) &= e^{-\frac{i}{\hbar}\mathcal{H}t}\psi(\mathbf{r}, 0) \\
&= e^{-\frac{i}{\hbar}\mathcal{H}t}\int_{\mathbb{R}^3} \tilde{\psi}(\mathbf{p})\psi_{\frac{\mathbf{p}}{|\mathbf{p}|}, |\mathbf{p}|} d\mathbf{p} \\
&= \int_{\mathbb{R}^3} \tilde{\psi}(\mathbf{p})e^{-\frac{i}{\hbar}\mathcal{H}t}\psi_{\frac{\mathbf{p}}{|\mathbf{p}|}, |\mathbf{p}|} d\mathbf{p} \\
&= \int_{\mathbb{R}^3} \tilde{\psi}(\mathbf{p})e^{-\frac{i|\mathbf{p}|^2 t}{2M\hbar}}\psi_{\frac{\mathbf{p}}{|\mathbf{p}|}, |\mathbf{p}|} d\mathbf{p}.
\end{aligned}
$$

8.5.2   The three-dimensional harmonic oscillator

Suppose that $V(\mathbf{r})$ is an arbitrary (analytic) potential describing a classical system with a stable equilibrium at the origin. Then a particle at the origin feels no force, so $0 = \mathbf{F}(0) = -\nabla V(0)$. Moreover, since this is a stable equilibrium, the particle must feel a restoring force near the origin, which implies that 0 is a local minimum for the potential V. By the second derivative test, it follows that the Hessian matrix of V is positive semi-definite at $\mathbf{r} = 0$:

$$\mathbb{H} := \left[\frac{\partial^2 V}{\partial_i \partial_j}(0)\right].$$

A Three-Dimensional World

This means that the eigenvalues k_j of $\mathbb{H}$ are non-negative—we will assume that they are all positive, so that $\mathbb{H}$ is actually positive definite. After shifting the potential by an additive constant (which does not affect the dynamics) we may also assume that $V(0) = 0$. Then the Taylor expansion of V at zero looks like

$$V(\mathbf{r}) = \frac{1}{2}\mathbf{r}^T \mathbb{H}\mathbf{r} + \text{higher-order terms.}$$

By the Spectral Theorem A.32, there exists a matrix $A \in SO(3)$ such that $A\mathbb{H}A^{-1} = \text{diag}(k_1, k_2, k_3)$. Hence, changing variables to $\mathbf{s} := A\mathbf{r}$ and ignoring the higher-order terms, we may approximate the potential V near 0 as

$$
\begin{aligned}
V(\mathbf{s}) &\approx \frac{1}{2}\mathbf{r}^T A^T \text{diag}(k_1, k_2, k_3) A\mathbf{r} \\
&= \frac{1}{2}\mathbf{s}^T \text{diag}(k_1, k_2, k_3)\mathbf{s} \\
&= \frac{k_1}{2}s_1^2 + \frac{k_2}{2}s_2^2 + \frac{k_3}{2}s_3^2.
\end{aligned}
$$

This is the sum of three one-dimensional simple harmonic oscillator potentials with spring constants $k_1, k_2, k_3 > 0$. Since we allow the "stiffness" to differ in the three directions, such a potential is called an *anisotropic harmonic oscillator*—the argument given above shows that it is ubiquitous, just like its one-dimensional counterpart.

We now turn to the quantum anisotropic harmonic oscillator, introducing the frequencies $\omega_j := \sqrt{\frac{k_j}{M}}$ to write the Hamiltonian:

$$
\begin{aligned}
\mathcal{H} &= \frac{\hbar^2}{2M}\Delta + \frac{k_1}{2}\hat{x}^2 + \frac{k_2}{2}\hat{y}^2 + \frac{k_3}{2}\hat{z}^2 \\
&= \frac{1}{2M}\left(\hat{p}_x^2 + \hat{p}_y^2 + \hat{p}_z^2\right) + \frac{M\omega_1^2}{2}\hat{x}^2 + \frac{M\omega_2^2}{2}\hat{y}^2 + \frac{M\omega_3^2}{2}\hat{z}^2 \\
&= \left(\frac{\hat{p}_x^2}{2M} + \frac{M\omega_1^2}{2}\hat{x}^2\right) + \left(\frac{\hat{p}_y^2}{2M} + \frac{M\omega_2^2}{2}\hat{y}^2\right) + \left(\frac{\hat{p}_z^2}{2M} + \frac{M\omega_3^2}{2}\hat{z}^2\right) \\
&=: \mathcal{H}_1 + \mathcal{H}_2 + \mathcal{H}_3.
\end{aligned}
$$

Thus, the Hamiltonian for the three-dimensional harmonic oscillator is just the sum of three one-dimensional harmonic oscillators. Since the Hamiltonians $\mathcal{H}_1$, $\mathcal{H}_2$, and $\mathcal{H}_3$ commute, we search for an orthonormal basis for position space consisting of simultaneous eigenfunctions for these three operators. Using the decomposition

$$L^2(\mathbb{R}^3) = L^2(\mathbb{R}, dx)\hat{\otimes}L^2(\mathbb{R}, dy)\hat{\otimes}L^2(\mathbb{R}, dz),$$

and the energy eigenfunctions ψ_n for the one-dimensional oscillator (see Section 7.4.3), we can write this basis as

$$\psi_{n_1, n_2, n_3}(x, y, z) = \psi_{n_1}(x)\psi_{n_2}(y)\psi_{n_3}(z) \qquad n_1, n_2, n_3 = 0, 1, 2, \ldots,$$

176   *Symmetry and Quantum Mechanics*

where $\mathcal{H}_j \psi_{n_j} = \hbar \omega_j \left(n_j + \frac{1}{2} \right) \psi_{n_j}$. It follows that

$$
\begin{aligned}
\mathcal{H}\psi_{n_1,n_2,n_3} &= (\mathcal{H}_1 + \mathcal{H}_2 + \mathcal{H}_3)\psi_{n_1,n_2,n_3} \\
&= \hbar \left(n_1\omega_1 + n_2\omega_2 + n_3\omega_3 + \frac{\omega_1 + \omega_2 + \omega_3}{2} \right) \psi_{n_x,n_y,n_x}.
\end{aligned}
$$

In the special case of the *isotropic harmonic oscillator*, we have $\omega_1 = \omega_2 = \omega_3$, so the energy levels are $E_n = \hbar\omega \left(n + \frac{3}{2} \right)$. But these levels are highly degenerate: the dimension of the eigenspace for E_n is given by the number of ways of writing the non-negative integer n as a sum of three non-negative integers n_1, n_2, n_3. This number is the dimension of the space of degree n homogeneous polynomials in three variables, so the degeneracy is $\frac{1}{2}(n+2)(n+1)$ by proposition 5.40.

8.5.3   Central potentials

We now restrict attention to *central potentials* that depend only on the radial distance to the origin: $V(\mathbf{r}) = V(r)$, where $r = |\mathbf{r}|$.

Exercise 8.17. *Show by direct computation that for a central potential $V(r)$, the resulting Hamiltonian $\mathcal{H} = \frac{1}{2M}\hat{\mathbf{p}} \cdot \hat{\mathbf{p}} + V(\hat{r})$ commutes with the angular-momentum operators $\hat{L}_x, \hat{L}_y, \hat{L}_z$.*

It follows from the previous exercise and Section 4.3.1 that the components of angular momentum are conserved quantities in a central potential. Moreover, the operators $\mathcal{H}, \hat{\mathbf{L}}^2$, and $\hat{L}_z$ commute with each other, so it makes sense to search for a basis for $L^2(\mathbb{R}^3)$ comprised of simultaneous eigenvectors for these three operators. In terms of observables, this means that we are looking for a basis for position space consisting of simultaneous eigenstates for energy, squared total orbital angular momentum, and z-angular momentum. We denote such a basis by $\{\psi_{E,l,m}\}$, where the observable energies $E \in \mathbb{R}$ will be determined by the potential $V(r)$, and the eigenvalues for $\hat{\mathbf{L}}^2$ and $\hat{L}_z$ are known from the representation theory of $SO(3)$ discussed in Section 8.3:

$$
\begin{aligned}
\mathcal{H}\psi_{E,l,m} &= E\psi_{E,l,m} \\
\hat{\mathbf{L}}^2 \psi_{E,l,m} &= l(l+1)\hbar^2 \psi_{E,l,m} \qquad l = 0, 1, 2, \ldots \\
\hat{L}_z \psi_{E,l,m} &= m\hbar \psi_{E,l,m} \qquad m = -l, -l+1, \ldots, l-1, l.
\end{aligned}
$$

In this context, the index l is called the *azimuthal quantum number* and m is called the *magnetic quantum number*. Note that the energy spectrum may not be discrete, but rather involve a continuum of values. Moreover, this labeling scheme may be insufficient, since the dimension of each simultaneous eigenspace may be greater than one.

In the search for such an eigenbasis, it will be convenient to exploit the rotational symmetry of the Hamiltonian and use spherical coordinates. Recall

A Three-Dimensional World

from exercise 5.47 that we may express the Laplacian in spherical coordinates as follows (note that we have changed notation from $\mathbf{S}$ to $\hat{\mathbf{L}}$ in order to emphasize orbital angular momentum):

$$\Delta = \frac{1}{r^2} \frac{\partial}{\partial r} \left(r^2 \frac{\partial}{\partial r} \right) - \frac{1}{\hbar^2 r^2} \hat{\mathbf{L}}^2.$$

The Hamiltonian is then

$$
\begin{aligned}
\mathcal{H} &= -\frac{\hbar^2}{2M} \Delta + V(\hat{r}) \\
&= -\frac{\hbar^2}{2Mr^2} \frac{\partial}{\partial r} \left(r^2 \frac{\partial}{\partial r} \right) + \frac{1}{2Mr^2} \hat{\mathbf{L}}^2 + V(\hat{r}) \\
&= -\frac{\hbar^2}{2M} \frac{\partial^2}{\partial r^2} - \frac{\hbar^2}{Mr} \frac{\partial}{\partial r} + \frac{1}{2Mr^2} \hat{\mathbf{L}}^2 + V(\hat{r}) \\
&= -\frac{\hbar^2}{2M} \left(\frac{\partial^2}{\partial r^2} + \frac{2}{r} \frac{\partial}{\partial r} \right) + \frac{1}{2Mr^2} \hat{\mathbf{L}}^2 + V(\hat{r}).
\end{aligned}
$$

From the discussion at the end of Section 8.3, we know that any separable function[2] of the form $R(r)Y_l^m(\theta, \phi) \in L^2(\mathbb{R}^3)$ is a simultaneous eigenvector for $\hat{\mathbf{L}}^2$ and $\hat{L}_z$. Hence, we will search for energy eigenfunctions of this form. Recalling that $\hat{\mathbf{L}}^2$ acts as $l(l+1)\hbar^2$ on such a function, the energy eigenfunction condition becomes a second-order ordinary differential equation for the radial function $R(r) = R_{E,l}(r)$:

$$\left[-\frac{\hbar^2}{2M} \left(\frac{d^2}{dr^2} + \frac{2}{r} \frac{d}{dr} \right) + \frac{l(l+1)\hbar^2}{2Mr^2} + V(r) \right] R_{E,l}(r) = E R_{E,l}(r). \quad (8.2)$$

Note in particular that this equation does not depend on the magnetic quantum number m. Hence, if $R_{E,l}$ is a solution for a particular E and l, then we obtain $2l + 1$ linearly independent eigenfunctions with energy E and squared orbital angular momentum $l(l + 1)\hbar^2$:

$$R_{E,l}(r)Y_l^m(\theta, \phi) \qquad m = -l, -l+1, \ldots, l.$$

This energy degeneracy is a consequence of the $SO(3)$-symmetry of central potentials: since the index m corresponds to the z-component of the angular momentum, any dependence of the energy on m would imply something special about the z-direction. But because central potentials are rotationally symmetric, all directions are on an equal footing, which forbids any such dependence.

---

[2] Here we allow the index m to take on both positive and negative values, so that we obtain all of the spherical harmonics $Y_l^{\pm m}$ described in Section 8.3.

8.5.4 The infinite spherical well

As an example of a non-trivial central potential, suppose that the particle is constrained to reside in a spherical region of radius $a > 0$. As in the one-dimensional square well from Section 7.4.2, we could model this with a central "potential function"

$$V(r) = \begin{cases} 0 & \text{if } 0 \le r < a \\ \infty & \text{if } r \ge a. \end{cases}$$

Instead, we restrict attention to the subspace

$$L^2([0, a], r^2 dr) \hat{\otimes} L^2(S^2, d\Omega)$$

of position space consisting of functions that are zero outside of the open ball of radius a centered at the origin. Inside this ball the particle is free, so the Hamiltonian is just $\mathcal{H} = -\frac{\hbar^2}{2M}\Delta$ considered as an operator on this subspace. The energy eigenfunction equation (8.2) is thus

$$\left[-\frac{\hbar^2}{2M}\left(\frac{d^2}{dr^2} + \frac{2}{r}\frac{d}{dr}\right) + \frac{l(l+1)\hbar^2}{2Mr^2} \right] R_{E,l}(r) = E R_{E,l}(r), \qquad (8.3)$$

and we are trying to solve it for $R_{E,l} \in L^2([0, a], r^2 dr)$. First assume that $E > 0$, so that the following quantity is real and positive: $k := \sqrt{\frac{2ME}{\hbar^2}}$. Then a bit of rearrangement (and replacing the index E with the index k) yields

$$\left[\frac{d^2}{dr^2} + \frac{2}{r}\frac{d}{dr} - \frac{l(l+1)}{r^2} + k^2 \right] R_{k,l}(r) = 0.$$

Setting $\rho = kr$ then gives

$$\left[\frac{d^2}{d\rho^2} + \frac{2}{\rho}\frac{d}{d\rho} + \left(1 - \frac{l(l+1)}{\rho^2}\right) \right] R_{k,l}\left(\frac{\rho}{k}\right) = 0.$$

This is the *spherical Bessel equation*. The solutions that are finite at $\rho = 0$ are the *spherical Bessel functions* $j_l(\rho)$, described as follows:

$$j_l(\rho) = (-\rho)^l \left(\frac{1}{\rho}\frac{d}{d\rho}\right)^l \frac{\sin(\rho)}{\rho}.$$

Thus, we must have $R_{k,l}(r) = j_l(kr)$, and the requirement that $R_{k,l}(a) = 0$ becomes $j_l(ka) = 0$. For each $l = 0, 1, 2, \ldots$, this condition is satisfied only for an infinite discrete set of values $k_{l,n} > 0$. Thus, for each l we get a discrete set of observable energies $E_{l,n} = \frac{k_{l,n}^2 \hbar^2}{2M}$, with corresponding radial eigenfunctions $R_{E_l,n}(r) = j_l(k_{l,n}r)$. For more details, see [9, Section 4.3.1] and [22, Section 10.4].

At the end of this section we will show that there are no physical solutions

A Three-Dimensional World

to the radial eigenfunction equation (8.3) for energies $E \leq 0$. In fact, we have found a Hilbert space basis for position space consisting of simultaneous eigenfunctions for energy, squared orbital angular momentum, and z-angular momentum given by

$$\psi_{n,l,m}(r,\theta,\phi) = j_l(k_{l,n}r)Y_l^m(\theta,\phi).$$

Now suppose that ψ is an arbitrary initial position-state:

$$\psi(r,\theta,\phi) = \sum_{n=1}^{\infty}\sum_{l=0}^{\infty}\sum_{m=-l}^{l} c_{n,l,m}\psi_{n,l,m}(r,\theta,\phi).$$

Then the time-evolution is given by

$$\begin{aligned}
\psi(r,\theta,\phi,t) &= e^{-\frac{i}{\hbar}\mathcal{H}t}\psi(r,\theta,\phi)\\
&= \sum_{n=1}^{\infty}\sum_{l=0}^{\infty}\sum_{m=-l}^{l} c_{n,l,m}e^{-\frac{i}{\hbar}\mathcal{H}t}\psi_{n,l,m}(r,\theta,\phi)\\
&= \sum_{n=1}^{\infty}\sum_{l=0}^{\infty}\sum_{m=-l}^{l} c_{n,l,m}e^{-\frac{i}{\hbar}E_{n,l}t}\psi_{n,l,m}(r,\theta,\phi).
\end{aligned}$$

As promised, we now wish to show that there are no physical solutions to the radial energy eigenfunction equation (8.3) for non-positive energies. First suppose that $E < 0$, and introduce the purely imaginary quantity $ik := \sqrt{\frac{2ME}{\hbar^2}}$. Then again setting $\rho = kr$, the same sequence of steps yields the *modified spherical Bessel equation*:

$$\left[\frac{d^2}{d\rho^2} + \frac{2}{\rho}\frac{d}{d\rho} - \left(1 + \frac{l(l+1)}{\rho^2}\right)\right] R_{k,l}\left(\frac{\rho}{k}\right) = 0.$$

The solutions that are finite at $\rho = 0$ are given by the *modified spherical Bessel functions*, which are essentially the spherical Bessel functions for purely imaginary arguments:

$$i_l(\rho) = (-i)^n j_l(i\rho).$$

These functions have no positive real zeros, and hence cannot satisfy the necessary boundary condition $i_l(ka) = 0$ for any value of $k > 0$. Thus, there are no eigenstates of negative energy.

All that remains is to show that there are no states of zero energy. For $E = 0$, we also have $k = 0$, and the radial equation becomes

$$\left[\frac{d^2}{dr^2} + \frac{2}{r}\frac{d}{dr} - \frac{l(l+1)}{r^2}\right] R_{0,l}(r) = 0.$$

This is a Cauchy-Euler equation with general solution

$$c_1 r^{\alpha_1} + c_2 r^{\alpha_2},$$

180   *Symmetry and Quantum Mechanics*

where
$$\alpha_1 = \frac{1}{2}\left(-1 + \sqrt{1 + 4l(l+1)}\right) \geq 0$$
and
$$\alpha_2 = \frac{1}{2}\left(-1 - \sqrt{1 + 4l(l+1)}\right) < 0.$$

The condition that $R_{0,l}$ be finite at $r = 0$ requires that $c_2 = 0$, but then the additional condition that $R_{0,l}(a) = 0$ requires that $c_1 = 0$ as well. Thus, there is no eigenstate of zero energy, as claimed.

8.6 Two-particle systems

We now want to model the interaction of two particles in three-dimensional space. We begin by assuming that the two particles are distinguishable (e.g., a proton and an electron), so that we can speak meaningfully of the "first" particle and the "second" particle. If $\mathbf{r}_1$ and $\mathbf{r}_2$ denote the classical position vectors of the two particles in $\mathbb{R}^3$, then the state space for the 2-particle system is the completed tensor product

$$L^2(\mathbb{R}^3, d\mathbf{r}_1)\hat{\otimes}L^2(\mathbb{R}^3, d\mathbf{r}_2) \cong L^2(\mathbb{R}^3 \times \mathbb{R}^3, d\mathbf{r}_1 d\mathbf{r}_2), \qquad (8.4)$$

and the state of the system is represented by a wavefunction $\psi(\mathbf{r}_1, \mathbf{r}_2)$. We consider a classical potential function V that depends only on the distance between the two particles: $V(\mathbf{r}_1, \mathbf{r}_2) = V(|\mathbf{r}_1 - \mathbf{r}_2|)$. If M_1 and M_2 denote the masses of the two particles, then the classical expression for the total energy is

$$E = \frac{\mathbf{p}_1 \cdot \mathbf{p}_1}{2M_1} + \frac{\mathbf{p}_2 \cdot \mathbf{p}_2}{2M_2} + V(|\mathbf{r}_1 - \mathbf{r}_2|).$$

Quantizing this expression yields the Hamiltonian

$$\mathcal{H} = \frac{\hat{\mathbf{p}}_1 \cdot \hat{\mathbf{p}}_1}{2M_1} + \frac{\hat{\mathbf{p}}_2 \cdot \hat{\mathbf{p}}_2}{2M_2} + V(|\hat{\mathbf{r}}_1 - \hat{\mathbf{r}}_2|). \qquad (8.5)$$

Exercise 8.18. *Check that the components of the total linear momentum operator $\hat{\mathbf{p}}_1 + \hat{\mathbf{p}}_2$ commute with the Hamiltonian $\mathcal{H}$. (Hint: Note that, in terms of the tensor product decomposition (8.4) the operator $\hat{\mathbf{p}}_1$ acts only on the first factor and $\hat{\mathbf{p}}_2$ acts only on the second, so the components of the two momentum operators commute.)*

Since the components of $\hat{\mathbf{p}}_1 + \hat{\mathbf{p}}_2$ commute with $\mathcal{H}$, total linear momentum is conserved in the sense of Section 4.3.1. To go further with the analysis, we change coordinates on $\mathbb{R}^3 \times \mathbb{R}^3$ from $(\mathbf{r}_1, \mathbf{r}_2)$ to $(\mathbf{r}, \mathbf{R})$, where

$$\mathbf{r} := \mathbf{r}_1 - \mathbf{r}_2 \qquad \text{(relative position)}$$

A Three-Dimensional World

181

and

$$\mathbf{R} := \frac{M_1 \mathbf{r}_1 + M_2 \mathbf{r}_2}{M_1 + M_2} \qquad \text{(center of mass)}.$$

The generators of translation along these new coordinate axes give rise to the linear momentum operators

$$\hat{\mathbf{p}} = \frac{M_2 \hat{\mathbf{p}}_1 - M_1 \hat{\mathbf{p}}_2}{M_1 + M_2} \qquad \text{(relative linear momentum)}$$

and

$$\hat{\mathbf{P}} = \hat{\mathbf{p}}_1 + \hat{\mathbf{p}}_2 \qquad \text{(total linear momentum)}.$$

To see this, set $M := M_1 + M_2$ and note that

$$(\mathbf{r}_1, \mathbf{r}_2) = \left(\mathbf{R} + \frac{M_2}{M}\mathbf{r}, \mathbf{R} - \frac{M_1}{M}\mathbf{r} \right).$$

Introduce components for the various position vectors as follows:

$$
\begin{aligned}
\mathbf{r}_j &= (x_j, y_j, z_j) \\
\mathbf{r} &= (x, y, z) \\
\mathbf{R} &= (X, Y, Z).
\end{aligned}
$$

Then writing $\Psi(\mathbf{r}, \mathbf{R}) = \psi(\mathbf{r}_1, \mathbf{r}_2)$ to denote the wavefunction in the new coordinates, we have

$$
\begin{aligned}
\hat{p}_x \Psi(\mathbf{r}, \mathbf{R}) &= i\hbar \frac{d}{ds} \Psi(\mathbf{r} - (s, 0, 0), \mathbf{R}) \Big|_{s=0} \\
&= i\hbar \frac{d}{ds} \psi\left(\mathbf{R} + \frac{M_2}{M}(\mathbf{r} - s\mathbf{e}_1), \mathbf{R} - \frac{M_1}{M}(\mathbf{r} - s\mathbf{e}_1) \right)\Big|_{s=0} \\
&= i\hbar \frac{\partial \psi}{\partial x_1}(\mathbf{r}_1, \mathbf{r}_2) \frac{-M_2}{M} + i\hbar \frac{\partial \psi}{\partial x_2}(\mathbf{r}_1, \mathbf{r}_2) \frac{M_1}{M} \\
&= \left[\frac{M_2}{M} \frac{\hbar}{i} \frac{\partial}{\partial x_1} - \frac{M_1}{M} \frac{\hbar}{i} \frac{\partial \psi}{\partial x_2} \right] \psi(\mathbf{r}_1, \mathbf{r}_2) \\
&= \left[\frac{M_2 \hat{p}_{x_1}}{M} - \frac{M_1 \hat{p}_{x_2}}{M} \right] \psi(\mathbf{r}_1, \mathbf{r}_2).
\end{aligned}
$$

Similarly, we have

$$
\begin{aligned}
\hat{P}_X \Psi(\mathbf{r}, \mathbf{R}) &= i\hbar \frac{d}{ds} \Psi(\mathbf{r}, \mathbf{R} - (s, 0, 0)) \Big|_{s=0} \\
&= i\hbar \frac{d}{ds} \psi\left(\mathbf{R} - s\mathbf{e}_1 + \frac{M_2}{M}\mathbf{r}, \mathbf{R} - s\mathbf{e}_1 - \frac{M_1}{M}\mathbf{r} \right)\Big|_{s=0} \\
&= i\hbar \frac{\partial \psi}{\partial x_1}(\mathbf{r}_1, \mathbf{r}_2)(-1) + i\hbar \frac{\partial \psi}{\partial x_2}(\mathbf{r}_1, \mathbf{r}_2)(-1) \\
&= \left[\frac{\hbar}{i} \frac{\partial}{\partial x_1} + \frac{\hbar}{i} \frac{\partial \psi}{\partial x_2} \right] \psi(\mathbf{r}_1, \mathbf{r}_2) \\
&= \left[\hat{p}_{x_1} + \hat{p}_{x_2} \right] \psi(\mathbf{r}_1, \mathbf{r}_2).
\end{aligned}
$$

182 *Symmetry and Quantum Mechanics*

In terms of the total and relative momentum operators, the Hamiltonian (8.5) becomes

$$\mathcal{H} = \frac{\hat{\mathbf{P}} \cdot \hat{\mathbf{P}}}{2M} + \frac{\hat{\mathbf{p}} \cdot \hat{\mathbf{p}}}{2\mu} + V(\hat{r}), \tag{8.6}$$

where $M = M_1 + M_2$ is the *total mass* and $\mu = \frac{M_1 M_2}{M_1 + M_2}$ is the *reduced mass*.

Exercise 8.19. *Show that the 2-particle Hamiltonian (8.5) takes the form (8.6) in the relative position/center of mass coordinates.*

Exercise 8.20. *Let* $\Phi(\mathbf{r}, \mathbf{R}) = (\mathbf{r}_1, \mathbf{r}_2) = \left(\mathbf{R} + \frac{M_2}{M}\mathbf{r}, \mathbf{R} - \frac{M_1}{M}\mathbf{r}\right)$ *denote the change of coordinate mapping. Show that the 6-by-6 Jacobian matrix of partial derivatives has the block form:*

$$D\Phi = \frac{1}{M} \begin{bmatrix} M_2 & M \\ -M_1 & M \end{bmatrix},$$

where $M = M_1 + M_2$ *and each entry represents a 3-by-3 scalar matrix. Conclude that* $\det(D\Phi) = 1$, *so that the change of variable theorem A.41 reads:*

$$\int_{\mathbb{R}^6} f(\mathbf{r}_1, \mathbf{r}_2) d\mathbf{r}_1 d\mathbf{r}_2 = \int_{\mathbb{R}^6} f(\Phi(\mathbf{r}, \mathbf{R})) d\mathbf{r} d\mathbf{R}.$$

By the previous exercise, the coordinates $(\mathbf{r}, \mathbf{R})$ lead to a corresponding decomposition of the position space $L^2(\mathbb{R}^3 \times \mathbb{R}^3, d\mathbf{r}_1 d\mathbf{r}_2)$ as the completed tensor product

$$L^2(\mathbb{R}^3, d\mathbf{R}) \hat{\otimes} L^2(\mathbb{R}^3, d\mathbf{r}).$$

Thus, finite linear combinations of separable functions $\psi_{CM}(\mathbf{R})\psi_{\mathrm{rel}}(\mathbf{r})$ are dense in the state space. In terms of this decomposition we can express the Hamiltonian (8.6) as

$$\mathcal{H} = \frac{\hat{\mathbf{P}} \cdot \hat{\mathbf{P}}}{2M} \otimes I + I \otimes \left(\frac{\hat{\mathbf{p}} \cdot \hat{\mathbf{p}}}{2\mu} + V(\hat{r})\right) = \mathcal{H}_{CM} \otimes I + I \otimes \mathcal{H}_{\mathrm{rel}}.$$

Hence, the dynamics splits into two pieces: the center of mass wavefunction $\psi_{CM}(\mathbf{R})$ evolves according to the free-particle Hamiltonian $\mathcal{H}_{CM} = \frac{\hat{\mathbf{P}} \cdot \hat{\mathbf{P}}}{2M}$, while the relative wavefunction $\psi_{\mathrm{rel}}(\mathbf{r})$ evolves according to the Hamiltonian $\mathcal{H}_{\mathrm{rel}} = \frac{\hat{\mathbf{p}} \cdot \hat{\mathbf{p}}}{2\mu} + V(\hat{r})$ describing the dynamics of a single particle of mass μ in the central potential $V(r)$. Since we already understand the free-particle energy eigenstates (Section 8.5.1), we focus on the relative Hamiltonian and drop the subscript "rel."

8.6.1 The Coulomb potential

The classical Coulomb potential describing the electrostatic interaction between a nucleus of positive charge Ze and a single electron of charge $-e$

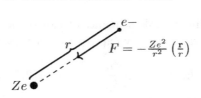

FIGURE 8.2: The Coulomb force on an electron due to a nucleus of positive charge Ze. The vector $\mathbf{r}$ (not shown) points *from* the nucleus *to* the electron.

is given[3] in relative coordinates by $V(r) = -\frac{Ze^2}{r}$. This potential leads to the attractive inverse-square Coulomb force between the nucleus and the electron (see Figure 8.2):

$$F(\mathbf{r}) = -\nabla V(r) = -\frac{Ze^2}{r^2}\left(\frac{\mathbf{r}}{r}\right).$$

Hence, to model the quantum dynamics of the electron in the Coulomb potential, we write down the (relative) Hamiltonian

$$\mathcal{H} = \frac{\hat{\mathbf{p}} \cdot \hat{\mathbf{p}}}{2\mu} - \frac{Ze^2}{\hat{r}} = -\frac{\hbar^2}{2\mu}\Delta - \frac{Ze^2}{\hat{r}}.$$

As with any central potential, this leads to the energy eigenfunction equation (8.2) for the radial functions $R_{E,l}(r) \in L^2(\mathbb{R}_{\geq 0}, r^2 dr)$:

$$\left[-\frac{\hbar^2}{2\mu}\left(\frac{d^2}{dr^2} + \frac{2}{r}\frac{d}{dr}\right) + \frac{l(l+1)\hbar^2}{2\mu r^2} - \frac{Ze^2}{r}\right]R_{E,l}(r) = ER_{E,l}(r).$$

By introducing the functions $u_{E,l}(r) := rR_{E,l}(r)$, this equation simplifies to

$$\left[-\frac{\hbar^2}{2\mu}\frac{d^2}{dr^2} + \frac{l(l+1)\hbar^2}{2\mu r^2} - \frac{Ze^2}{r}\right]u_{E,l}(r) = Eu_{E,l}(r), \tag{8.7}$$

which is the one-dimensional Schrödinger equation with the *effective potential* $V_{\text{eff}}(r) := \frac{l(l+1)\hbar^2}{2\mu r^2} - \frac{Ze^2}{r}$.

Exercise 8.21. Derive the differential equation (8.7) for $u_{E,l}$ starting with the one for $R_{E,l}$. Moreover, use the relation $rR(r) = u(r)$ to show that $R \in L^2(\mathbb{R}_{\geq 0}, r^2 dr)$ if and only if $u \in L^2(\mathbb{R}_{\geq 0}, dr)$.

For negative energies $E < 0$ (corresponding to bound states), the standard way to solve the equation (8.7) is via power series methods (see [9, Section 4.2] or [22, Section 10.2]). Here we simply quote the results. The requirement that $u_{E,l}$ be square-integrable leads to a quantization of the allowable energy levels E_n:

$$E_n = -\frac{\mu Z^2 e^4}{2\hbar^2 n^2} \qquad n = 1, 2, 3, \ldots. \tag{8.8}$$

---

[3]Here we follow [22] and use Gaussian units to avoid numerical factors. In SI units (as in [9]), the potential would be $V(r) = -\frac{Ze^2}{4\pi\epsilon_0 r}$.

184 Symmetry and Quantum Mechanics

If we introduce a dimensionless quantity[4] called the *fine structure constant* defined by $\alpha := \frac{e^2}{\hbar c}$, then we may express the energies as

$$E_n = -\frac{\mu Z^2 c^2 \alpha^2}{2n^2} \qquad n = 1, 2, 3, \ldots.$$

For each $n \geq 1$, the energy E_n occurs as an eigenvalue for the operator in (8.7) when $l = 0, 1, \ldots, n - 1$. If we introduce the dimensionless position variable

$$\rho = r\sqrt{\frac{8\mu|E_n|}{\hbar^2}},$$

then the eigenfunction $u_{E_n,l} = u_{n,l}$ may be written

$$u_{n,l}(\rho) = \rho^{l+1} e^{-\frac{\rho}{2}} F^l_{n-1-l}(\rho),$$

where F^l_{n-1-l} is a polynomial of degree $n - 1 - l$ obtained by appropriate normalization of the *associated Laguerre polynomial* L^{2l+1}_{n-1-l}. These are defined as follows:

$$L^p_{q-p}(x) := (-1)^p \left(\frac{d}{dx}\right)^p L_q(x),$$

where

$$L_q(x) := e^x \left(\frac{d}{dx}\right)^q (e^{-x} x^q)$$

is the qth *Laguerre polynomial*.

To return to the original position variable r, note that

$$\begin{aligned}
\rho &= r\sqrt{\frac{8\mu|E_n|}{\hbar^2}} \\
&= r\sqrt{\frac{8\mu^2 Z^2 c^2 \alpha^2}{2\hbar^2 n^2}} \\
&= r\frac{2Z}{n}\frac{\mu c \alpha}{\hbar} \\
&= \frac{2Z}{n}\frac{r}{a},
\end{aligned}$$

where $a := \frac{\hbar}{\mu c \alpha}$ is a length known as the *Bohr radius*.

Hence, the eigenfunction of energy $E_n < 0$, squared total orbital angular momentum $l(l+1)\hbar^2$, and z-angular momentum $m\hbar$ is given by

$$\begin{aligned}
\psi_{n,l,m}(r,\theta,\phi) &= R_{n,l}(r)Y^m_l(\theta,\phi) \\
&= \frac{1}{r}u_{n,l}\left(\frac{2Z}{n}\frac{r}{a}\right)Y^m_l(\theta,\phi) \\
&= \left(\frac{2Z}{na}\right)^{l+1} r^l e^{-\frac{Zr}{na}} F^l_{n-1-l}\left(\frac{2Z}{n}\frac{r}{a}\right)Y^m_l(\theta,\phi).
\end{aligned}$$

---

[4]Here, c denotes the speed of light, which is the same in all inertial reference frames (see Chapter 9). The numerical value of α is approximately $\frac{1}{137}$.

A Three-Dimensional World

For a fixed $n \geq 1$, the allowable values for l are $l = 0, 1, \ldots, n - 1$, and for each such l, the index m can take on the $2l + 1$ values $m = -l, -l + 1, \ldots, l$. Hence, the total dimension of the eigenspace for the energy $E_n < 0$ is given by

$$\sum_{l=0}^{n-1}(2l + 1) = 2\sum_{l=0}^{n-1}(l + \frac{1}{2}) = 2\left(\frac{n(n-1)}{2} + \frac{n}{2}\right) = n^2.$$

The striking thing here is the unexpectedly large degeneracy: while the independence of the energies E_n from the magnetic quantum number m is explained by the rotational invariance of central potentials, the independence from the azimuthal quantum number l is special to the Coulomb potential. This extra degeneracy is explained by the fact that the Coulomb potential enjoys an $SO(4)$-symmetry extending the $SO(3)$-symmetry shared by all central potentials. This extra symmetry stems for a quantum version of the classical Runge-Lenz vector, which is a conserved quantity in the 2-body Kepler problem for planetary orbits (see [10, Section 2.6]). In fact, by exploiting the $SO(4)$-symmetry, one can prove that the negative energy eigenspaces are irreducible representations of the Lie algebra $\mathfrak{so}(4)$ and even determine the possible energies E_n via a purely algebraic argument very much in the spirit of this book. The algebraic manipulations are cumbersome, however, so we will simply provide three references for the details: [10, Section 18.4]; [15, Chapters 8–9]; [20, Chapter 3, Section 5.3].

Finally, for a discussion of the positive energy solutions to equation (8.7)—which correspond to scattering states—see [20, Chapter 3, Section 5.2].

8.7 Particles with spin

We now want to incorporate spin into our 3-dimensional model. First of all, recall the group $G = H_3 \rtimes SO(3)$ from Section 8.4 and its action on position space $L^2(\mathbb{R}^3)$:

$$(\mathbf{v}, \mathbf{w}, \alpha, A) \star \psi(\mathbf{r}) = e^{-\frac{i}{\hbar}(\alpha + \mathbf{v} \cdot (\mathbf{r} - \mathbf{w}))} \psi(A^{-1}(\mathbf{r} - \mathbf{w})).$$

Using the double cover $f \colon SU(2) \to SO(3)$ from theorem 2.39, we can extend this to an action of the group $\widetilde{G} := H_3 \rtimes SU(2)$ on position space, defined by

$$(\mathbf{v}, \mathbf{w}, \alpha, B) \star \psi(\mathbf{r}) = e^{-\frac{i}{\hbar}(\alpha + \mathbf{v} \cdot (\mathbf{r} - \mathbf{w}))} \psi(f(B^{-1})(\mathbf{r} - \mathbf{w})).$$

Now recall from Section 5.5 that spinor space for a spin-s particle is the irreducible $SU(2)$-representation π_s of dimension $2s + 1$. Denote by

$$\{|k\rangle : k = -2s, -2s + 2, \ldots, 2s\}$$

186   *Symmetry and Quantum Mechanics*

the z-basis for spinor space representing the states of definite spin $\frac{k\hbar}{2}$ in the z-direction:

$$\hat{S}_z|k\rangle = \frac{k\hbar}{2}|k\rangle.$$

Note that we may consider π_s as a representation of $\widetilde{G}$ in which the subgroup H_3 acts trivially: $\pi_s(\mathbf{v}, \mathbf{w}, \alpha, B) := \pi_s(B)$. Thus, both position space and spinor space are representations of $\widetilde{G}$, and the total state space of a spin-s particle is their tensor product

$$L^2(\mathbb{R}^3) \otimes \pi_s = L^2(\mathbb{R}^3) \otimes \mathbb{C}^{2s+1} \cong L^2(\mathbb{R}^3, \mathbb{C}^{2s+1}).$$

So a particle state is modeled by a spinor-valued wavefunction $\boldsymbol{\psi} \colon \mathbb{R}^3 \to \mathbb{C}^{2s+1}$. The components $\psi_k(\mathbf{r}) \in L^2(\mathbb{R}^3)$ are given by the expansion in the z-basis of spinor space:

$$\boldsymbol{\psi}(\mathbf{r}) = \sum_k \psi_k(\mathbf{r})|k\rangle.$$

Finally, the action of $\widetilde{G}$ is given by:

$$(\mathbf{v}, \mathbf{w}, \alpha, B) \star \boldsymbol{\psi}(\mathbf{r}) = e^{-\frac{i}{\hbar}(\alpha + \mathbf{v}\cdot(\mathbf{r}-\mathbf{w}))} \pi_s(B) \boldsymbol{\psi}(f(B^{-1})(\mathbf{r} - \mathbf{w})).$$

Exercise 8.22. *Extend (again) the interpretation of the G-action developed in exercise 8.15 to the group $\widetilde{G}$: suppose that observer M is rotated with respect to P according to $A \in SO(3)$, and choose $B = f^{-1}(A) \in SU(2)$. Moreover, suppose that M is also moving through P's reference frame with velocity $-\mathbf{v}$ so that M is located at position $\mathbf{r} = \mathbf{w}$ at time $t = 0$. If at time $t = 0$ observer P describes a spin-s particle with mass m via the wavefunction $\boldsymbol{\psi}(\mathbf{r})$, then M would describe the same particle (at time $t = 0$) via the wavefunction*

$$\boldsymbol{\psi}'(\mathbf{r}) = (m\mathbf{v}, \mathbf{w}, \alpha, B)^{-1} \star \boldsymbol{\psi}(\mathbf{r}).$$

Suppose that $\mathcal{H} = -\frac{\hbar^2}{2M}\Delta + V(\hat{\mathbf{r}})$ is a (spin-independent) Hamiltonian as in Section 8.5. We have been thinking of $\mathcal{H}$ as an operator on $L^2(\mathbb{R}^3)$, but it extends to an operator on $L^2(\mathbb{R}^3) \otimes \pi_s$ by acting trivially on the spinor space factor: $\mathcal{H} = \mathcal{H} \otimes I$. Let $\psi(\mathbf{r}) \in L^2(\mathbb{R}^3)$ be an eigenfunction with energy E for the original operator $\mathcal{H}$. Then for each basis spinor $|k\rangle$, we obtain a spinor-valued wavefunction $\boldsymbol{\psi}(\mathbf{r}) := \psi(\mathbf{r})|k\rangle$ that is also an $\mathcal{H}$-eigenfunction with the same energy E. Varying the index k, we obtain $2s+1$ linearly independent eigenfunctions in $L^2(\mathbb{R}^3) \otimes \pi_s$. It follows that if the eigenspace for energy E has dimension n in the absence of spin, then the corresponding energy eigenspace for a spin-s particle will have dimension $(2s+1)n$. In particular, for a given potential V, a spin-$\frac{1}{2}$ particle has energy eigenspaces of twice the dimension of a spin-0 particle. Indeed, for each position state of the spin-0 particle, the spin-$\frac{1}{2}$ particle can be either spin up or spin down along the z-axis (or in a superposition of the two); since the Hamiltonian is spin-independent, the spin-state does not affect the energy of the particle.

A Three-Dimensional World

The space $L^2(\mathbb{R}^3) \otimes \pi_s$ is clearly irreducible as a representation of $\widetilde{G} = H_3 \rtimes SU(2)$, but we can ask for its decomposition as a representation of the subgroup $SU(2)$. For this, we work in spherical coordinates as in Section 8.3 and use the decomposition of $L^2(S^2)$ into $SO(3)$-irreducibles together with the Clebsch-Gordan rules from Section 6.2:

$$
\begin{aligned}
L^2(\mathbb{R}^3) \otimes \pi_s &\cong L^2(\mathbb{R}_{\geq 0}, r^2 dr) \hat{\otimes} L^2(S^2, d\Omega) \otimes \pi_s \\
&= L^2(\mathbb{R}_{\geq 0}, r^2 dr) \hat{\otimes} \left(\widehat{\bigoplus_{l=0}^{\infty}} \mathcal{Y}_l \right) \otimes \pi_s \\
&= L^2(\mathbb{R}_{\geq 0}, r^2 dr) \hat{\otimes} \left(\widehat{\bigoplus_{l=0}^{\infty}} \pi_l \right) \otimes \pi_s \\
&= L^2(\mathbb{R}_{\geq 0}, r^2 dr) \hat{\otimes} \left(\widehat{\bigoplus_{l=0}^{\infty}} \bigoplus_{j=|l-s|}^{l+s} \pi_j \right).
\end{aligned}
$$

The finite direct sums inside the parentheses indicate the mixing of orbital angular momentum with spin. To explain this physical interpretation, recall that we have realized the representation π_l as the space of spherical harmonics $\mathcal{Y}_l \subset L^2(S^2)$, with basis functions $Y_l^m(\theta, \phi)$ representing states of squared orbital momentum $l(l+1)\hbar^2$ and z-angular momentum $m\hbar$:

$$
\hat{\mathbf{L}}^2 Y_l^m = l(l+1)\hbar^2 Y_l^m \qquad \text{and} \qquad \hat{L}_z Y_l^m = m\hbar Y_l^m.
$$

Similarly, the representation π_s has basis spinors $|k\rangle$ representing states of squared spin $s(s+1)\hbar^2$ and z-spin $\frac{k\hbar}{2}$:

$$
\hat{\mathbf{S}}^2 |k\rangle = s(s+1)\hbar^2 |k\rangle \qquad \text{and} \qquad \hat{S}_z |k\rangle = \frac{k\hbar}{2} |k\rangle.
$$

On the tensor product $\mathcal{Y}_l \otimes \pi_s$, the orbital angular momentum operators act only on the first factor, while the spin operators act only on the second. But as a representation of $\mathfrak{su}(2)_{\mathbb{C}} = \mathfrak{sl}_2(\mathbb{C})$, the action of the scaled Pauli basis $\frac{\hbar}{2}\sigma_j$ is given by the components of the *total angular momentum operator*:

$$
\hat{\mathbf{J}} := \hat{\mathbf{L}} + \hat{\mathbf{S}} = \hat{\mathbf{L}} \otimes I + I \otimes \hat{\mathbf{S}}.
$$

From Section 6.2, we also know that this representation decomposes into a direct sum of irreducibles as indicated:

$$
\mathcal{Y}_l \otimes \pi_s = \bigoplus_{j=|l-s|}^{l+s} \pi_j.
$$

But then $\hat{\mathbf{J}}^2$ acts as $j(j+1)\hbar^2$ on π_j, which has a basis consisting of eigenvectors for $\hat{J}_z$. Hence, for each triple (l, s, j) with $|l - s| \leq j \leq l + s$, there is a basis

188　　　　　　　　　　*Symmetry and Quantum Mechanics*

of the subrepresentation $\pi_j \subset \mathcal{Y}_l \otimes \pi_s \subset L^2(S^2) \otimes \pi_s$ given by the $2j+1$ spinor-valued wavefunctions denoted

$$\psi_{l,s,j,m_j}(\mathbf{r}) \qquad m_j = -j, j+1, \ldots, j.$$

These represent particle states with the following definite values of our chosen observables:

$\hat{\mathbf{L}}^2$ 　　　　squared orbital angular momentum $l(l+1)\hbar^2$

$\hat{\mathbf{S}}^2$ 　　　　squared spin $s(s+1)\hbar^2$

$\hat{\mathbf{J}}^2$ 　　　　squared total angular momentum $j(j+1)\hbar^2$

$\hat{J}_z$ 　　　　z-component of total angular momentum $m_j\hbar$.

Now suppose that we have two distinguishable particles as in Section 8.6, but now with spins s_1 and s_2 respectively. Then the total state space for the 2-particle system is given by the completed tensor product

$$\left(L^2(\mathbb{R}^3, d\mathbf{r}_1) \otimes \pi_{s_1}\right) \hat{\otimes} \left(L^2(\mathbb{R}^3, d\mathbf{r}_2) \otimes \pi_{s_2}\right) \cong L^2(\mathbb{R}^6, d\mathbf{r}_1 d\mathbf{r}_2) \otimes \pi_{s_1} \otimes \pi_{s_2}.$$

Changing to the relative/center-of-mass coordinates $(\mathbf{r}, \mathbf{R})$, the state space may be written as

$$L^2(\mathbb{R}^3, d\mathbf{R}) \hat{\otimes} L^2(\mathbb{R}^3, d\mathbf{r}) \otimes \pi_{s_1} \otimes \pi_{s_2}.$$

If the potential $V(\mathbf{r})$ depends only on the separation $\mathbf{r} = \mathbf{r}_1 - \mathbf{r}_2$, then the Hamiltonian decomposes into a free part describing the center of mass and a relative Hamiltonian:

$$\mathcal{H} = \frac{\hat{\mathbf{P}} \cdot \hat{\mathbf{P}}}{2M} \otimes I + I \otimes \left(\frac{\hat{\mathbf{p}} \cdot \hat{\mathbf{p}}}{2\mu} + V(\hat{r})\right).$$

Since the vector $\mathbf{r}$ points from the second particle to the first, we should consider the relative Hamiltonian $\mathcal{H}_{\text{rel}} = \frac{\hat{\mathbf{p}} \cdot \hat{\mathbf{p}}}{2\mu} + V(\hat{r})$ as an operator on $L^2(\mathbb{R}^3, d\mathbf{r}) \otimes \pi_{s_1}$, thereby accounting for the spin of the first particle. On the other hand, we have a choice as to whether we should also include the second spinor space, or instead associate π_{s_2} with the center of mass Hamiltonian $\mathcal{H}_{CM} = \frac{\hat{\mathbf{P}} \cdot \hat{\mathbf{P}}}{2M}$, which would then act on $L^2(\mathbb{R}^3, d\mathbf{R}) \otimes \pi_{s_2}$. The latter option is especially reasonable when the second particle is much more massive than the first, so that the center of mass is essentially the location of the second particle. However, we might want to alter the relative Hamiltonian to account for the interaction between the spins of the two particles (as in example 6.10 in Chapter 6, revisited below), in which case we must consider $\mathcal{H}_{\text{rel}}$ as an operator on $L^2(\mathbb{R}^3, d\mathbf{r}) \otimes \pi_{s_1} \otimes \pi_{s_2}$.

8.7.1 The hydrogen atom

We now return to the Coulomb potential $V(r) = -\frac{Ze^2}{r}$ describing the interaction between a nucleus of positive charge Ze and an electron with charge $-e$. As in Section 8.6.1, we have the (relative) Hamiltonian

$$\mathcal{H} = -\frac{\hbar^2}{2\mu}\Delta - \frac{Ze^2}{\hat{r}},$$

now viewed as an operator on $L^2(\mathbb{R}^3, d\mathbf{r}) \otimes \pi_{\frac{1}{2}}$ since the electron is a spin-$\frac{1}{2}$ particle. (For now, we are ignoring the spin of the nucleus by associating it with the center of mass wavefunction.) As described in the previous section, tensoring with $\pi_{\frac{1}{2}}$ has the effect of doubling the dimension of each energy eigenspace. So, for every $n = 1, 2, 3, \ldots$, there are now $2n^2$ linearly independent states of energy E_n. As we will discuss in the next section, this explains the number of elements in each row of the periodic table of the elements!

Taking $Z = 1$ in the potential V corresponds to the hydrogen atom, with nucleus a single proton of charge e. From (8.8), we see that the energy levels of hydrogen are

$$E_n = -\frac{\mu c^2 \alpha^2}{2n^2} \qquad n = 1, 2, 3, \ldots.$$

So far the spin of the electron has only served to double the dimension of each energy eigenspace. But the fact that the electron has spin actually changes the dynamics, and we can model these effects by introducing various correction terms to the Hamiltonian. As a result, the energy levels change slightly and the energy degeneracy is broken—this is called the *fine structure*[5] and *hyperfine structure* of hydrogen. The actual computation of the energy shifts requires perturbation theory, so we will content ourselves with a brief description of the correction terms in the Hamiltonian and the magnitudes of the resulting energy shifts (see [9, Sections 6.3, 6.5] for details).

The first correction term accounts for *spin-orbit coupling*: the interaction between the spin of the electron and the magnetic field generated by the motion of the proton. Roughly (and classically) speaking, from the point of view of the electron, the orbiting proton (being charged) generates a magnetic field $\mathbf{B} = \frac{e}{M_e c^2 r^3}\mathbf{L}$ pointing in the same direction as the electron's orbital angular momentum $\mathbf{L}$ (computed in the rest frame of the proton, which is essentially the center of mass frame, since $M_p >> M_e$.) As discussed in example 4.1 of Chapter 4, the electron has a magnetic dipole moment $\boldsymbol{\mu}_e$ producing an interaction energy of $-\boldsymbol{\mu}_e \cdot \mathbf{B}$ with the magnetic field. Quantizing this expression amounts to replacing $\boldsymbol{\mu}_e$ with the correct multiple (the *gyromagnetic ratio*) of the electron spin operator to yield the *spin-orbit Hamiltonian*:

$$\mathcal{H}_{\mathrm{so}} := \frac{e^2}{2M_e^2 c^2}\frac{\hat{\mathbf{L}} \cdot \hat{\mathbf{S}}_e}{\hat{r}^3}.$$

---

[5]The fine structure of hydrogen also includes a relativistic correction; see [9, Section 6.3].

190   *Symmetry and Quantum Mechanics*

Adding this correction term to the Coulomb Hamiltonian yields

$$\mathcal{H}' := \mathcal{H} + \mathcal{H}_{\text{so}} = -\frac{\hbar^2}{2\mu}\Delta - \frac{e^2}{\hat{r}} + \frac{e^2}{2M_e^2 c^2}\frac{\hat{\mathbf{L}} \cdot \hat{\mathbf{S}}_e}{\hat{r}^3}.$$

The extra term destroys the l-degeneracy of the energy levels E_n: the new energies $E'_{n,l,j}$ now depend on the squared orbital angular momentum $l(l+1)\hbar^2$ as well as the squared total[6] angular momentum $j(j+1)\hbar^2$. But the shifts are very small: the differences $E_n - E'_{n,l,j}$ are on the order of $\alpha^2 E_n$, where $\alpha \approx \frac{1}{137}$ is the fine structure constant.

The hyperfine structure of hydrogen arises from *spin-spin coupling*: the interaction between the magnetic dipole moments of the electron and proton. The necessary correction term $\mathcal{H}_{\text{ss}}$ is complicated (see [9, equation 6.86]), but in the case of the lowest energy state the relevant portion is simply

$$\frac{b}{\hbar^2}\hat{\mathbf{S}}_e \cdot \hat{\mathbf{S}}_p,$$

where b is a constant with the dimensions of energy (this Hamiltonian was the subject of example 6.10 in Chapter 6). Adding the correction to our growing Hamiltonian yields:

$$\mathcal{H}'' = \mathcal{H}' + \mathcal{H}_{\text{ss}} = -\frac{\hbar^2}{2\mu}\Delta - \frac{e^2}{\hat{r}} + \frac{e^2}{2M_e^2 c^2}\frac{\hat{\mathbf{L}} \cdot \hat{\mathbf{S}}_e}{\hat{r}^3} + \mathcal{H}_{\text{ss}}.$$

In order for this to make sense, we need to consider $\mathcal{H}'$ as an operator on $L^2(\mathbb{R}^3, d\mathbf{r}) \otimes \pi_{\frac{1}{2}} \otimes \pi_{\frac{1}{2}}$, where the first copy of $\pi_{\frac{1}{2}}$ represents the electron spin and the second copy represents the proton spin. As before, tensoring with $\pi_{\frac{1}{2}}$ has the effect of doubling the dimension of each energy eigenspace for $\mathcal{H}'$. In particular, the lowest energy E_1 for $\mathcal{H}'$ now has a 4-dimensional eigenspace given by combining the ground state wavefunction $\psi_{1,0,0}(r, \theta, \phi)$ for the spin-zero Hamiltonian $\mathcal{H}$ with the 4-dimensional space of spin-states for the proton-electron system. But as we saw in example 6.10 in Chapter 6, the operator $\frac{b}{\hbar^2}\hat{\mathbf{S}}_e \cdot \hat{\mathbf{S}}_p$ breaks this energy degeneracy, yielding a one-dimensional eigenspace for the energy $E_1 - 3b$ and a three-dimensional eigenspace for the energy $E_1 + b$. In general, the spin-spin term $\mathcal{H}_{\text{ss}}$ produces tiny energy shifts on the order of $\frac{M_e}{M_p}\alpha^2 E_n$, much smaller than the shifts due to spin-orbit coupling, which are on the order of $\alpha^2 E_n$.

When a hydrogen atom transitions from the $E_1 + b$ energy level to the $E_1 - 3b$ energy level, it emits electromagnetic radiation having the corresponding energy difference of $4b$. This energy corresponds to a wavelength of approximately 21 centimeters, putting it in the microwave portion of the electromagnetic spectrum. This 21-centimeter radiation is commonly observed in radio astronomy.

---

[6]Miraculously, once one makes the relativistic correction as well, a bit of degeneracy is restored: the energies $E'_{n,j}$ depend on j but not on $l = j \pm \frac{1}{2}$.

8.8 Identical particles

We return now to the question of identical particles, first considered in Section 6.3. We wish to extend that discussion to the context of two identical spin-s particles in three-dimensional space. The state of the pair of particles is described by a $\pi_s \otimes \pi_s$-valued wavefunction $\psi(\mathbf{r}_1, \mathbf{r}_2)$ in the space

$$(L^2(\mathbb{R}^3, d\mathbf{r}_1) \otimes \pi_s)\hat{\otimes}(L^2(\mathbb{R}^3, d\mathbf{r}_2) \otimes \pi_s) \cong L^2(\mathbb{R}^6, d\mathbf{r}_1 d\mathbf{r}_2) \otimes \pi_s \otimes \pi_s.$$

But since the particles are indistinguishable, switching the labels on the two particles must result in the same physical state. That is, physical states must be eigenfunctions of the label switching operator Q, which has eigenvalues ± 1 since $Q^2 = I$. On a decomposable wavefunction of the form $\psi(\mathbf{r}_1, \mathbf{r}_2)|\phi_1\rangle \otimes |\phi_2\rangle$, the operator Q acts as

$$Q(\psi(\mathbf{r}_1, \mathbf{r}_2)|\phi_1\rangle \otimes |\phi_2\rangle) = \psi(\mathbf{r}_2, \mathbf{r}_1)|\phi_2\rangle \otimes |\phi_1\rangle.$$

Hence, for a general state we require that

$$\sum_j \psi_j(\mathbf{r}_1, \mathbf{r}_2)|\phi_{1,j}\rangle \otimes |\phi_{2,j}\rangle = \pm \sum_j \psi_j(\mathbf{r}_2, \mathbf{r}_1)|\phi_{2,j}\rangle \otimes |\phi_{1,j}\rangle.$$

In fact, the Spin-Statistics Theorem states that particles with integral spin (bosons) are symmetric under exchange, while those with half-integral spin (fermions) are antisymmetric. Note the subtlety here: fermion states may be formed by combining symmetric scalar wavefunctions with antisymmetric spinors, or by combining antisymmetric scalar wavefunctions with symmetric spinors. Similarly, bosons may be formed by combining scalar wavefunctions and spinors of the same parity. See [20, Section 4.3] for further discussion.

Now suppose that $\boldsymbol{\psi}_1(\mathbf{r}_1)$ and $\boldsymbol{\psi}_2(\mathbf{r}_2)$ in $L^2(\mathbb{R}^3) \otimes \pi_s$ are two individual particle states. Then the state describing two identical particles, one in each state, is given by the normalization of

$$\boldsymbol{\psi}_1 \otimes \boldsymbol{\psi}_2 \pm Q(\boldsymbol{\psi}_1 \otimes \boldsymbol{\psi}_2),$$

where we use the $+$ sign if the particles are bosons and the $-$ sign if the particles are fermions. Note that in the case of fermions, this expression becomes zero if we take $\boldsymbol{\psi}_1 = \boldsymbol{\psi}_2$. This simple observation implies the *Pauli exclusion principle*: two identical fermions cannot occupy the same state.

The exclusion principle has major implications for the structure of atoms, which are composed of a nucleus of positive charge Ze together with $Z \geq 1$ electrons, each of charge $-e$. For $Z = 1$ we get hydrogen, for $Z = 2$ we get helium, for $Z = 3$ we get lithium, and so on through the periodic table of the elements (see Figure 8.3). But what is the meaning of the traditional organization of the periodic table into rows of varying lengths? To discover the

s		d										p					
1 H																	2 He
3 Li	4 Be											5 B	6 C	7 N	8 O	9 F	10 Ne
11 Na	12 Mg											13 Al	14 Si	15 P	16 S	17 Cl	18 Ar
19 K	20 Ca	21 Sc	22 Ti	23 V	24 Cr	25 Mn	26 Fe	27 Co	28 Ni	29 Cu	30 Zn	31 Ga	32 Ge	33 As	34 Se	35 Br	36 Kr
37 Rb	38 Sr	39 Y	40 Zr	41 Nb	42 Mo	43 Tc	44 Ru	45 Rh	46 Pd	47 Ag	48 Cd	49 In	50 Sn	51 Sb	52 Te	53 I	54 Xe
55 Cs	56 Ba		72 Hf	73 Ta	74 W	75 Re	76 Os	77 Ir	78 Pt	79 Au	80 Hg	81 Tl	82 Pb	83 Bi	84 Po	85 At	86 Rn
87 Fr	88 Ra		104 Rf	105 Db	106 Sg	107 Bh	108 Hs	109 Mt	110 Ds	111 Rg	112 Cn	113 Uut	114 Uuq	115 Uup	116 Uuh	117 Uus	118 Uuo
Lanthanides			57 La	58 Ce	59 Pr	60 Nd	61 Pm	62 Sm	63 Eu	64 Gd	65 Tb	66 Dy	67 Ho	68 Er	69 Tm	70 Yb	71 Lu
Actinides			89 Ac	90 Th	91 Pa	92 U	93 Np	94 Pu	95 Am	96 Cm	97 Bk	98 Cf	99 Es	100 Fm	101 Md	102 No	103 Lr
			f														

FIGURE 8.3: The Periodic Table of the Elements. The groups of columns marked with letters indicate elements with a common value of azimuthal quantum number l for their highest-energy electron in the ground state: s corresponds to $l = 0$, p to $l = 1$, d to $l = 2$, f to $l = 3$.

A Three-Dimensional World
193

answer, let's think about constructing the lowest possible energy state (the *ground state*) of each element by starting with a nucleus of charge Ze and adding Z electrons one at a time according to the Pauli exclusion principle, each in the lowest available energy eigenstate determined by the Coulomb Hamiltonian $\mathcal{H} = -\frac{\hbar^2}{2\mu}\Delta - \frac{Ze^2}{r}$ acting on $L^2(\mathbb{R}^3) \otimes \pi_{\frac{1}{2}}$. Here we are ignoring any interaction between the electrons (except for the exclusion principle), so that each additional electron is governed by the same Hamiltonian $\mathcal{H}$.

Recall from the beginning of Section 8.7.1 that (for a fixed value of $Z \geq 1$) the Hamiltonian $\mathcal{H}$ has an infinite sequence of negative energies E_n given by formula (8.8). Moreover, for each $n \geq 1$, the E_n-eigenspace has dimension $2n^2$, where the factor of 2 comes from the electron spin; a basis of eigenfunctions is denoted by $\psi_{n,l,m}|\pm z\rangle$ where $l = 0, 1, \ldots, n-1$ and $m = -l, -l+1, \ldots, l$. Here $\psi_{n,l,m}$ denotes the state with energy E_n, squared orbital angular momentum $l(l+1)\hbar^2$, and z-angular momentum $m\hbar$.

Note that, although the actual energy values E_n depend on Z, the dimension of the eigenspaces is independent of Z: for each value of $Z \geq 1$, there is always a two-dimensional space of lowest energy E_1, then an 8-dimensional space of energy E_2, followed by an 18-dimensional space of energy E_3, and so on. Hence, for the purposes of the present discussion, we can imagine constructing the elements one-by-one by adding electrons to the eigenspaces in order, simultaneously increasing the value of Z to ensure that the resulting elements are electrically neutral. When we add electrons within a given energy level, we will always make use of the lowest available azimuthal quantum number l. Let's get started:

(H) For $Z = 1$ we have a single electron in the state $\psi_{1,0,0}|+z\rangle$. This is the ground state of the hydrogen atom.

(He) For $Z = 2$, we add a second electron with the same energy but opposite spin: $\psi_{1,0,0}|-z\rangle$. Antisymmetrizing, we get the ground state of helium:

$$\frac{1}{\sqrt{2}}\psi_{1,0,0}(\mathbf{r}_1)\psi_{1,0,0}(\mathbf{r}_2)(|+z\rangle \otimes |-z\rangle - |-z\rangle \otimes |+z\rangle).$$

At this point we have exhausted the 2-dimensional eigenspace for energy E_1, and so we move on to E_2, starting a new row of the periodic table.

(Li) For $Z = 3$, we add a third electron in the state $\psi_{2,0,0}|+z\rangle$. This is lithium.

(Be) For $Z = 4$, we add a fourth electron with the same energy as the third, but with the opposite spin: $\psi_{2,0,0}|-z\rangle$. This is beryllium.

We have now exhausted all the states with energy E_2 and $l = 0$, the so-called *2s-orbitals*. We now move on to the *2p-orbitals* with energy E_2 and $l = 1$, for which there is a 6-dimensional space: $m = -1, 0, 1$ and spin up or spin down for each: boron, carbon, nitrogen, oxygen, fluorine, and neon. We have now

194   *Symmetry and Quantum Mechanics*

filled up the 8-dimensional eigenspace for energy E_2, and we move on to E_3, starting a new row of the table.

For E_3, we begin with the two $3s$-orbital states ($l = m = 0$; Na and Mg) and then the six $3p$-orbital states ($l = 1, m = -1, 0, 1$; Al, Si, P, S, Cl, Ar). Then we move on to the $3d$-*orbitals* ($l = 2, m = -2, -1, 0, 1, 2$) of which there are ten, thus accounting for the entire 18-dimensional eigenspace for E_3. But the reader will notice that in the periodic table, those ten $3d$-orbital states occur in a new row, *after* the pair of $4s$-orbital states for the next energy E_4: K and Ca. This deviation from the "expected order" is a result of the fact that we have completely ignored the interaction between electrons in the atoms. Nevertheless, it is clear that the Coulomb Hamiltonian eigenspaces go a long way toward explaining the basic pattern.

Moreover, the structure displayed in the periodic table is the basis for all of chemistry! In fact, the chemical properties of elements are largely determined by the "outermost" electrons of the atoms—i.e., those with the highest energies. For this reason, elements in the same column of the periodic table have similar chemical properties. For instance, elements in the final column are those with completely filled p-orbitals[7] and empty higher orbitals. Such elements do not like to lose/gain/share electrons, and hence exhibit low chemical reactivity; they are known as the *noble gases* since they generally refuse to interact with other elements. Likewise, fluorine (F) and chlorine (Cl) are each only one electron away from having full p-orbitals, while sodium (Na) has one "extra" electron in a $3s$-orbital on top of completely filled $2p$-orbitals. Hence, both F and Cl like to join with Na to form NaF and the more familiar NaCl (table salt), in which the Na atom gives its "extra" electron to account for the "missing" electron in its companion. This is a special case of a rule of thumb called the *Octet Rule*: elements in the s- and p-columns of the periodic table like to form molecules in which the outer electrons are transferred or shared so that each atom in the molecule has full p-orbitals. Another example is carbon dioxide (CO_2), in which each oxygen atom shares two of its six outer electrons with the carbon atom, while the carbon atom shares two of its four outer electrons with each oxygen atom. As a result of this sharing, each atom in the CO_2 molecule has 8 outer electrons, forming a complete octet comprised of two occupied $2s$-orbitals and six occupied $2p$-orbitals.

As our friends M and P conclude their discussion of the periodic table, they marvel at how far they've come since Chapter 1—all the way from rotations of physical space to the structure of table salt. The mathematician M wonders about the origin of the Spin-Statistics Theorem. Observer P says something about a relativistic version of quantum mechanics called *Quantum Field Theory*, but that is a whole other story—for an introduction see [5], [14], or the growing series of volumes [24, 25, 26], which has the noble goal of providing "A Bridge between Mathematicians and Physicists." In the final

---

[7]There are no p-orbitals for the lowest energy E_1, but helium is traditionally placed in the final column because it has completely filled s-orbitals and empty higher orbitals.

chapter of our text, we will simply point the way toward a relativistic version of quantum mechanics for individual particles, without taking on a quantum treatment of fields as ultimately required by the special theory of relativity.

Chapter 9

Toward a Relativistic Theory

In which M and P discover the central extension of the Galilean group, the restricted Lorentz group $SO^+(1,3)$, and the Dirac equation.

Let's return to our observers M and P as they were at the beginning, located together, looking at empty space. In Chapter 1, we studied the consequences of their choosing different coordinate axes for the space around their common location, which led us to the consideration of the rotation group $SO(3)$. In Chapter 2 we studied the universal double cover $f\colon SU(2) \to SO(3)$, and found that the defining representation of $SU(2)$ provides a model for electron spin. Later, in Chapters 7 and 8, we wondered about the consequences of M and P choosing different locations to plant their feet, and we were led to the translation action of $(\mathbb{R}^3, +)$ and the Schrödinger representation of the Heisenberg group H_3 on position space $L^2(\mathbb{R}^3)$. In all of these cases, we assumed that M and P were at rest with respect to each other. In this chapter, we ask about the consequences of relative, uniform motion in a straight line.

9.1 Galilean relativity

Suppose that P has chosen an orthonormal basis for $(V, \langle,\rangle)$ as in Chapter 1, thereby identifying the physical space around him with $(\mathbb{R}^3, \cdot)$. In addition to having identical meter sticks and protractors (when compared to each other at rest), we now assume that M and P have identically constructed watches (a tacit assumption throughout this book). Just as in Chapter 1, M has chosen a different orthonormal basis, obtained from P's by a rotation $A \in SO(3)$. But in addition, we now assume that M is traveling with constant velocity $\mathbf{v}$ with respect to P, so that at time t, M is located at $\mathbf{x}(t) = t\mathbf{v}$ (see Figure 9.1). If t' denotes the time on M's watch, we assume that $t' = 0$ when $t = 0$ (i.e., when M and P are at the same location). Our question is simple: what is the relationship between P's space-time coordinates $(t, \mathbf{x})$ and M's coordinates $(t', \mathbf{x}')$? In the next section, we will briefly review Einstein's radical answer to this question (in the form of special relativity) and investigate some immediate consequences for quantum mechanics. But in this section, we stick with the non-relativistic framework, in which the passage of time is assumed to be common to all observers, regardless of their states of motion, so $t' = t$. Furthermore, space is assumed to be an immutable stage on which events transpire.

197

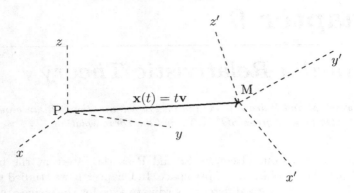

FIGURE 9.1: Observer M's rotated coordinate system moving away from observer P at a constant velocity **v**.

Perhaps surprisingly, it will turn out that the non-relativistic story is more complicated than the relativistic tale developed in the next section. The reason is essentially group-theoretic: as shown below, we will need to construct a *central extension* of the Galilean group $\mathcal{G}$ in order to describe the relationship between M and P's wavefunctions in the non-relativistic setting. In the context of special relativity, we will see that the relevant group is the restricted Poincaré group $\mathcal{P}_0$, and we only need to pass to its universal cover in order to obtain an action on wavefunctions. Although the comparison is interesting and instructive, the reader wishing to skip the algebraic complications of the Galilean group may jump to Section 9.2 after reading example 9.1.

In non-relativistic mechanics, the connection between the coordinates $(t, \mathbf{x})$ and $(t', \mathbf{x}')$ is called a *Euclidean transformation*. At time $t = t' = 0$, observers M and P are at the same location, and by assumption, their spatial coordinates differ by a rotation $A \in SO(3)$. Recall that A sends P's basis $\mathbf{u}_j$ to M's basis $\mathbf{u}'_j$, so M's coordinates $\mathbf{x}'$ are obtained from P's coordinates $\mathbf{x}$ via the inverse matrix:

$$(0, \mathbf{x}') = (0, A^{-1}\mathbf{x}).$$

Since at each time t, observer P would describe M's location as $t\mathbf{v}$, we must subtract the time-dependent vector $t\mathbf{v}$ from P's coordinates and then rotate to obtain M's coordinates:

$$(t', \mathbf{x}') = (t, A^{-1}(\mathbf{x} - t\mathbf{v})).$$

This coordinate transformation is the inverse of the linear transformation $\mathcal{E}(\mathbf{v}, A) \colon \mathbb{R}^4 \to \mathbb{R}^4$ given by

$$\mathcal{E}(\mathbf{v}, A)(t, \mathbf{x}) = (t, A\mathbf{x} + t\mathbf{v}).$$

In this way, we obtain a representation of the *Euclidean group* $\mathbb{R}^3 \rtimes SO(3)$ on $\mathbb{R}^4$ defined by $(\mathbf{v}, A) \mapsto \mathcal{E}(\mathbf{v}, A)$. This action extends the $SO(3)$-action on

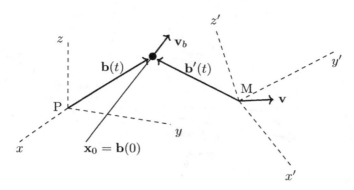

FIGURE 9.2: The motion $\mathbf{b}(t)$ of a ball traveling with velocity $\mathbf{v}_b$ with respect to observer P. The vector $\mathbf{b}'(t)$ describes the motion in observer M's rotated coordinate system, which is moving away from P with velocity $\mathbf{v}$.

the physical space $\mathbb{R}^3 = \{0\} \times \mathbb{R}^3$ from Chapter 1. We refer to $\mathcal{E}(\mathbf{v}, I)$ as a *Euclidean boost* with velocity $\mathbf{v}$.

Example 9.1. *Suppose that P observes a ball moving through space with constant velocity $\mathbf{v}_b$. If at time $t = 0$ the ball is located at the point $\mathbf{x}_0$, then P observes the straight-line trajectory (see Figure 9.2)*

$$\mathbf{b}(t) = \mathbf{x}_0 + t\mathbf{v}_b.$$

In terms of space-time coordinates, P writes the trajectory as

$$(t, \mathbf{b}(t)) = (t, \mathbf{x}_0 + t\mathbf{v}_b).$$

Let's determine the trajectory $\mathbf{b}'(t') = \mathbf{b}'(t)$ observed by M according to the classical, non-relativistic story described above. Applying the inverse of the Euclidean transformation $\mathcal{E}(\mathbf{v}, A)$ yields:

$$(t', \mathbf{b}'(t')) = \mathcal{E}(\mathbf{v}, A)^{-1}(t, \mathbf{x}_0 + t\mathbf{v}_b) = (t, A^{-1}(\mathbf{x}_0 + t\mathbf{v}_b - t\mathbf{v})),$$

so that $\mathbf{b}'(t) = A^{-1}\mathbf{x}_0 + tA^{-1}(\mathbf{v}_b - \mathbf{v})$. Note that M observes the ball to be traveling with the different constant velocity $A^{-1}(\mathbf{v}_b - \mathbf{v})$.

We may combine the Euclidean group with the group of time- and space-translations $(\mathbb{R}, +) \times (\mathbb{R}^3, +)$ to obtain the Galilean group.

Definition 9.2. *The* Galilean group *is $\mathcal{G} := (\mathbb{R} \times \mathbb{R}^3) \rtimes (\mathbb{R}^3 \rtimes SO(3))$ with group law*

$$(s_2, \mathbf{w}_2, \mathbf{v}_2, A_2)(s_1, \mathbf{w}_1, \mathbf{v}_1, A_1) :=$$
$$(s_2 + s_1, \mathbf{w}_2 + A_2\mathbf{w}_1 + s_1\mathbf{v}_2, \mathbf{v}_2 + A_2\mathbf{v}_1, A_2 A_1).$$

Its action on Euclidean space-time $\mathbb{R}^4$ is given by

$$(s, \mathbf{w}, \mathbf{v}, A) \star (t, \mathbf{x}) = (t + s, A\mathbf{x} + \mathbf{w} + t\mathbf{v}). \tag{9.1}$$

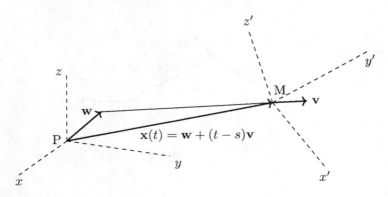

FIGURE 9.3: Observer M's rotated coordinate system moving away from observer P at constant velocity $\mathbf{v}$, starting at the location $\mathbf{x} = \mathbf{w}$ at time $t = s$.

Exercise 9.3. *Show that the given composition law does endow $\mathcal{G}$ with the structure of a group, and verify that (9.1) yields a group action on $\mathbb{R}^4$. Show that the inverse of an element of $\mathcal{G}$ is given by*

$$(s, \mathbf{w}, \mathbf{v}, A)^{-1} = (-s, A^{-1}(s\mathbf{v} - \mathbf{w}), -A^{-1}\mathbf{v}, A^{-1}).$$

We interpret the action (9.1) as follows: suppose that $(t, \mathbf{x})$ are P's coordinates for a space-time event. Moreover, suppose that M has set her watch *behind* P's by s seconds (so $t' = t - s$) and is moving with respect to P as follows (see Figure 9.3):

- M is rotated according to $A \in SO(3)$;
- at time $t = s$ (corresponding to $t' = 0$) M is located at $\mathbf{x} = \mathbf{w}$;
- M is moving with constant velocity $\mathbf{v}$.

Then M's space-time coordinates $(t', \mathbf{x}')$ are obtained by acting on P's with the inverse of the group element $(s, \mathbf{w}, \mathbf{v}, A)$:

$$\begin{aligned}
(t', \mathbf{x}') &= (s, \mathbf{w}, \mathbf{v}, A)^{-1} \star (t, \mathbf{x}) \\
&= (-s, A^{-1}(s\mathbf{v} - \mathbf{w}), -A^{-1}\mathbf{v}, A^{-1}) \star (t, \mathbf{x}) \\
&= (t - s, A^{-1}\mathbf{x} + A^{-1}(s\mathbf{v} - \mathbf{w}) - tA^{-1}\mathbf{v}) \\
&= (t - s, A^{-1}(\mathbf{x} - \mathbf{w} - (t - s)\mathbf{v})).
\end{aligned}$$

We would now like to reformulate the quantum mechanics of a single (spinless) particle in terms of the space-time $\mathbb{R}^4$. Recall from Chapter 8 that the position state of such a particle at a given instant t is modeled by a unit vector $\psi(t, \mathbf{x}) \in L^2(\mathbb{R}^3)$. Hence, we may consider the time-evolution of an initial state $\psi(0, \mathbf{x})$ as a function $\mathbb{R} \to L^2(\mathbb{R}^3)$ given by $t \mapsto \psi(t, \mathbf{x})$. Changing

Toward a Relativistic Theory

viewpoints only slightly, we may put the time- and space-dependence on an equal footing and think of $\psi \colon \mathbb{R}^4 \to \mathbb{C}$ as a complex-valued function of four variables. As described in Section 8.3, the spatial-translation and rotation subgroup $G_0 = \{0\} \times \mathbb{R}^3 \rtimes (\{\mathbf{0}\} \rtimes SO(3))$ of the Galilean group acts on position space $L^2(\mathbb{R}^3)$, providing the wavefunction transformations connecting observers whose reference frames differ by a spatial translation $\mathbf{w}$ and rotation A. In detail, the G_0-action on wavefunctions is given by

$$(\mathbf{w}, A) \star \psi(t, \mathbf{x}) = \psi((\mathbf{w}, A)^{-1} \star (t, \mathbf{x})) = \psi(t, A^{-1}(\mathbf{x} - \mathbf{w})).$$

Then if P describes a particle with the wavefunction $\psi(t, \mathbf{x})$, the rotated and translated observer M would describe the same particle with the wavefunction

$$\psi'(t, \mathbf{x}) = (\mathbf{w}, A)^{-1} \star \psi(t, \mathbf{x}) = \psi(t, A\mathbf{x} + \mathbf{w}).$$

Since M's coordinates $\mathbf{x}'$ are related to P's as $\mathbf{x}' = A^{-1}(\mathbf{x} - \mathbf{w})$, this new wavefunction has the property that $\psi'(t', \mathbf{x}') = \psi(t, \mathbf{x})$, so that M's amplitude density for the particle to be at the space-time location $(t', \mathbf{x}')$ is the same as P's amplitude density for the particle to be at the *same* space-time location $(t, \mathbf{x})$, only expressed in different coordinates. We would like to extend this G_0-action (and its interpretation) to an action of the full Galilean group acting on the wavefunctions $\psi(t, \mathbf{x})$. We will see that this isn't such a simple matter.

We begin by considering a pure time-translation $s = (s, \mathbf{0}, \mathbf{0}, I)$. In terms of the preceding discussion, we imagine that the only difference between the observers M and P is that M's watch is set s seconds behind P's. Then M's wavefunction should be obtained from P's by acting via the inverse time-translation, which has the effect of evolving P's wavefunction *forward* in time:

$$\psi'(t, \mathbf{x}) = (-s) \star \psi(t, \mathbf{x}) = \psi(s \star (t, \mathbf{x})) = \psi(t + s, \mathbf{x}).$$

In order to put time-translation together with the space-translation and rotation action, note that according to the group law in $\mathcal{G}$ from definition 9.2, we have $(s, \mathbf{w}, \mathbf{0}, A) = s(0, \mathbf{w}, \mathbf{0}, A)$, so a zero-velocity group element acts on wavefunctions as

$$
\begin{aligned}
(s, \mathbf{w}, \mathbf{0}, A) \star \psi(t, \mathbf{x}) = \quad &= \quad s \star ((0, \mathbf{w}, \mathbf{0}, A) \star \psi(t, \mathbf{x})) \\
&= \quad s \star \psi(t, A^{-1}(\mathbf{x} - \mathbf{w})) \\
&= \quad \psi(t - s, A^{-1}(\mathbf{x} - \mathbf{w})).
\end{aligned}
$$

Hence, we have discovered the action of the space-time-translation and rotation subgroup $\mathbb{R} \times G_0 = \mathbb{R} \times \mathbb{R}^3 \rtimes (\{\mathbf{0}\} \rtimes SO(3)) \subset \mathcal{G}$ on wavefunctions.

Example 9.4. *Before moving on to Euclidean boosts, let's see how the $\mathbb{R} \times G_0$-action works out for a free particle of mass m with definite momentum $\mathbf{p} \in \mathbb{R}^3$. Setting $p = |\mathbf{p}|$, such a particle has definite energy $\frac{p^2}{2m}$ (since momentum eigenstates are also energy eigenstates; see Section 8.5.1), so observer P describes the particle via the wavefunction*

$$\psi(t, \mathbf{x}) = e^{-\frac{i}{\hbar} \frac{p^2 t}{2m}} \psi_{\mathbf{p}, p}(\mathbf{x}) = e^{\frac{i}{\hbar}(\mathbf{p} \cdot \mathbf{x} - \frac{p^2 t}{2m})}.$$

202   *Symmetry and Quantum Mechanics*

Observer M would then use the wavefunction

$$\begin{aligned}
\psi'(t,\mathbf{x}) &= (s,\mathbf{w},\mathbf{0},A)^{-1} \star \psi(t,\mathbf{x}) \\
&= (-s,-A^{-1}\mathbf{w},\mathbf{0},A^{-1}) \star \psi(t,\mathbf{x}) \\
&= \psi(t+s,A\mathbf{x}+\mathbf{w}) \\
&= e^{\frac{i}{\hbar}(\mathbf{p}\cdot(A\mathbf{x}+\mathbf{w})-\frac{p^2(t+s)}{2m})} \\
&= e^{\frac{i}{\hbar}((\mathbf{p}\cdot\mathbf{w})-\frac{p^2 s}{2m})} e^{\frac{i}{\hbar}((A^{-1}\mathbf{p})\cdot\mathbf{x}-\frac{p^2 t}{2m})}.
\end{aligned}$$

Since A is an orthogonal matrix, we have $|A^{-1}\mathbf{p}|^2 = p^2$ so ψ' describes a particle with definite momentum $A^{-1}\mathbf{p}$ (together with a global phase out front). This is exactly what we should expect, since $A^{-1}\mathbf{p}$ gives M's rotated description of the free particle's momentum.

Now consider a general element $(s,\mathbf{w},\mathbf{v},A) \in \mathcal{G}$. We wish to determine its action on a wavefunction, so that acting by its inverse yields M's wavefunction from P's when M is rotated via A, moving with velocity $\mathbf{v}$, and displaced to $\mathbf{x} = \mathbf{w}$ at time $t = s$ (see Figure 9.3). Actually, we have already answered a portion of this question (for $s = 0$) in Chapter 8. In particular, exercise 8.15 shows that M's wavefunction (at time $t' = 0$) is obtained from P's wavefunction (at the same instant $t = 0$) as follows:

$$\psi'(0,\mathbf{x}) = e^{\frac{i\alpha}{\hbar}} e^{-\frac{i}{\hbar}m\mathbf{v}\cdot A\mathbf{x}} \psi(0,A\mathbf{x}+\mathbf{w}). \tag{9.2}$$

Here, $e^{\frac{i\alpha}{\hbar}}$ is a global phase, m is the mass of the particle under observation, and we have made allowance for the fact that exercise 8.15 assumed M to be moving with velocity $-\mathbf{v}$, whereas now M is moving with velocity $\mathbf{v}$. Our job is to determine the relationship between the time-evolutions of these initial wavefunctions.

The formula (9.2) for $\psi'(0,\mathbf{x})$ comes from the action of the group $G = H_3 \rtimes SO(3)$ on wavefunctions described in Section 8.4. To review, the group G consists[1] of elements $(m\mathbf{v},\mathbf{w},\alpha,A)$, with group law:

$$(m\mathbf{v}_2,\mathbf{w}_2,\alpha_2,A_2)(m\mathbf{v}_1,\mathbf{w}_1,\alpha_1,A_1) =$$
$$(m(\mathbf{v}_2 + A_2\mathbf{v}_1),\mathbf{w}_2 + A_2\mathbf{w}_1,\alpha_2 + \alpha_1 + m\mathbf{v}_2 \cdot A_2\mathbf{w}_1,A_2 A_1).$$

By exercise 8.14, the inverse in G is given by the formula:

$$(m\mathbf{v},\mathbf{w},\alpha,A)^{-1} = (-mA^{-1}\mathbf{v},-A^{-1}\mathbf{w},-\alpha + m\mathbf{v}\cdot\mathbf{w},A^{-1}). \tag{9.3}$$

Finally, the G-action on $L^2(\mathbb{R}^3)$ is defined by

$$(m\mathbf{v},\mathbf{w},\alpha,A) \star \psi(\mathbf{x}) = e^{-\frac{i}{\hbar}(\alpha+m\mathbf{v}\cdot(\mathbf{x}-\mathbf{w}))} \psi(A^{-1}(\mathbf{x}-\mathbf{w})). \tag{9.4}$$

---

[1] We now explicitly incorporate the particle mass m into the notation of the group elements. Essentially, this is because we wish to interpret $\mathbf{v}$ as a velocity, while the Schrödinger representation of the Heisenberg group requires that the first component of H_3 should have the dimensions of momentum.

Toward a Relativistic Theory

Acting via the inverse of the element $(-m\mathbf{v}, \mathbf{w}, \alpha, A)$ yields $\psi'(0, \mathbf{x})$ as described in (9.2):

$$(-m\mathbf{v}, \mathbf{w}, \alpha, A)^{-1} \star \psi(0, \mathbf{x})$$
$$= (mA^{-1}\mathbf{v}, -A^{-1}\mathbf{w}, -\alpha - m\mathbf{v} \cdot \mathbf{w}, A^{-1}) \star \psi(0, \mathbf{x})$$
$$= e^{-\frac{i}{\hbar}(-\alpha - m\mathbf{v} \cdot \mathbf{w} + mA^{-1}\mathbf{v} \cdot (\mathbf{x} + A^{-1}\mathbf{w}))} \psi(0, A(\mathbf{x} + A^{-1}\mathbf{w}))$$
$$= e^{\frac{i}{\hbar}(\alpha - m\mathbf{v} \cdot A\mathbf{x})} \psi(0, A\mathbf{x} + \mathbf{w}).$$

Now, in order to discover the relationship between the time-evolutions $\psi'(t, \mathbf{x})$ and $\psi(t, \mathbf{x})$, we would like to act via the inverse time-evolution operator $-t = (-t, 0, 0, I) \in \mathcal{G}$. But there is a subtlety here, since before we can safely mix the G-action described above with the time-evolution $(\mathbb{R}, +)$-action, we need to realize these two groups as subgroups of some common group acting on wavefunctions in a way that extends the given subgroup actions. Note that the Galilean group $\mathcal{G}$ cannot serve as this common group, since it does not contain $G = H_3 \rtimes SO(3)$ as a subgroup due to the central elements α in the Heisenberg group. In fact, the group we are looking for is called a *central extension* of the Galilean group, defined as $\mathcal{G}^\theta := \mathbb{R} \times \mathcal{G}$ with group law of the form

$$(\theta_2, g_2)(\theta_1, g_1) = (\theta_2 + \theta_1 + \theta(g_2, g_1), g_2 g_1), \tag{9.5}$$

where we must determine an appropriate function $\theta \colon \mathcal{G} \times \mathcal{G} \to \mathbb{R}$.

Note that the group $G = H_3 \rtimes SO(3)$ is itself a central extension of $(\mathbb{R}^3 \times \mathbb{R}^3) \rtimes SO(3)$: writing $(m\mathbf{v}, \mathbf{w}, \alpha, A) = (\alpha, f)$, where $f = (m\mathbf{v}, \mathbf{w}, A) \in (\mathbb{R}^3 \times \mathbb{R}^3) \rtimes SO(3)$, we may express the group law in G by

$$(\alpha_2, f_2)(\alpha_1, f_1) = (\alpha_2 + \alpha_1 + \alpha(f_2, f_1), f_2 f_1),$$

where $\alpha(f_2, f_1) := m\mathbf{v}_2 \cdot A_2 \mathbf{w}_1$. This gives us a hint for the function $\theta(g_2, g_1)$, but we must be careful, because we have some choice about how to embed G as a subgroup of $\mathcal{G}^\theta$. In fact, for our purposes, it will be convenient to work with an isomorphic copy of the group G, where the inversion formula involves a simple negation for the central component ($\alpha \mapsto -\alpha$), rather than the more complicated $\alpha \mapsto -\alpha + m\mathbf{v} \cdot \mathbf{w}$ from (9.3). This may be achieved by redefining the composition law on the underlying set of G in such a way that the isomorphism type of the resulting group is unchanged. To accomplish this, we consider the set $G_{\mathrm{Gal}} := \mathbb{R} \times \mathbb{R}^3 \times \mathbb{R}^3 \times SO(3)$ with group law

$$(\theta_2, \mathbf{w}_2, \mathbf{v}_2, A_2)(\theta_1, \mathbf{w}_1, \mathbf{v}_1, A_1) =$$
$$\left(\theta_2 + \theta_1 + \frac{1}{2}(\mathbf{w}_2 \cdot A_2 \mathbf{v}_1 - \mathbf{v}_2 \cdot A_2 \mathbf{w}_1), \mathbf{w}_2 + A_2 \mathbf{w}_1, \mathbf{v}_2 + A_2 \mathbf{v}_1, A_2 A_1\right).$$

Exercise 9.5. *Check that, with this composition law, G_{Gal} is in fact a group, and that the inverse of an element is given by*

$$(\theta, \mathbf{w}, \mathbf{v}, A)^{-1} = (-\theta, -A^{-1}\mathbf{w}, -A^{-1}\mathbf{v}, A^{-1}).$$

204   *Symmetry and Quantum Mechanics*

Consider the map $\varphi\colon G \to G_{\text{Gal}}$ *defined by*

$$\varphi(m\mathbf{v}, \mathbf{w}, \alpha, A) = \left(\frac{\alpha}{m} - \frac{1}{2}\mathbf{v}\cdot\mathbf{w}, \mathbf{w}, -\mathbf{v}, A\right).$$

Show that φ is an isomorphism of groups, with inverse

$$\varphi^{-1}(\theta, \mathbf{w}, \mathbf{v}, A) = \left(-m\mathbf{v}, \mathbf{w}, m\left(\theta - \frac{1}{2}\mathbf{v}\cdot\mathbf{w}\right), A\right).$$

Using the isomorphism of the previous exercise, we may view the G-action (9.4) on $L^2(\mathbb{R}^3)$ as a G_{Gal}-action instead:

$$
\begin{aligned}
(\theta, \mathbf{w}, \mathbf{v}, A) \star \psi(\mathbf{x}) \quad &:= \quad \varphi^{-1}(\theta, \mathbf{w}, \mathbf{v}, A) \star \psi(\mathbf{x}) \\
&= \quad \left(-m\mathbf{v}, \mathbf{w}, m\left(\theta - \frac{1}{2}\mathbf{v}\cdot\mathbf{w}\right), A\right) \star \psi(\mathbf{x}) \\
&= \quad e^{-\frac{i}{\hbar}(m(\theta - \frac{1}{2}\mathbf{v}\cdot\mathbf{w}) - m\mathbf{v}\cdot(\mathbf{x}-\mathbf{w}))}\psi(A^{-1}(\mathbf{x}-\mathbf{w})) \\
&= \quad e^{-\frac{im}{\hbar}(\theta - \mathbf{v}\cdot(\mathbf{x}-\frac{1}{2}\mathbf{w}))}\psi(A^{-1}(\mathbf{x}-\mathbf{w})).
\end{aligned}
$$

We now view G_{Gal} as a subset of $\mathbb{R} \times G$ via the inclusion

$$(\theta, \mathbf{w}, \mathbf{v}, A) \mapsto (\theta, \mathbf{0}, \mathbf{w}, \mathbf{v}, A).$$

Moreover, $\mathbb{R} \times G_0 \subset G$ is also a subset of $\mathbb{R} \times G$ via the inclusion

$$(s, \mathbf{w}, \mathbf{0}, A) \mapsto (0, s, \mathbf{w}, \mathbf{0}, A).$$

We now wish to specify a function $\theta(g_2, g_1)$ so that the resulting group law (9.5) on $G^\theta = \mathbb{R} \times G$ restricts to the group laws of G_{Gal} and $\mathbb{R} \times G_0$. In fact, we will now show that the function θ is completely determined by the following additional requirement:

- the function $\theta(g_2, g_1)$ is zero if either input is a pure time-translation $s = (s, \mathbf{0}, \mathbf{0}, I) \in G$:

$$\theta(g, s) = \theta(s, g) = 0 \text{ for all } g \in G.$$

Indeed, if the function θ satisfies this requirement and defines the correct group laws on G_{Gal} and $\mathbb{R} \times G_0$, we find that the group law on G^θ must be

$$
\begin{aligned}
(\theta_2, &s_2, \mathbf{w}_2, \mathbf{v}_2, A_2)(\theta_1, s_1, \mathbf{w}_1, \mathbf{v}_1, A_1) \\
&= (\theta_2, s_2, \mathbf{w}_2, \mathbf{v}_2, A_2)s_1(\theta_1, 0, \mathbf{w}_1, \mathbf{v}_1, A_1) \\
&= (\theta_2, s_2 + s_1, \mathbf{w}_2 + s_1\mathbf{v}_2, \mathbf{v}_2, A_2)(\theta_1, 0, \mathbf{w}_1, \mathbf{v}_1, A_1) \\
&= (s_2 + s_1)(\theta_2, 0, \mathbf{w}_2 + s_1\mathbf{v}_2, \mathbf{v}_2, A_2)(\theta_1, 0, \mathbf{w}_1, \mathbf{v}_1, A_1) \\
&= (s_2 + s_1)(\theta_2 + \theta_1 + \frac{1}{2}((\mathbf{w}_2 + s_1\mathbf{v}_2)\cdot A_2\mathbf{v}_1 - \mathbf{v}_2\cdot A_2\mathbf{w}_1), 0, \\
&\qquad\qquad \mathbf{w}_2 + A_2\mathbf{w}_1, \mathbf{v}_2 + A_2\mathbf{v}_1, A_2A_1) \\
&= (\theta_2 + \theta_1 + \frac{1}{2}((\mathbf{w}_2 + s_1\mathbf{v}_2)\cdot A_2\mathbf{v}_1 - \mathbf{v}_2\cdot A_2\mathbf{w}_1), s_2 + s_1, \\
&\qquad\qquad \mathbf{w}_2 + A_2\mathbf{w}_1, \mathbf{v}_2 + A_2\mathbf{v}_1, A_2A_1).
\end{aligned}
$$

Toward a Relativistic Theory   205

Hence, the function $\theta(g_2, g_1)$ defining the group law on $\mathcal{G}^\theta$ is

$$\theta(g_2, g_1) = \frac{1}{2}(\mathbf{w}_2 \cdot A_2 \mathbf{v}_1 - \mathbf{v}_2 \cdot A_2 \mathbf{w}_1 + s_1 \mathbf{v}_2 \cdot A_2 \mathbf{v}_1). \tag{9.6}$$

Exercise 9.6. *Verify that with the definition (9.6) of $\theta(g_2, g_1)$, the set $\mathcal{G}^\theta :=$ $\mathbb{R} \times \mathcal{G}$ becomes a group with operation*

$$(\theta_2, g_2)(\theta_1, g_1) = (\theta_2 + \theta_1 + \theta(g_2, g_1), g_2 g_1).$$

Check that the following formula defines a $\mathcal{G}^\theta$-action on wavefunctions $\psi(t, \mathbf{x})$, extending the actions of G_{Gal} and $\mathbb{R} \times G_0$ defined earlier:

$$(\theta, s, \mathbf{w}, \mathbf{v}, A) \star \psi(t, \mathbf{x}) = e^{-\frac{im}{\hbar}(\theta - \mathbf{v} \cdot (\mathbf{x} - \frac{1}{2}\mathbf{w}))} \psi(t - s, A^{-1}(\mathbf{x} - \mathbf{w})).$$

Now recall our problem (see Figure 9.3): observer M has set her watch behind P's by s seconds, is rotated via $A \in SO(3)$, and is moving with velocity $\mathbf{v}$ so that at time $t = s$ she is located at $\mathbf{x} = \mathbf{w}$. If P describes the state of a mass-m particle with the wavefunction $\psi(t, \mathbf{x})$, what is M's wavefunction $\psi'(t, \mathbf{x})$? To answer this question, we proceed in three stages:

1. First, act on P's initial wavefunction $\psi(0, \mathbf{x})$ by the inverse time-translation $-s$ to account for the fact that M's watch is set s seconds behind P's:

$$(-s) \star \psi(0, \mathbf{x}) = \psi(s, \mathbf{x}).$$

2. Second, act by the inverse of the element $\varphi(-m\mathbf{v}, \mathbf{w}, \frac{1}{2}m\mathbf{v} \cdot \mathbf{w}, A) = (0, 0, \mathbf{w}, \mathbf{v}, A) \in G_{\mathrm{Gal}}$ to account for the rotation, relative motion, and initial displacement, thereby obtaining M's initial wavefunction $\psi'(0, \mathbf{x})$:

$$\psi'(0, \mathbf{x}) = (0, 0, \mathbf{w}, \mathbf{v}, A)^{-1} \star \psi(s, \mathbf{x}).$$

(Here, we have chosen $\alpha = \frac{1}{2}m\mathbf{v} \cdot \mathbf{w}$ so that the corresponding element of G_{Gal} has $\theta = 0$.)

3. Third, act by the inverse time-translation $-t$ to evolve M's initial wavefunction to time t:

$$\psi'(t, \mathbf{x}) = (-t) \star \psi'(0, \mathbf{x}) = [(-t)(0, 0, \mathbf{w}, \mathbf{v}, A)^{-1}(-s)] \star \psi(0, \mathbf{x}).$$

Computing in the group $\mathcal{G}^\theta$, we see that we wish to act on $\psi(0, \mathbf{x})$ by the group element

$$
\begin{aligned}
(-t)(0, 0, \mathbf{w}, \mathbf{v}, A)^{-1}(-s) &= [s(0, 0, \mathbf{w}, \mathbf{v}, A)t]^{-1} \\
&= [s(0, t, \mathbf{w} + t\mathbf{v}, \mathbf{v}, A)]^{-1} \\
&= (0, s + t, \mathbf{w} + t\mathbf{v}, \mathbf{v}, A)^{-1} \\
&= (0, -s - t, -A^{-1}(\mathbf{w} + t\mathbf{v}), -A^{-1}\mathbf{v}, A^{-1}).
\end{aligned}
$$

206   *Symmetry and Quantum Mechanics*

We thus find that

$$\begin{aligned}
\psi'(t,\mathbf{x}) &= (0, -s-t, -A^{-1}(\mathbf{w}+t\mathbf{v}), -A^{-1}\mathbf{v}, A^{-1}) \star \psi(0,\mathbf{x}) \\
&= e^{-\frac{im}{\hbar}(A^{-1}\mathbf{v}\cdot(\mathbf{x}+\frac{1}{2}A^{-1}(\mathbf{w}+t\mathbf{v})))}\psi(t+s, A(\mathbf{x}+(\mathbf{w}+t\mathbf{v}))) \\
&= e^{-\frac{im}{\hbar}\mathbf{v}\cdot(A\mathbf{x}+\frac{1}{2}(\mathbf{w}+t\mathbf{v})))}\psi(t+s, A\mathbf{x}+\mathbf{w}+t\mathbf{v}).
\end{aligned}$$

This is the wavefunction that M would use to describe the same particle as P describes by $\psi(t,\mathbf{x})$. The next exercise demonstrates that this really is the correct transformation law for physical states.

Exercise 9.7. *Suppose that $\psi(t,\mathbf{x})$ is a solution to the Schrödinger equation with potential $V(\mathbf{x})$:*

$$i\hbar\frac{\partial\psi}{\partial t} = -\frac{\hbar^2}{2m}\Delta\psi + V(\mathbf{x})\psi.$$

Show that the transformed wavefunction $\psi'(t,\mathbf{x})$ described above is a solution to the Schrödinger equation with transformed (and time-dependent!) potential $V'(\mathbf{x},t) = V(A\mathbf{x}+\mathbf{w}+t\mathbf{v})$. (Hint: see exercise 8.16 and the discussion in Section 7.4.)

9.2   Special relativity

The fundamental postulate of special relativity is that the speed of light, $c = 2.99792458 \times 10^8 \frac{m}{s}$, is the same for all inertial observers. In example 9.1, we saw that the Euclidean transformations lead to different velocities for a ball as observed by M and P. If we think naively about a ball traveling at (or near) the speed of light, this shows that the Euclidean transformations cannot be correct in the relativistic context. To determine the correct transformation laws, we give up on the notion of absolute time and space, and instead insist on the universality of the speed of light.

So now suppose that P observes a flash of light at his origin at time $t = 0$ (see Figure 9.4). At a later time, t, the light will have expanded to a sphere of radius ct, so that (t, x_1, x_2, x_3) describes the space-time coordinates[2] of a point on the light-sphere if and only if:

$$x_1^2 + x_2^2 + x_3^2 = (ct)^2 \quad \text{or} \quad (ct)^2 - x_1^2 - x_2^2 - x_3^2 = 0.$$

We are assuming that M travels with constant velocity $\mathbf{v}$ with respect to P, so that P observes her trajectory to be $\mathbf{x}(t) = t\mathbf{v}$. Moreover, we assume that $t' = 0$ when $t = 0$, so that their space-time origins coincide. With these assumptions, M also observes a flash of light at her origin at time $t' = 0$, so

---

[2]It is traditional (and convenient) in the relativistic context to shift notation from x, y, z for the spatial coordinates to x_1, x_2, x_3.

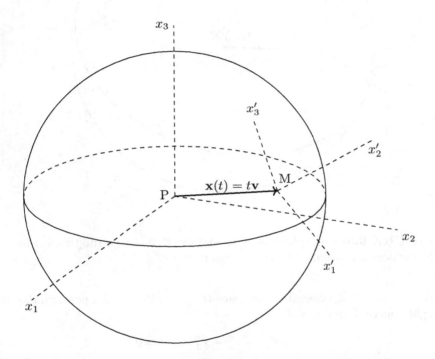

FIGURE 9.4: The sphere of light at time $t > 0$ resulting from a flash at P's space-time origin. The sphere has radius ct.

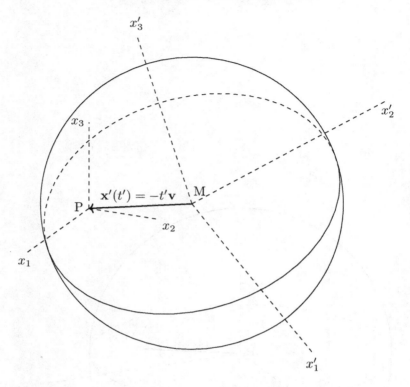

FIGURE 9.5: The sphere of light at time $t' > 0$ resulting from a flash at M's space-time origin. The sphere has radius ct'.

that (t', x'_1, x'_2, x'_3) describes the space-time coordinates of a point on the same light-sphere if and only if:

$$(ct')^2 - x'^2_1 - x'^2_2 - x'^2_3 = 0.$$

Figure 9.5 shows the situation from M's point of view, in which P is moving with velocity $-\mathbf{v}$. Note that we have explicitly used our assumption that the speed c is the same for both observers.

To reveal the consequences of these observations, we begin by replacing the time coordinates t, t' by the coordinates $x_0 := ct$ and $x'_0 := ct'$ having the dimensions of length. Then we define a function $\langle , \rangle_\mathrm{M} \colon \mathbb{R}^4 \times \mathbb{R}^4 \to \mathbb{R}^4$ by

$$\langle x, \tilde{x} \rangle_\mathrm{M} := x_0 \tilde{x}_0 - x_1 \tilde{x}_1 - x_2 \tilde{x}_2 - x_3 \tilde{x}_3.$$

By the next exercise, $\langle , \rangle_\mathrm{M}$ is a *non-degenerate indefinite inner product*, which means that it satisfies the following properties for $\lambda \in \mathbb{R}$ and $x, \tilde{x}, y \in \mathbb{R}^4$:

i) $\langle \lambda x + y, \tilde{x} \rangle_\mathrm{M} = \lambda \langle x, \tilde{x} \rangle_\mathrm{M} + \langle y, \tilde{x} \rangle_\mathrm{M}$ (*linearity*);

Toward a Relativistic Theory

ii) $\langle x, \tilde{x} \rangle_{\mathrm{M}} = \langle \tilde{x}, x \rangle_{\mathrm{M}}$ (*symmetry*);

iii) If $\langle x, y \rangle_{\mathrm{M}} = 0$ for all y, then $x = 0$.   (*non-degeneracy*).

The resulting structure $(\mathbb{R}^4, \langle, \rangle_{\mathrm{M}})$ is called *Minkowski space*.

Exercise 9.8. *Check that $\langle, \rangle_{\mathrm{M}}$ is a non-degenerate indefinite inner product on $\mathbb{R}^4$.*

Denote by $\Lambda \colon \mathbb{R}^4 \to \mathbb{R}^4$ the transformation that sends M's coordinates $x' = (x_0', x_1', x_2', x_3')$ for an event to P's coordinates $x = (x_0, x_1, x_2, x_3)$ for the same event. Then the discussion above implies that $\langle x, x \rangle_{\mathrm{M}} = 0$ if and only if x describes a point on P's light-sphere. Similarly, $\langle x', x' \rangle_{\mathrm{M}} = 0$ if an only if x' describes a point on M's light-sphere. But these two light-spheres are the same, so we see that $\langle x', x' \rangle_{\mathrm{M}} = 0$ if and only if $\langle \Lambda(x'), \Lambda(x') \rangle_{\mathrm{M}} = 0$. The simplest way to ensure this is to require that Λ be a linear transformation that preserves the indefinite inner product, i.e., an automorphism of Minkowski space. The next exercise introduces the Lorentz group, which is the automorphism group of Minkowski space; the reader should compare the development with the discussion of the orthogonal groups in Section 1.2 and the unitary groups in Section 2.3.

Exercise 9.9. *Suppose that $\Lambda \colon \mathbb{R}^4 \to \mathbb{R}^4$ is a linear operator, and let $\varepsilon := \{\mathbf{e}_0, \mathbf{e}_1, \mathbf{e}_2, \mathbf{e}_3\}$ denote the standard basis of $\mathbb{R}^4$. Then Λ is represented (with respect to ε) by a 4×4 matrix of real numbers, which we also denote by Λ:*

$$\Lambda = [\Lambda_{ij}] \quad \textit{where} \quad \Lambda(\mathbf{e}_j) = \sum_{i=0}^{3} \Lambda_{ij} \mathbf{e}_i.$$

a) Show that Λ preserves the Minkowski inner product on $\mathbb{R}^4$ if and only if the matrix Λ satisfies

$$\Lambda^T g \Lambda = g \quad \textit{where} \quad g := \begin{bmatrix} 1 & 0 & 0 & 0 \\ 0 & -1 & 0 & 0 \\ 0 & 0 & -1 & 0 \\ 0 & 0 & 0 & -1 \end{bmatrix}.$$

Denote this set of matrices by $O(1,3)$, and show that it forms a subgroup of $GL(4, \mathbb{R})$, called the Lorentz group. *It is the symmetry group of Minkowski space $(\mathbb{R}^4, \langle, \rangle_{\mathrm{M}})$. Show that every element of $O(1,3)$ has determinant ± 1. The subgroup of Lorentz matrices with determinant 1 is called the* special Lorentz group, *denoted $SO(1,3)$.*

b) Use part a) to show that the upper left diagonal matrix element Λ_{00} satisfies $|\Lambda_{00}| \geq 1$ for all $\Lambda \in O(1,3)$. Denote by $O^+(1,3)$ the subset of Lorentz matrices with $\Lambda_{00} \geq 1$. Show that $O^+(1,3)$ is a subgroup of $O(1,3)$, called the orthochronous Lorentz group. *If we impose the additional requirement that the determinant is 1, we obtain the* special orthochronous Lorentz group, *denoted $SO^+(1,3)$.*

210 *Symmetry and Quantum Mechanics*

c) Suppose that $\Lambda \in O(1,3)$ is an arbitrary Lorentz transformation. Verify that

> *i) if $\det(\Lambda) = -1$ and $\Lambda_{00} \geq 1$, then $\Lambda = (\Lambda g)g$ where $\Lambda g \in SO^+(1,3)$;*
>
> *ii) if $\det(\Lambda) = 1$ and $\Lambda_{00} \leq -1$, then $\Lambda = (-\Lambda)(-I)$ where $-\Lambda \in SO^+(1,3)$;*
>
> *iii) if $\det(\Lambda) = -1$ and $\Lambda_{00} \leq -1$, then $\Lambda = (-\Lambda g)(-g)$ where $-\Lambda g \in SO^+(1,3)$.*

Hence, every Lorentz transformation may be written uniquely in one of the forms $\pm\Lambda$ or $\pm\Lambda g$ for some $\Lambda \in SO^+(1,3)$. Conclude that the Lorentz group may be written as a semi-direct product:

$$O(1,3) \cong SO^+(1,3) \rtimes (\{I,g\} \times \{I,-g\}).$$

Here, the group law on the semi-direct product is given by

$$(\Lambda_1, \tau_1)(\Lambda_2, \tau_2) = (\Lambda_1\tau_1\Lambda_2\tau_1, \tau_1\tau_2),$$

for all $\Lambda_1, \Lambda_2 \in SO^+(1,3)$ and $\tau_1, \tau_2 \in \{\pm I, \pm g\}$. Since the determinant is continuous, as is the function $\Lambda \mapsto \Lambda_{00}$, we see that $O(1,3)$ is not connected. In fact, we will later see (corollary 9.14) that $SO^+(1,3)$ is the connected component of the identity in $O(1,3)$.

d) Show that the rotation group $SO(3)$ is a subgroup of $SO^+(1,3)$ via the embedding

$$SO(3) \ni A \mapsto \begin{bmatrix} 1 & 0 \\ 0 & A \end{bmatrix} \in SO^+(1,3).$$

We now assume that the transformation Λ that sends M's coordinates x' to P's coordinates x is an element of the special orthochronous Lorentz group $SO^+(1,3)$. To explain this restriction to a subgroup of $O(1,3)$, we use part c) of the previous exercise. Note that the element g corresponds to a spatial reflection, $\mathbf{x}' = -\mathbf{x}$, which we forbid according to M and P's agreement from Chapter 1 to use right-handed coordinate systems. Similarly, $-g$ corresponds to an observer whose time runs backward compared to P's: $t' = -t$. Since we are assuming that M and P's watches both run forward (although perhaps at different rates), we forbid this transformation as well. Finally, $-I = (-g)g$ corresponds to a reflection in both space and time. Note that if the relative velocity is $\mathbf{v} = 0$, then M and P are at rest together as in Chapter 1, and the transformation Λ is simply a rotation in $SO(3) \subset SO^+(1,3)$ as described in part d) of the previous exercise. From now on, we will refer to $SO^+(1,3)$ as the *restricted Lorentz group*.

Since $SO^+(1,3)$ is a closed subgroup of $GL(4,\mathbb{C})$, it follows from [11, Corollary 3.45] that it is a Lie group. In order to better understand its structure, we determine its Lie algebra $\mathfrak{so}^+(1,3)$.

Toward a Relativistic Theory

Proposition 9.10. *The Lie algebra of the restricted Lorentz group* $SO^+(1,3)$ *is the space of 4-by-4 real g-skew-symmetric matrices:*

$$\mathfrak{so}^+(1,3) = \left\{ X \in M(4,\mathbb{R}) \mid X^T = -gXg^{-1} \right\}.$$

Proof. Suppose that $b \colon \mathbb{R} \to SO^+(1,3)$ is a one-to-one differentiable curve satisfying $b(0) = I$. Then differentiating the Lorentz condition $b(s)^T g b(s) = g$ at $s = 0$ yields:

$$\dot{b}(0)^T g + g\dot{b}(0) = 0,$$

so that the tangent vector $\dot{b}(0)$ is a 4×4 real g-skew-symmetric matrix:

$$\dot{b}(0)^T = -g\dot{b}(0)g^{-1}.$$

Conversely, suppose that X is any 4×4 real g-skew-symmetric matrix, and consider the curve in $GL(4,\mathbb{R})$ defined by $s \mapsto \exp(sX)$. Then compute

$$
\begin{aligned}
\exp(sX)^T g \exp(sX) &= \exp(sX^T) g \exp(sX) \\
&= \exp(-sgXg^{-1}) g \exp(sX) \\
&= g \exp(-sX) g^{-1} g \exp(sX) \\
&= g \exp(-sX) \exp(sX) \\
&= g,
\end{aligned}
$$

which shows that $\exp(sX)$ is a curve in $O(1,3)$. Since it passes through the identity at $s = 0$, it follows by continuity that $\exp(sX)$ is actually a curve in $SO^+(1,3)$. Since $X = \frac{d}{ds}\exp(sX)|_{s=0}$, we see that X is a tangent vector to $SO^+(1,3)$ at the identity as required. $\square$

The Lie algebra of $\mathfrak{so}^+(1,3)$ is 6-dimensional, and the following matrices provide a basis:

$$
\mathcal{L}_1 = \begin{bmatrix} 0 & 0 & 0 & 0 \\ 0 & 0 & 0 & 0 \\ 0 & 0 & 0 & -1 \\ 0 & 0 & 1 & 0 \end{bmatrix}
\qquad
\mathcal{K}_1 = \begin{bmatrix} 0 & 1 & 0 & 0 \\ 1 & 0 & 0 & 0 \\ 0 & 0 & 0 & 0 \\ 0 & 0 & 0 & 0 \end{bmatrix}
$$

$$
\mathcal{L}_2 = \begin{bmatrix} 0 & 0 & 0 & 0 \\ 0 & 0 & 0 & 1 \\ 0 & 0 & 0 & 0 \\ 0 & -1 & 0 & 0 \end{bmatrix}
\qquad
\mathcal{K}_2 = \begin{bmatrix} 0 & 0 & 1 & 0 \\ 0 & 0 & 0 & 0 \\ 1 & 0 & 0 & 0 \\ 0 & 0 & 0 & 0 \end{bmatrix}
$$

$$
\mathcal{L}_3 = \begin{bmatrix} 0 & 0 & 0 & 0 \\ 0 & 0 & -1 & 0 \\ 0 & 1 & 0 & 0 \\ 0 & 0 & 0 & 0 \end{bmatrix}
\qquad
\mathcal{K}_3 = \begin{bmatrix} 0 & 0 & 0 & 1 \\ 0 & 0 & 0 & 0 \\ 0 & 0 & 0 & 0 \\ 1 & 0 & 0 & 0 \end{bmatrix}.
$$

Exercise 9.11. *Check that these matrices form a basis of* $\mathfrak{so}^+(1,3)$ *and write down the commutation relations.*

212 *Symmetry and Quantum Mechanics*

The matrices $\mathcal{L}_j$ generate the subalgebra $\mathfrak{so}(3) \subset \mathfrak{so}^+(1,3)$ corresponding to the rotation subgroup $SO(3) \subset SO^+(1,3)$. In particular, if $\mathbf{u}$ is a unit vector in physical space, then $\exp(\theta \mathbf{u} \cdot \mathcal{L}) \in SO(3)$ is a rotation through the angle θ around the $\mathbf{u}$-axis (here, $\mathcal{L} := (\mathcal{L}_1, \mathcal{L}_2, \mathcal{L}_3)$ is the vector of rotation generators). We claim that the matrices $\mathcal{K}_j$ are the generators of *Lorentz boosts* in the coordinate directions, yielding the space-time coordinate transformations when M is moving along one of P's coordinate axes.

To justify this interpretation of the elements $\mathcal{K}_j$, suppose that M is traveling along P's first coordinate axis with speed v, so that $\mathbf{v} = (v, 0, 0)$. We claim that the transformation sending P's coordinates x to M's coordinates x' is given by $\exp(-\varphi \mathcal{K}_1)$ for an appropriately chosen value of the parameter φ.

Exercise 9.12. *Show by explicit computation of the infinite series that*

$$\exp(-\varphi \mathcal{K}_1) = \begin{bmatrix} \cosh(\varphi) & -\sinh(\varphi) & 0 & 0 \\ -\sinh(\varphi) & \cosh(\varphi) & 0 & 0 \\ 0 & 0 & 1 & 0 \\ 0 & 0 & 0 & 1 \end{bmatrix}.$$

Here, the hyperbolic trigonometric functions *are defined as*

$$\cosh(\varphi) \quad := \quad \frac{1}{2}(e^\varphi + e^{-\varphi}) = \cos(i\varphi)$$

$$\sinh(\varphi) \quad := \quad \frac{1}{2}(e^\varphi - e^{-\varphi}) = -i\sin(i\varphi).$$

Applying the transformation $\exp(-\varphi \mathcal{K}_1)$ to the column vector x representing P's space-time coordinates yields the system of equations

$$\begin{aligned} x_0' &= \cosh(\varphi)x_0 - \sinh(\varphi)x_1 \\ x_1' &= -\sinh(\varphi)x_0 + \cosh(\varphi)x_1 \\ x_2' &= x_2 \\ x_3' &= x_3. \end{aligned}$$

To understand the physical meaning of these equations, set $x_1' = 0$ in the second equation and divide by $\cosh(\varphi)x_0$ to obtain

$$\tanh(\varphi) = \frac{\sinh(\varphi)}{\cosh(\varphi)} = \frac{x_1}{x_0} = \frac{x_1}{ct} = \frac{v}{c},$$

since at time t, observer M (at $x_1' = 0$) is at a distance $x_1 = vt$ from P. Returning with this information to the first two equations and using the identity $\cosh^2(\varphi) - \sinh^2(\varphi) = 1$, we find:

$$t' = \frac{x_0'}{c} = \cosh(\varphi)\left(\frac{x_0}{c} - \frac{\tanh(\varphi)}{c}x_1\right) = \frac{1}{\sqrt{1 - \frac{v^2}{c^2}}}\left(t - \frac{v}{c^2}x_1\right)$$

$$x_1' = \cosh(\varphi)(x_1 - \tanh(\varphi)x_0) = \frac{1}{\sqrt{1 - \frac{v^2}{c^2}}}(x_1 - vt).$$

The factor $\cosh(\varphi) = (1 - \frac{v^2}{c^2})^{-\frac{1}{2}}$ is responsible for the phenomena of *time dilation* and *Lorentz contraction* that are explained in all elementary accounts of special relativity. For our purposes, the key point to observe is that for $\frac{v}{c}$ small (i.e., non-relativistic velocities), these equations reduce to the Euclidean transformation

$$t' \approx t \qquad \text{and} \qquad x_1' \approx x_1 - vt.$$

The parameter φ defined by $\tanh(\varphi) = \frac{v}{c}$ is called the *rapidity* of the corresponding boost. A similar analysis reveals that the boosts $\exp(-\varphi \mathcal{K}_2)$ and $\exp(-\varphi \mathcal{K}_3)$ provide the transformations from P's coordinates to M's in the cases when M is traveling along P's second or third coordinate axes with speed $v = c \tanh(\varphi)$. In general, if $\mathbf{u} \in (\mathbb{R}^3, \cdot)$ is a unit vector and P observes M to be traveling with speed $v = c \tanh(\varphi)$ in the direction $\mathbf{u}$, then the corresponding coordinate transformation is given by the boost $\exp(-\varphi \mathbf{u} \cdot \mathcal{K})$, where $\mathcal{K} := (\mathcal{K}_1, \mathcal{K}_2, \mathcal{K}_3)$ is the vector of boost generators. The minus sign in the argument of the exponential comes about for the same reason that the inverse of the rotation matrix A appeared in Chapter 1: if $\Lambda := \exp(\varphi \mathbf{u} \cdot \mathcal{K})$ is the boost that sends P's *basis* for Minkowski space to M's basis, then the inverse boost $\Lambda^{-1} = \exp(-\varphi \mathbf{u} \cdot \mathcal{K})$ sends P's *coordinates* for Minkowski space to M's coordinates.

We have now identified two types of Lorentz transformations:

- (Pure Rotations) $A = \exp(\theta \mathbf{u} \cdot \mathcal{L}) \in SO(3)$ is a rotation through the angle θ around the $\mathbf{u}$-axis;

- (Pure Boosts) $\Phi = \exp(\varphi \mathbf{u} \cdot \mathcal{K})$ is a boost with rapidity φ in the $\mathbf{u}$-direction.

We were led to the Lorentz transformations based on our questions about a rotated observer in uniform motion, so we expect that every restricted Lorentz transformation Λ should be obtained as a combination of Lorentz boosts and spatial rotations. In order to confirm this hunch, we prove the following proposition which provides an explicit decomposition of Λ as a product of rotations and boosts. Our proof is adapted from the treatment in [8, Chapter 6], which studies the higher-dimensional Lorentz groups.

Proposition 9.13. *Every restricted Lorentz transformation* $\Lambda \in SO^+(1,3)$ *may be written as a product of the form* $\Lambda(A_1, \varphi, A_2)$, *where* $A_1, A_2 \in SO(3)$:

$$
\begin{bmatrix} 1 & 0 & 0 & 0 \\ 0 & & & \\ 0 & & A_1 & \\ 0 & & & \end{bmatrix}
\begin{bmatrix} \cosh(\varphi) & 0 & 0 & \sinh(\varphi) \\ 0 & 1 & 0 & 0 \\ 0 & 0 & 1 & 0 \\ \sinh(\varphi) & 0 & 0 & \cosh(\varphi) \end{bmatrix}
\begin{bmatrix} 1 & 0 & 0 & 0 \\ 0 & & & \\ 0 & & A_2 & \\ 0 & & & \end{bmatrix}.
$$

Proof. Begin by writing Λ in block form as follows, where $\mathbf{v}, \mathbf{w} \in \mathbb{R}^3$ are column vectors and M is a 3×3 matrix.

$$\Lambda = \begin{bmatrix} \Lambda_{00} & \mathbf{v}^T \\ \mathbf{w} & M \end{bmatrix}.$$

Symmetry and Quantum Mechanics

Writing out the Lorentz conditions $\Lambda^T g \Lambda = g = \Lambda g \Lambda^T$ yields the following conditions on the blocks of Λ:

$$\begin{aligned}
\Lambda_{00}^2 &= 1 + |\mathbf{w}|^2 = 1 + |\mathbf{v}|^2 \\
\Lambda_{00} \mathbf{v} &= M^T \mathbf{w} \\
\Lambda_{00} \mathbf{w} &= M \mathbf{v} \\
M^T M &= I + \mathbf{v}\mathbf{v}^T.
\end{aligned}$$

If $\mathbf{v} = \mathbf{0}$, then $\mathbf{w} = \mathbf{0}$, $\Lambda_{00} = 1$, and $M^T M = I$. It follows that $\Lambda \in SO(3)$ is a pure rotation. So for the remainder of the proof, we assume that $\mathbf{v} \neq \mathbf{0}$. In this case, the 3×3 matrix $\mathbf{v}\mathbf{v}^T$ has a 2-dimensional kernel equal to the plane orthogonal to $\mathbf{v}$; choose an orthonormal basis $\mathbf{u}_1, \mathbf{u}_2$ for this space. Also, the vector $\mathbf{v}$ is an eigenvector for $\mathbf{v}\mathbf{v}^T$ with eigenvalue $|\mathbf{v}|^2$. It follows that $\mathbf{u}_1$ and $\mathbf{u}_2$ are eigenvectors for $M^T M$ with eigenvalue 1, and $\mathbf{v}$ is an eigenvector for $M^T M$ with eigenvalue $1 + |\mathbf{v}|^2 = \Lambda_{00}^2$. Let Q be the orthogonal matrix with columns $\mathbf{u}_1, \mathbf{u}_2, \frac{1}{|\mathbf{v}|}\mathbf{v}$. By swapping $\mathbf{u}_1$ and $\mathbf{u}_2$ if necessary, we may ensure that $Q \in SO(3)$. Then Q diagonalizes $M^T M$, so that we have $Q^{-1} M^T M Q = \mathrm{diag}(1, 1, \Lambda_{00}^2)$. Setting $P = Q\mathrm{diag}(1, 1, \Lambda_{00})Q^{-1}$, we see that P is a symmetric and positive definite matrix such that $P^2 = M^T M$ (here we use our assumption that $\Lambda_{00} \geq 1$, so that P has positive eigenvalues). Now define $A = MP^{-1}$, and note that A is orthogonal:

$$A^T A = (MP^{-1})^T MP^{-1} = P^{-1} M^T M P^{-1} = P^{-1} P^2 P^{-1} = I.$$

Hence, we have $M = AP$, the polar decomposition of the matrix M.

Since $\mathbf{v}$ is an eigenvector for P with eigenvalue Λ_{00}, we have

$$\Lambda_{00} \mathbf{w} = M\mathbf{v} = AP\mathbf{v} = A\Lambda_{00}\mathbf{v} = \Lambda_{00} A\mathbf{v},$$

so that $\mathbf{w} = A\mathbf{v}$. Now return to the full matrix Λ:

$$\Lambda = \begin{bmatrix} \Lambda_{00} & \mathbf{v}^T \\ \mathbf{w} & M \end{bmatrix} = \begin{bmatrix} \Lambda_{00} & \mathbf{v}^T \\ A\mathbf{v} & AP \end{bmatrix} = \begin{bmatrix} 1 & \mathbf{0}^T \\ \mathbf{0} & A \end{bmatrix} \begin{bmatrix} \Lambda_{00} & \mathbf{v}^T \\ \mathbf{v} & P \end{bmatrix}.$$

Denote the final matrix in the previous display by N. Note that the vectors $(0, \mathbf{u}_1)$ and $(0, \mathbf{u}_2)$ are orthonormal eigenvectors for N with eigenvalue 1. In addition, the vectors $(\pm|\mathbf{v}|, \mathbf{v})$ are also eigenvectors for N:

$$\begin{bmatrix} \Lambda_{00} & \mathbf{v}^T \\ \mathbf{v} & P \end{bmatrix} \begin{bmatrix} \pm|\mathbf{v}| \\ \mathbf{v} \end{bmatrix} = \begin{bmatrix} \pm\Lambda_{00}|\mathbf{v}| + |\mathbf{v}|^2 \\ \pm|\mathbf{v}|\mathbf{v} + \Lambda_{00}\mathbf{v} \end{bmatrix} = (\Lambda_{00} \pm |\mathbf{v}|) \begin{bmatrix} \pm|\mathbf{v}| \\ \mathbf{v} \end{bmatrix}.$$

Since $\Lambda_{00}^2 = 1 + |\mathbf{v}|^2$, both of the eigenvalues are positive, and in fact they are reciprocals:

$$(\Lambda_{00} + |\mathbf{v}|)(\Lambda_{00} - |\mathbf{v}|) = \Lambda_{00}^2 - |\mathbf{v}|^2 = 1.$$

Hence, we may take logarithms to obtain $\pm\varphi := \log(\Lambda_{00} \pm |\mathbf{v}|)$, so that the

Toward a Relativistic Theory

215

eigenvalues may be expressed as $e^{\pm\varphi}$. The following is thus an orthonormal eigenbasis for the matrix N, with eigenvalues $e^{\varphi}, 1, 1, e^{-\varphi}$:

$$\frac{1}{\sqrt{2}|\mathbf{v}|}\begin{bmatrix}|\mathbf{v}|\\\mathbf{v}\end{bmatrix}, \begin{bmatrix}0\\\mathbf{u}_1\end{bmatrix}, \begin{bmatrix}0\\\mathbf{u}_2\end{bmatrix}, \frac{1}{\sqrt{2}|\mathbf{v}|}\begin{bmatrix}-|\mathbf{v}|\\\mathbf{v}\end{bmatrix}.$$

Let R be the orthogonal matrix with these vectors as columns. Then $R^{-1}NR = \mathrm{diag}(e^{\varphi}, 1, 1, e^{-\varphi})$. We now conjugate further with the orthogonal matrix

$$S := \frac{1}{\sqrt{2}}\begin{bmatrix} 1 & 0 & 0 & 1 \\ 0 & \sqrt{2} & 0 & 0 \\ 0 & 0 & \sqrt{2} & 0 \\ -1 & 0 & 0 & 1 \end{bmatrix}.$$

Direct computation reveals that

$$S^{-1}\mathrm{diag}(e^{\varphi}, 1, 1, e^{-\varphi})S = \begin{bmatrix} \cosh(\varphi) & 0 & 0 & \sinh(\varphi) \\ 0 & 1 & 0 & 0 \\ 0 & 0 & 1 & 0 \\ \sinh(\varphi) & 0 & 0 & \cosh(\varphi) \end{bmatrix}$$

and that

$$RS = \begin{bmatrix} 1 & \mathbf{0}^T \\ \mathbf{0} & Q \end{bmatrix},$$

where $Q \in SO(3)$ is the matrix that diagonalizes P. Returning once more to the full matrix Λ, we have

$$
\begin{aligned}
\Lambda &= \begin{bmatrix} 1 & \mathbf{0}^T \\ \mathbf{0} & A \end{bmatrix} N \\
&= \begin{bmatrix} 1 & \mathbf{0}^T \\ \mathbf{0} & A \end{bmatrix} RS \begin{bmatrix} \cosh(\varphi) & 0 & 0 & \sinh(\varphi) \\ 0 & 1 & 0 & 0 \\ 0 & 0 & 1 & 0 \\ \sinh(\varphi) & 0 & 0 & \cosh(\varphi) \end{bmatrix} S^{-1}R^{-1} \\
&= \begin{bmatrix} 1 & \mathbf{0}^T \\ \mathbf{0} & A \end{bmatrix}\begin{bmatrix} 1 & \mathbf{0}^T \\ \mathbf{0} & Q \end{bmatrix} \begin{bmatrix} \cosh(\varphi) & 0 & 0 & \sinh(\varphi) \\ 0 & 1 & 0 & 0 \\ 0 & 0 & 1 & 0 \\ \sinh(\varphi) & 0 & 0 & \cosh(\varphi) \end{bmatrix}\begin{bmatrix} 1 & \mathbf{0}^T \\ \mathbf{0} & Q^{-1} \end{bmatrix} \\
&= \begin{bmatrix} 1 & \mathbf{0}^T \\ \mathbf{0} & A_1 \end{bmatrix} \begin{bmatrix} \cosh(\varphi) & 0 & 0 & \sinh(\varphi) \\ 0 & 1 & 0 & 0 \\ 0 & 0 & 1 & 0 \\ \sinh(\varphi) & 0 & 0 & \cosh(\varphi) \end{bmatrix}\begin{bmatrix} 1 & \mathbf{0}^T \\ \mathbf{0} & A_2 \end{bmatrix},
\end{aligned}
$$

where we have set $A_1 = AQ$ and $A_2 = Q^{-1} \in SO(3)$. Taking the determinant, we have

$$1 = \det(\Lambda) = \det(A_1)(\cosh^2(\varphi) - \sinh^2(\varphi))\det(A_2) = \det(A_1),$$

so that $A_1 \in SO(3)$ as desired. $\qquad\square$

216   *Symmetry and Quantum Mechanics*

Corollary 9.14. *The restricted Lorentz group $SO^+(1,3)$ is connected.*

Proof. The proposition shows that there is a continuous and surjective mapping of topological spaces $SO(3) \times \mathbb{R} \times SO(3) \to SO^+(1,3)$ defined by $(A_1, \varphi, A_2) \mapsto \Lambda(A_1, \varphi, A_2)$. Since continuous maps preserve connectedness, it follows that $SO^+(1,3)$ is the connected component of the identity of $O(1,3)$. $\qquad\square$

9.3 $SL_2(\mathbb{C})$ is the universal cover of $SO^+(1,3)$

In Chapter 2, we saw that the special unitary group $SU(2)$ is the universal double cover of the rotation group $SO(3)$. Moreover, we exhibited the double cover $f \colon SU(2) \to SO(3)$ explicitly via the conjugation action of $SU(2)$ on its Lie algebra, $\mathfrak{su}(2) = iH_0(2)$, the real vector space of 2×2 traceless, skew-Hermitian matrices. Recall that we identified $iH_0(2)$ with the Euclidean space $(\mathbb{R}^3, \cdot)$ via the mapping $\mathbf{e}_j \mapsto \frac{1}{2i}\sigma_j$. Explicitly:

$$(x_1, x_2, x_3) \mapsto \frac{1}{2i} \left[\begin{array}{cc} x_3 & x_1 - ix_2 \\ x_1 + ix_2 & -x_3 \end{array} \right].$$

In terms of this identification, the conjugation action of $B \in SU(2)$ on $iH_0(2)$ yields an element $f(B) \in SO(3)$, so that we obtain the rotation action of $SO(3)$ on Euclidean space from the conjugation action of $SU(2)$ on its Lie algebra (see Section 2.4).

In the previous section, we saw that $SO(3)$ is a subgroup of $SO^+(1,3)$, so it is natural to ask whether the double cover f extends to a double cover $\tilde{f}$ of the restricted Lorentz group? It would be especially nice to describe $\tilde{f}$ in such a way as to recover the $SO^+(1,3)$-action on Minkowski space from a natural action of the covering group. To this end, we observe that $SU(2)$ is a 3-dimensional subgroup of the 6-dimension special linear group $SL(2, \mathbb{C})$ studied in Section 3.4.2. Since $SO^+(1,3)$ is also 6-dimensional, this is a promising start.

Note that as a subgroup of $SO^+(1,3)$, the rotation group $SO(3)$ acts on the spatial part of Minkowski space, which is isomorphic to $\mathbb{R}^3$ with the *negative dot product*:

$$\langle (0, x_1, x_2, x_3), (0, \tilde{x}_1, \tilde{x}_2, \tilde{x}_3) \rangle_{\mathrm{M}} = -(x_1\tilde{x}_1 + x_2\tilde{x}_2 + x_3\tilde{x}_3).$$

To account for this change of signs, we consider the conjugation action of $SU(2)$ on $H_0(2)$, the space of traceless Hermitian matrices rather than skew-Hermitian. Since $H_0(2)$ is obtained from $iH_0(2)$ through multiplication by i, the positive definite inner product on $iH_0(2)$ is transformed into a negative definite inner product on $H_0(2)$, just as we would like. Explicitly, the negative

Toward a Relativistic Theory

definite inner product on $H_0(2)$ is determined by taking the matrices $\frac{1}{2}\sigma_j$ to be orthogonal basis vectors, each with squared "length" -1.

Now $H_0(2)$ is a 3-dimensional subspace of the 4-dimensional vector space $H(2)$ consisting of all 2×2 Hermitian matrices. However, the conjugation action of $SU(2)$ on $H_0(2)$ does not obviously extend to an $SL(2,\mathbb{C})$-action on $H(2)$, since if $C \in SL(2,\mathbb{C})$ and $X \in H(2)$, it does not follow that CXC^{-1} is Hermitian. But note that for $B \in SU(2)$, we have $B^{-1} = B^\dagger$, so that $BXB^{-1} = BXB^\dagger$, and the latter formula does extend to an $SL(2,\mathbb{C})$-action on $H(2)$:

$$(CXC^\dagger)^\dagger = (C^\dagger)^\dagger X^\dagger C^\dagger = CXC^\dagger.$$

To establish the connection with Minkowski space, we alter our earlier identification of $(\mathbb{R}^3, \cdot)$ with $iH_0(2)$ to obtain an identification of $(\mathbb{R}^4, \langle, \rangle_{\mathrm{M}})$ with $H(2)$:

$$x = (x_0, x_1, x_2, x_3) \mapsto \frac{1}{2}\left[\begin{array}{cc} x_0 + x_3 & x_1 - ix_2 \\ x_1 + ix_2 & x_0 - x_3 \end{array}\right] =: H(x). \tag{9.7}$$

Exercise 9.15. *Show that for all x in Minkowski space, we have* $4\det(H(x)) = \langle x, x\rangle_{\mathrm{M}}$.

Now observe that the $SL(2,\mathbb{C})$-action on $H(2)$ preserves the determinant:

$$\det(CXC^\dagger) = \det(C)\det(X)\det(C^\dagger) = \det(X).$$

Via the identification with Minkowski space, we see that each $C \in SL(2,\mathbb{C})$ determines a linear transformation $\tilde{f}(C)\colon \mathbb{R}^4 \to \mathbb{R}^4$ that preserves the Minkowski inner product, so that $\tilde{f}(C) \in O(1,3)$. Explicitly:

$$\tilde{f}(C) = \Lambda \iff H(\Lambda x) = CH(x)C^\dagger \quad \text{for all } x \in \mathbb{R}^4. \tag{9.8}$$

Thus, we obtain a continuous homomorphism $\tilde{f}\colon SL(2,\mathbb{C}) \to O(1,3)$. Since $SL(2,\mathbb{C})$ is connected (see the end of Section 3.4.2), it follows that the image of $\tilde{f}$ must be contained in the connected component of the identity, which by corollary 9.14 is the restricted Lorentz group $SO^+(1,3)$. Moreover, exercise 2.37 from Chapter 2 shows that the kernel of $\tilde{f}$ is $\pm I$, so $\tilde{f}$ induces an isomorphism of $SL(2,\mathbb{C})/\{\pm I\}$ with a subgroup of $SO^+(1,3)$. In fact, $\tilde{f}$ is surjective. To show this, it suffices by Proposition 9.13 to prove that the image contains the spatial rotations and the boosts in the z-direction. The subgroup $SU(2) \subset SL(2,\mathbb{C})$ maps onto the group of spatial rotations $SO(3) \subset SO^+(1,3)$, and the next example shows how to obtain the boosts as well.

Example 9.16. *Consider the matrix*

$$C = \left[\begin{array}{cc} e^a & 0 \\ 0 & e^{-a} \end{array}\right] \in SL(2,\mathbb{C}).$$

218 *Symmetry and Quantum Mechanics*

To determine $\tilde{f}(C) \in SO^+(1,3)$, we make the computation

$$C \begin{bmatrix} x_0 + x_3 & x_1 - ix_2 \\ x_1 + ix_2 & x_0 - x_3 \end{bmatrix} C^\dagger = \begin{bmatrix} e^{2a}(x_0 + x_3) & x_1 - ix_2 \\ x_1 + ix_2 & e^{-2a}(x_0 - x_3) \end{bmatrix}$$

$$=: \begin{bmatrix} y_0 + y_3 & y_1 - iy_2 \\ y_1 + iy_2 & y_0 - x_3 \end{bmatrix},$$

where $y_1 = x_1, y_2 = x_2$ and y_0 and y_3 are obtained by solving the system

$$y_0 + y_3 = e^{2a}(x_0 + x_3)$$
$$y_0 - y_3 = e^{-2a}(x_0 + x_3).$$

Adding and subtracting the two equations yields the solution

$$y_0 = \cosh(2a)x_0 + \sinh(2a)x_3$$
$$y_3 = \sinh(2a)x_0 + \cosh(2a)x_3.$$

Thus, we see that the map sending x to y is

$$\tilde{f}(C) = \begin{bmatrix} \cosh(2a) & 0 & 0 & \sinh(2a) \\ 0 & 1 & 0 & 0 \\ 0 & 0 & 1 & 0 \\ \sinh(2a) & 0 & 0 & \cosh(2a) \end{bmatrix},$$

which describes a boost with rapidity $\varphi = 2a$ in z-direction.

Thus, we have a surjective homomorphism $\tilde{f} \colon SL(2, \mathbb{C}) \to SO^+(1,3)$ with kernel $\{\pm I\}$, showing that the simply connected group $SL(2, \mathbb{C})$ is the universal double cover of the restricted Lorentz group, extending the double cover $f \colon SU(2) \to SO(3)$ from Chapter 2. Taking the derivative at the identity of $SL(2, \mathbb{C})$, we obtain an isomorphism of Lie algebras $D\tilde{f} \colon \mathfrak{sl}_2(\mathbb{C}) \to \mathfrak{so}^+(1,3)$.

Exercise 9.17. *Show that the isomorphism $D\tilde{f}$ is given by:*

$$D\tilde{f}\left(\frac{1}{2i}\sigma_j\right) = \mathcal{L}_j \quad \text{and} \quad D\tilde{f}\left(\frac{1}{2}\sigma_j\right) = \mathcal{K}_j.$$

Conclude that $D\tilde{f}$ preserves adjoints in the sense that $D\tilde{f}(X^\dagger) = (D\tilde{f}(X))^T$ for all $X \in \mathfrak{sl}_2(\mathbb{C})$.

Proposition 9.18. *The universal double cover $\tilde{f} \colon SL(2, \mathbb{C}) \to SO^+(1,3)$ preserves adjoints: if $\tilde{f}(C) = \Lambda$, then $\tilde{f}(C^\dagger) = \Lambda^T$.*

Proof. We first prove that the claim holds if $C = \exp(X)$ for some $X \in \mathfrak{sl}_2(\mathbb{C})$.

Toward a Relativistic Theory

By proposition 5.20 we know that $\tilde{f}(\exp(X)) = \exp(D\tilde{f}(X))$. Hence,

$$
\begin{aligned}
\tilde{f}(C^\dagger) &= \tilde{f}(\exp(X)^\dagger) \\
&= \tilde{f}(\exp(X^\dagger)) \\
&= \exp(D\tilde{f}(X^\dagger)) \\
&= \exp((D\tilde{f}(X))^T) \\
&= \exp(D\tilde{f}(X))^T \\
&= \tilde{f}(\exp(X))^T \\
&= \tilde{f}(C)^T.
\end{aligned}
$$

But since $SL(2,\mathbb{C})$ is connected, it follows from [11, corollary 3.47] that every $C \in SL(2,\mathbb{C})$ may be written in the form[3]

$$
C = \exp(X_1)\exp(X_2)\cdots\exp(X_m)
$$

for some $X_j \in \mathfrak{sl}_2(\mathbb{C})$. We then have

$$
\begin{aligned}
\tilde{f}(C^\dagger) &= \tilde{f}(\exp(X_m)^\dagger \exp(X_{m-1})^\dagger \cdots \exp(X_1)^\dagger) \\
&= \tilde{f}(\exp(X_m)^\dagger)\tilde{f}(\exp(X_{m-1})^\dagger)\cdots\tilde{f}(\exp(X_1)^\dagger) \\
&= \tilde{f}(\exp(X_m))^T\tilde{f}(\exp(X_{m-1}))^T\cdots\tilde{f}(\exp(X_1))^T \\
&= (\tilde{f}(\exp(X_1))\tilde{f}(\exp(X_2))\cdots\tilde{f}(\exp(X_m)))^T \\
&= (\tilde{f}(\exp(X_1)\exp(X_2))\cdots\exp(X_m)))^T \\
&= \tilde{f}(C)^T.
\end{aligned}
$$

$\square$

As a first attempt at understanding the physical significance of the double cover $\tilde{f}$, consider the following idea. Let $\Lambda \in SO^+(1,3)$ be the Lorentz transformation that sends P's basis for Minkowski space to M's (thus, Λ^{-1} is the matrix sending P's coordinates x to M's coordinates x'). Then there exists a matrix $C \in SL(2,\mathbb{C})$, unique up to sign, such that $\tilde{f}(C) = \Lambda$. Now suppose that $|\phi\rangle \in (\mathbb{C}^2, \cdot)$ represents the spin-state of an electron at rest at P's origin, as in Chapter 2. Then it is tempting to suppose that $C|\phi\rangle$ represents the spin-state of the electron for M. After all, when $\Lambda = A$ is a rotation in $SO(3)$, then $C = B \in SU(2)$, and $B|\phi\rangle$ *does* represent the spin-state for M. However, if Λ is not a rotation, then C is not unitary, and $C|\phi\rangle$ does not have unit norm. Thus, it does not represent a spin-state for M.

This actually shouldn't surprise us: in the context of Chapter 2, the electron was at rest with respect to both M and P, and that is no longer the case. In fact, the electron is moving away from M at constant velocity $-\mathbf{v}$, so M's description of the electron will necessarily involve position and momentum in addition to spin. Hence, the correct question to ask is: suppose that P

---

[3]In fact, the exponential map $\exp\colon \mathfrak{sl}_2(\mathbb{C}) \to SL(2,\mathbb{C})$ is almost surjective: for every $C \in SL(2,\mathbb{C})$, either C or $-C$ is in the image (see [8, proposition 6.14]).

220 *Symmetry and Quantum Mechanics*

describes an electron via a spinor-valued wavefunction $\psi(t, \mathbf{x})$. What is M's wavefunction for the same particle if she is rotated and moving uniformly with respect to P according to a restricted Lorentz transformation Λ? We will answer this question for the case of a free electron in the next section.

The restricted Lorentz group $SO^+(1,3)$ is the relativistic analogue of the Euclidean group $\mathbb{R}^3 \rtimes SO(3)$, consisting of Euclidean boosts and spatial rotations. We can obtain a relativistic analogue of the Galilean group $\mathcal{G}$ by enlarging $SO^+(1,3)$ to include the space- and time-translations:

$$\mathcal{P}_0 := \mathbb{R}^4 \rtimes SO^+(1,3).$$

This is the *restricted Poincaré group*, which is the connected component of the full *Poincaré group* $\mathcal{P} := \mathbb{R}^4 \rtimes O(1,3)$. The restricted Poincaré group acts on Minkowski space as follows:

$$(w, \Lambda) \star x := \Lambda x + w.$$

For $w = (cs, \mathbf{w})$, acting by the inverse of (w, Λ) yields M's coordinates x' from P's coordinates x when M is rotated and moving away from P with constant velocity starting at $\mathbf{x} = \mathbf{w}$ at times $t = s$ and $t' = 0$ (see Figure 9.3):

$$x' = (w, \Lambda)^{-1} \star x = \Lambda^{-1}(x - w).$$

In order to make sense of the statement that $t = s$ when $t' = 0$, imagine that the point with coordinates $\mathbf{w}$ in P's reference frame is visibly marked. Then M sets her watch to $t' = 0$ and flashes a light just as she passes the marked point. P sets his watch to $t = \frac{|\mathbf{w}|}{c} + s$ just as the light from M reaches his eye, thereby accounting for the finite time it took the light signal to reach him.

9.4   The Dirac equation

We now try to formulate the quantum mechanics of a single relativistic particle in terms of Minkowski space. To that end, we consider the time-evolution of an initial wavefunction $\psi(0, \mathbf{x}) \in L^2(\mathbb{R}^3)$ as a complex-valued function of four variables $\psi(x_0, \mathbf{x}) = \psi(x)$, where $x_0 = ct$.

In the non-relativistic setting, the time-evolution is governed by the Schrödinger equation

$$i\hbar \frac{\partial \psi}{\partial t} = \mathcal{H}_S \psi = -\frac{\hbar^2}{2m} \Delta \psi + V(\mathbf{x})\psi,$$

where the Schrödinger Hamiltonian $\mathcal{H}_S$ is obtained by quantizing the non-relativistic relationship between energy and momentum: $E = \frac{|\mathbf{p}|^2}{2m} + V(\mathbf{x})$. To

Toward a Relativistic Theory 221

discover the correct time-evolution in the relativistic context, we should instead start with the relativistic relationship between energy and momentum.

In special relativity, the energy-momentum of a particle is represented by a 4-vector $p = (\frac{E}{c}, \mathbf{p})$ in Minkowski space. For a particle of mass m, the squared Minkowski length is given by the constant $m^2 c^2$:

$$m^2 c^2 = \langle p, p \rangle_{\mathrm{M}} = \frac{E^2}{c^2} - |\mathbf{p}|^2.$$

This leads to the following relativistic relationship between energy and momentum for a free particle of mass m:

$$E^2 = c^2 |\mathbf{p}|^2 + m^2 c^4, \tag{9.9}$$

so that a particle has a *rest-energy* of $E = mc^2$ in its own rest-frame, where its ordinary spatial momentum is $\mathbf{p} = 0$. Note that (using the binomial theorem), we do recover the ordinary kinetic energy (shifted by mc^2) in the non-relativistic regime where $|\mathbf{p}| = m|\mathbf{v}|$ is small compared to mc:

$$
\begin{aligned}
E &= \sqrt{m^2 c^4 + c^2 |\mathbf{p}|^2} \\
&= mc^2 \left(1 + \left(\frac{|\mathbf{p}|}{mc} \right)^2 \right)^{\frac{1}{2}} \\
&= mc^2 \left(1 + \frac{1}{2} \left(\frac{|\mathbf{p}|}{mc} \right)^2 - \frac{1}{8} \left(\frac{|\mathbf{p}|}{mc} \right)^4 + \cdots \right) \\
&\approx mc^2 + \frac{|\mathbf{p}|^2}{2m}.
\end{aligned}
$$

To quantize the relation (9.9), we replace E with $\mathcal{H}$ and $\mathbf{p}$ with $\hat{\mathbf{p}} = \frac{\hbar}{i} \nabla$ to obtain

$$\mathcal{H}^2 = -c^2 \hbar^2 \Delta + m^2 c^4. \tag{9.10}$$

Recalling that $\mathcal{H} = i\hbar \frac{\partial}{\partial t} = ic\hbar \frac{\partial}{\partial x_0}$, we are led to the *Klein-Gordan equation*:

$$\frac{\partial^2 \psi}{\partial x_0^2} = \left(\Delta - \frac{m^2 c^2}{\hbar^2} \right) \psi. \tag{9.11}$$

This equation is Lorentz invariant, but it is second order in time, so that (among other problems) an initial state $\psi(0, \mathbf{x})$ does not uniquely determine the time-evolution. Dirac's brilliant insight was to retain a first-order equation by explicitly constructing a square root of the right-hand side of equation (9.10).

To that end, write $\mathcal{H}_D = \frac{c\hbar}{i} (\alpha_1 \partial_1 + \alpha_2 \partial_2 + \alpha_3 \partial_3) + mc^2 \beta$, where $\partial_j := \frac{\partial}{\partial x_j}$. We wish to determine the quantities α_j and β so that $\mathcal{H}_D^2 = -c^2 \hbar^2 \Delta + m^2 c^4$.

222 *Symmetry and Quantum Mechanics*

Exercise 9.19. *Show there is no solution to this problem with $\alpha_j, \beta \in \mathbb{C}$.*

But suppose that we allow for the possibility that the α_j and β are non-commuting quantities that commute with the partial derivatives? Then squaring $\frac{i}{c\hbar}\mathcal{H}_D$ yields

$$
\left(\alpha_1\partial_1 + \alpha_2\partial_2 + \alpha_3\partial_3 + \frac{imc}{\hbar}\beta\right)^2 = \alpha_1^2\partial_1^2 + \alpha_2^2\partial_2^2 + \alpha_3^2\partial_3^2 - \frac{m^2c^2}{\hbar^2}\beta^2
$$
$$
+(\alpha_1\alpha_2 + \alpha_2\alpha_1)\partial_1\partial_2
$$
$$
+(\alpha_2\alpha_3 + \alpha_3\alpha_2)\partial_2\partial_3
$$
$$
+(\alpha_1\alpha_3 + \alpha_3\alpha_1)\partial_1\partial_3
$$
$$
+\frac{imc}{\hbar}(\alpha_1\beta + \beta\alpha_1)\partial_1
$$
$$
+\frac{imc}{\hbar}(\alpha_2\beta + \beta\alpha_2)\partial_2
$$
$$
+\frac{imc}{\hbar}(\alpha_3\beta + \beta\alpha_3)\partial_3.
$$

Hence, our problem will be solved if the α_j and β satisfy the relations

$$
\alpha_j^2 = 1, \qquad \alpha_i\alpha_j = -\alpha_j\alpha_i \quad (i \neq j) \tag{9.12}
$$
$$
\beta^2 = 1, \qquad \beta\alpha_j = -\alpha_j\beta. \tag{9.13}
$$

Note that the 2×2 Pauli matrices σ_j satisfy the relations (9.12) involving only the α's. But there is no way to choose an additional 2×2 matrix β so as to satisfy the remaining relations (9.13). To make room for β without moving too far away from the Pauli matrices, consider the 4×4 block matrices

$$
\alpha_j := \begin{bmatrix} -\sigma_j & 0 \\ 0 & \sigma_j \end{bmatrix}, \qquad \beta := \begin{bmatrix} 0 & I \\ I & 0 \end{bmatrix}. \tag{9.14}
$$

Exercise 9.20. *Check that these α_j and β satisfy the necessary relations (9.12–9.13).*

By specifying the *Dirac Hamiltonian*:

$$
\mathcal{H}_D = \frac{c\hbar}{i}(\alpha_1\partial_1 + \alpha_2\partial_2 + \alpha_3\partial_3) + mc^2\beta,
$$

we are forced to consider $\mathbb{C}^4$-valued wavefunctions $\psi(x) \in L^2(\mathbb{R}^4, \mathbb{C}^4)$, although the physical interpretation of the space $\mathbb{C}^4$ is not yet clear. In any case, the time-evolution of an initial wavefunction $\psi(0, \mathbf{x})$ is now governed by the *Dirac equation*:

$$
ic\hbar\partial_0\psi(x) = \mathcal{H}_D\psi(x).
$$

Note that if $\psi(x)$ is a solution to the Dirac equation, then each of its four components is a solution to the Klein-Gordan equation (9.11):

$$\begin{aligned}
\partial_0^2 \psi &= \partial_0(-\frac{i}{c\hbar}\mathcal{H}_D\psi) \\
&= -\frac{i}{c\hbar}\mathcal{H}_D(\partial_0\psi) \\
&= \left(-\frac{i}{c\hbar}\right)^2 \mathcal{H}_D^2\psi \\
&= -\frac{1}{c^2\hbar^2}(-c^2\hbar^2\Delta + m^2c^4)\psi \\
&= \left(\Delta - \frac{m^2c^2}{\hbar^2}\right)\psi.
\end{aligned}$$

It will be helpful to reformulate the Dirac equation so that the time and space variables appear in a more symmetric way. We begin by re-expressing the Dirac equation as a kernel condition $\widetilde{\mathcal{H}}_D\psi(x) = 0$, where

$$\begin{aligned}
\widetilde{\mathcal{H}}_D &:= ic\hbar\partial_0 - \mathcal{H}_D \\
&= ic\hbar\partial_0 - \frac{c\hbar}{i}(\alpha_1\partial_1 + \alpha_2\partial_2 + \alpha_3\partial_3) - mc^2\beta \\
&= ic\hbar(\partial_0 + \alpha_1\partial_1 + \alpha_2\partial_2 + \alpha_3\partial_3) - mc^2\beta.
\end{aligned}$$

In order to put the time and space variables on an even more equal footing, multiply on the left by β and use the fact that $\beta^2 = I$:

$$\begin{aligned}
\beta\widetilde{\mathcal{H}}_D &= ic\hbar(\beta\partial_0 + \beta\alpha_1\partial_1 + \beta\alpha_2\partial_2 + \beta\alpha_3\partial_3) - mc^2 \\
&= ic\hbar(\gamma^0\partial_0 + \gamma^1\partial_1 + \gamma^2\partial_2 + \gamma^3\partial_3) - mc^2.
\end{aligned}$$

Here we have introduced the *gamma matrices*

$$\gamma^0 := \beta = \begin{bmatrix} 0 & I \\ I & 0 \end{bmatrix}, \qquad \gamma^j := \beta\alpha_j = \begin{bmatrix} 0 & \sigma_j \\ -\sigma_j & 0 \end{bmatrix}.$$

Since the matrix β is invertible, we haven't changed the kernel, so we may write the Dirac equation (and hence the time-evolution of a $\mathbb{C}^4$-valued wavefunction) as follows:

$$i\hbar\sum_{\mu=0}^{3}\gamma^\mu\partial_\mu\psi = mc\psi. \qquad (9.15)$$

In order to interpret the Dirac equation, we need to connect the copy of $\mathbb{C}^4$ on which the gamma matrices act to something that we have seen before. The appearance of the Pauli matrices suggests that a representation of the Lie algebra $\mathfrak{sl}_2(\mathbb{C})$ may be lurking about. Note that $\mathfrak{sl}_2(\mathbb{C})$ may be viewed in two ways:

224   *Symmetry and Quantum Mechanics*

($\mathbb{R}$) as the 6-dimensional *real* Lie algebra of the Lie group $SL(2,\mathbb{C})$:

$$\mathfrak{sl}_2(\mathbb{C}) = \text{span}_{\mathbb{R}} \left\{ \frac{1}{2i}\sigma_j, \frac{1}{2}\sigma_j \right\}$$

($\mathbb{C}$) as the 3-dimensional *complex* Lie algebra obtained by complexifying the Lie algebra of $SU(2)$:

$$\mathfrak{sl}_2(\mathbb{C}) = \mathfrak{su}(2)_{\mathbb{C}} = \text{span}_{\mathbb{C}} \left\{ \frac{1}{2i}\sigma_j \right\}.$$

Viewed as a complex Lie algebra, $\mathfrak{sl}_2(\mathbb{C})$ has a unique 2-dimensional irreducible representation[4]:

$$\pi_{\frac{1}{2}}\left((a+bi)\frac{1}{2i}\sigma_j \right) = (a+bi)\frac{1}{2i}\sigma_j = \frac{a}{2i}\sigma_j + \frac{b}{2}\sigma_j.$$

This is the representation of $\mathfrak{sl}_2(\mathbb{C})$ (now viewed as a real Lie algebra) induced by the defining representation of $SL(2,\mathbb{C})$ on $\mathbb{C}^2$. But we obtain a distinct irreducible representation of $SL(2,\mathbb{C})$ by acting on $\mathbb{C}^2$ via the inverse conjugate transpose:

$$\pi_{\frac{1}{2}}^*(C) := C^{\dagger-1} \qquad \text{for } C \in SL(2,\mathbb{C}).$$

To determine the corresponding representation of $\mathfrak{sl}_2(\mathbb{C})$, we differentiate the 1-parameter subgroup generated by an element $(a+bi)\frac{\sigma_j}{2i}$:

$$\frac{d}{d\theta}\left(\exp\left((a+bi)\frac{\theta}{2i}\sigma_j \right)^{\dagger-1} \right) |_{\theta=0} = \frac{d}{d\theta}\exp\left((a-bi)\frac{\theta}{2i}\sigma_j \right) |_{\theta=0}$$

$$= (a-bi)\frac{\sigma_j}{2i}.$$

Hence, as a representation of the real Lie algebra $\mathfrak{sl}_2(\mathbb{C})$ we have

$$\pi_{\frac{1}{2}}^*\left((a+bi)\frac{1}{2i}\sigma_j \right) = (a-bi)\frac{1}{2i}\sigma_j = \frac{a}{2i}\sigma_j - \frac{b}{2}\sigma_j.$$

Note that the representations $\pi_{\frac{1}{2}}$ and $\pi_{\frac{1}{2}}^*$ agree when restricted to the real Lie subalgebra $\mathfrak{su}(2) = \text{span}_{\mathbb{R}}\left\{ \frac{1}{2i}\sigma_j \right\} \subset \mathfrak{sl}_2(\mathbb{C})$, but they differ by a sign on the Hermitian generators $\frac{1}{2}\sigma_j$.

Now observe that the matrices α_j from (9.14) are the images of the Pauli matrices $\sigma_j \in \mathfrak{sl}_2(\mathbb{C})$ under the direct sum of our representations:

$$\pi_{\frac{1}{2}}^* \oplus \pi_{\frac{1}{2}}(\sigma_j) = \begin{bmatrix} -\sigma_j & 0 \\ 0 & \sigma_j \end{bmatrix} = \alpha_j.$$

---

[4]In order to simplify the notation, we write π_s for the spin-s representation of $SU(2)$ as well as the induced representation of $\mathfrak{sl}_2(\mathbb{C})$ obtained by differentiation and complexification. In Chapter 5 we denoted this representation by $(D\pi_s)_{\mathbb{C}}$.

Toward a Relativistic Theory

This suggests that we should view the copy of $\mathbb{C}^4$ in the Dirac equation (9.15) as the $SL(2,\mathbb{C})$-representation $\Pi := \pi^*_{\frac{1}{2}} \oplus \pi_{\frac{1}{2}}$. When restricted to the subgroup $SU(2) \subset SL(2,\mathbb{C})$, we obtain two copies of the spin-$\frac{1}{2}$ representation of $SU(2)$ describing the spin of a non-relativistic electron as in Chapter 2. For this reason, the 4-dimensional representation Π of $SL(2,\mathbb{C})$ is called the *Dirac spinor representation*. The role of the matrix β is to change basis on $\mathbb{C}^4$, yielding an isomorphic representation:

$$\begin{array}{ccc}
SL(2,\mathbb{C}) \times \mathbb{C}^4 & \xrightarrow{\;\;\Pi\;\;} & \mathbb{C}^4 \\
{\scriptstyle \mathrm{id} \times \beta} \downarrow & & \downarrow {\scriptstyle \beta} \\
SL(2,\mathbb{C}) \times \mathbb{C}^4 & \xrightarrow{\;\beta\Pi\beta^{-1}\;} & \mathbb{C}^4.
\end{array}$$

The top row corresponds to our original form of the Dirac equation with the alpha matrices:

$$0 = \widetilde{\mathcal{H}_D}\psi.$$

Applying the isomorphism β yields the Dirac equation (9.15) with the gamma matrices, corresponding to the bottom row:

$$0 = (\beta\widetilde{\mathcal{H}_D}\beta^{-1})\beta\psi = \beta\widetilde{\mathcal{H}_D}\psi.$$

To understand the meaning of the Dirac equation, write $\psi = (\psi_L, \psi_R)$ where $\psi_L, \psi_R \colon \mathbb{R}^4 \to \mathbb{C}^2$. As mentioned above, if we restrict the Dirac spinor representation to the subgroup $SU(2)$, we obtain the direct sum of two copies of the spin-$\frac{1}{2}$ representation. So it seems that we might be describing two spin-$\frac{1}{2}$ particles instead of just one! But the situation becomes a bit clearer if we write out the Dirac equation (9.15) explicitly in terms of ψ_L and ψ_R. Writing $\boldsymbol{\sigma} := (\sigma_1, \sigma_2, \sigma_3)$, we have

$$\begin{aligned}
\frac{mc}{i\hbar}\psi_L &= \partial_0\psi_R + (\boldsymbol{\sigma}\cdot\nabla)\psi_R \\
\frac{mc}{i\hbar}\psi_R &= \partial_0\psi_L - (\boldsymbol{\sigma}\cdot\nabla)\psi_L.
\end{aligned}$$

Hence, ψ_R completely determines ψ_L and vice-versa, so that there are really only two independent components in a solution to the Dirac equation. The two-component spinors ψ_L and ψ_R are called *Weyl spinors*, and they have a physical interpretation in terms of "left-handed" and "right-handed" particles (see [5, Section 4.3] and [14, Chapters 36–37]).

We are now ready to tackle the main question of this chapter. Suppose that P describes an electron via the Dirac spinor wavefunction $\psi(x)$. Also, suppose that observer M is rotated and moving uniformly according to a Lorentz transformation Λ, in such a way that at corresponding times $t = s$ and $t' = 0$ she is located at $\mathbf{x} = \mathbf{w}$. Then M's space-time coordinates x' are obtained from P's via the transformation $x' = \Lambda^{-1}(x - w)$, where $w = (cs, \mathbf{w})$. What is the wavefunction $\psi'(x)$ that M uses to describe the same electron?

226 *Symmetry and Quantum Mechanics*

This is the relativistic analogue of our question from section 9.1, which we answered by constructing a $\mathcal{G}^\theta$-action on wavefunctions, where $\mathcal{G}^\theta$ is a central extension of the Galilean group $\mathcal{G}$. Since the restricted Poincaré group $\mathcal{P}_0 = \mathbb{R}^4 \rtimes SO^+(1,3)$ is the relativistic analogue of the Galilean group, this suggests that we should try to find a $\mathcal{P}_0$-action on Dirac spinor wavefunctions. Luckily, this is easier than in the Galilean case, since we do not need to pass to a central extension, but only to the universal cover $\widetilde{\mathcal{P}_0} := \mathbb{R}^4 \rtimes SL(2,\mathbb{C})$.

In fact, we will see that the following $\widetilde{\mathcal{P}_0}$-action is the one we want:

$$(w, C) \star \boldsymbol{\psi}(x) := \Pi(C)\boldsymbol{\psi}(\tilde{f}(C^{-1})(x - w)),$$

where $\tilde{f} \colon SL(2,\mathbb{C}) \to SO^+(1,3)$ is the universal double cover from Section 9.3. Using this action, we may describe M's wavefunction $\boldsymbol{\psi}'(x)$ as follows: choose an element $C \in SL(2,\mathbb{C})$ such that $\tilde{f}(C) = \Lambda \in SO^+(1,3)$, noting that C is well defined up to a sign. Then $\boldsymbol{\psi}'$ is obtained from $\boldsymbol{\psi}$ by acting with the inverse of (w, C):

$$\boldsymbol{\psi}'(x) = (w, C)^{-1} \star \boldsymbol{\psi}(x) = \Pi(C^{-1})\boldsymbol{\psi}(\Lambda x + w).$$

To provide some justification for this assertion, we will show in Theorem 9.22 that if $\boldsymbol{\psi}(x)$ is a solution to the Dirac equation, then so is $\boldsymbol{\psi}'(x)$.

Following the strategy in [5, Section 4.2], we will need to make use of a certain matrix identity expressing the effect of letting $\Lambda \in SO^+(1,3)$ act on the vector of gamma matrices $\gamma := (\gamma^0, \gamma^1, \gamma^2, \gamma^3)$ to obtain a new vector of matrices:

$$(\Lambda\gamma)^\rho := \sum_{\nu=0}^{3} \Lambda_{\rho\nu}\gamma^\nu \quad \text{for } \rho = 0, 1, 2, 3.$$

Beware of a possible confusion: $\Lambda\gamma^\nu$ would denote the product of two 4×4 matrices, while $(\Lambda\gamma)^\rho$ denotes a linear combination of the four gamma matrices, with coefficients taken from the row of Λ indexed by ρ. The matrix identity that we need is stated in the following proposition.

Proposition 9.21. *Suppose that* $C \in SL(2,\mathbb{C})$ *satisfies* $\tilde{f}(C) = \Lambda \in SO^+(1,3)$. *Then*

$$(\Lambda\gamma)^\rho = \Pi(C^{-1})\gamma^\rho\Pi(C). \tag{9.16}$$

Here, $\Pi = \pi_{\frac{1}{2}}^* \oplus \pi_{\frac{1}{2}}$ *is the Dirac spinor representation of* $SL(2,\mathbb{C})$.

Proof. We begin by looking explicitly at an arbitrary linear combination of gamma matrices,

$$\sum_{\mu=0}^{3} x_\mu\gamma^\mu = \begin{bmatrix} 0 & x_0 I \\ x_0 I & 0 \end{bmatrix} + \sum_{j=1}^{3} \begin{bmatrix} 0 & x_j\sigma_j \\ -x_j\sigma_j & 0 \end{bmatrix}$$

$$= \begin{bmatrix} 0 & H(x) \\ H(gx) & 0 \end{bmatrix},$$

Toward a Relativistic Theory

where $g = \mathrm{diag}(1, -1, -1, -1)$ is the spatial inversion element of the Lorentz group and $H(x)$ is the matrix (9.7) corresponding to $x = (x_0, x_1, x_2, x_3)$ under our identification of Minkowski space with 2×2 Hermitian matrices.

Recall from (9.8) that $\tilde{f}(C) = \Lambda$ means that $H(\Lambda x) = CH(x)C^\dagger$. Since $\tilde{f}(C^\dagger) = \Lambda^T$ by proposition 9.18, we also have $H(\Lambda^T x) = C^\dagger H(x)C$. Finally, using the Lorentz condition $g\Lambda^T = \Lambda^{-1}g$, we see that

$$H(g\Lambda^T x) = H(\Lambda^{-1}gx) = C^{-1}H(gx)(C^{-1})^\dagger.$$

Bringing these relations to our linear combination of gamma matrices, we have

$$
\begin{aligned}
\sum_{\mu=0}^{3} x_\mu \gamma^\mu &= \begin{bmatrix} 0 & H(x) \\ H(gx) & 0 \end{bmatrix} \\
&= \begin{bmatrix} 0 & C^{\dagger-1}H(\Lambda^T x)C^{-1} \\ CH(g\Lambda^T x)C^\dagger & 0 \end{bmatrix} \\
&= \begin{bmatrix} C^{\dagger-1} & 0 \\ 0 & C \end{bmatrix} \begin{bmatrix} 0 & H(\Lambda^T x) \\ H(g\Lambda^T x) & 0 \end{bmatrix} \begin{bmatrix} C^\dagger & 0 \\ 0 & C^{-1} \end{bmatrix} \\
&= \Pi(C) \sum_{\nu=0}^{3} (\Lambda^T x)_\nu \gamma^\nu \Pi(C^{-1}) \\
&= \Pi(C) \sum_{\nu=0}^{3} \left(\sum_{\mu=0}^{3} \Lambda^T_{\nu\mu} x_\mu \right) \gamma^\nu \Pi(C^{-1}) \\
&= \Pi(C) \sum_{\mu=0}^{3} x_\mu \left(\sum_{\nu=0}^{3} \Lambda_{\mu\nu} \gamma^\nu \right) \Pi(C^{-1}) \\
&= \Pi(C) \sum_{\mu=0}^{3} x_\mu (\Lambda\gamma)^\mu \Pi(C^{-1}).
\end{aligned}
$$

Rearranging a bit, we find that

$$\sum_{\mu=0}^{3} x_\mu (\Lambda\gamma)^\mu = \Pi(C^{-1}) \sum_{\mu=0}^{3} x_\mu \gamma^\mu \Pi(C).$$

For each $\rho = 0, 1, 2, 3$ in turn, setting $x_\mu = \delta_{\mu\rho}$ (the Kronecker delta) yields our desired matrix identity:

$$(\Lambda\gamma)^\rho = \Pi(C^{-1})\gamma^\rho \Pi(C).$$

$\square$

Theorem 9.22. *Suppose that $\psi(x)$ is a solution to the Dirac equation:*

$$\sum_{\mu=0}^{3} \gamma^\mu \partial_\mu \psi(x) = \frac{mc}{i\hbar} \psi(x).$$

228   *Symmetry and Quantum Mechanics*

Then for all $(w, C) \in \mathcal{P}_0 = \mathbb{R}^4 \rtimes SO^+(1,3)$, the transformed wavefunction $\psi'(x)$ is also a solution to the Dirac equation, where

$$\psi'(x) = (w, C)^{-1} \star \psi(x) = \Pi(C^{-1})\psi(\Lambda x + w).$$

Proof. We start by computing a single partial derivative using the chain rule:

$$
\begin{aligned}
\partial_\mu \psi'(x) &= \partial_\mu(\psi(\Lambda x + w)) \\
&= \sum_{\rho=0}^{3}(\partial_\rho \psi)(\Lambda x + w)\partial_\mu \left(\sum_{\nu=0}^{3} \Lambda_{\rho\nu} x_\nu + w_\rho \right) \\
&= \sum_{\rho=0}^{3}(\partial_\rho \psi)(\Lambda x + w)\Lambda_{\rho\mu}.
\end{aligned}
$$

We then find that

$$
\begin{aligned}
\sum_{\mu=0}^{3} \gamma^\mu \partial_\mu \psi'(x) &= \sum_{\mu=0}^{3} \gamma^\mu \partial_\mu \left(\Pi(C^{-1})\psi(\Lambda x + w) \right) \\
&= \sum_{\mu=0}^{3} \gamma^\mu \Pi(C^{-1}) \sum_{\rho=0}^{3}(\partial_\rho \psi)(\Lambda x + w)\Lambda_{\rho\mu} \\
&= \sum_{\mu,\rho=0}^{3} \Lambda_{\rho\mu} \gamma^\mu \Pi(C^{-1})(\partial_\rho \psi)(\Lambda x + w) \\
&= \sum_{\rho=0}^{3} (\Lambda\gamma)^\rho \Pi(C^{-1})(\partial_\rho \psi)(\Lambda x + w) \\
&= \sum_{\rho=0}^{3} \Pi(C^{-1})\gamma^\rho \Pi(C)\Pi(C^{-1})(\partial_\rho \psi)(\Lambda x + w) \\
&= \Pi(C^{-1}) \sum_{\rho=0}^{3} \gamma^\rho (\partial_\rho \psi)(\Lambda x + w) \\
&= \Pi(C^{-1})\frac{mc}{i\hbar}\psi(\Lambda x + w) \\
&= \frac{mc}{i\hbar}\psi'(x).
\end{aligned}
$$

Hence, $\psi'(x)$ also satisfies the Dirac equation, as claimed. $\square$

As explained in [5, Section 4.3], the Dirac equation reduces to the Schrödinger equation in the non-relativistic limit of small velocities. Moreover, one can incorporate a Coulomb potential into the Dirac Hamiltonian in order to model the relativistic hydrogen atom, and the resulting energy levels include the fine structure that was put in "by hand" in Section 8.7.1.

Toward a Relativistic Theory

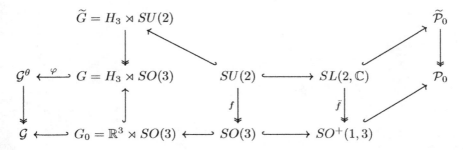

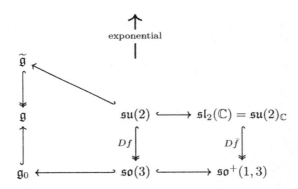

FIGURE 9.6: M and P's farewell diagram, illustrating the relationship between some of the various Lie groups (top) and Lie algebras (bottom) that have featured in the book.

Here we take leave of our observers M and P as they marvel at the success of the Dirac equation, and at the elegance of the story they have discovered about symmetry and quantum mechanics. As a collaborative farewell, they draw Figure 9.6, which serves as a reminder of the Lie groups and algebras they have met in the course of their exploration.

Appendix A

Appendices

A.1 Linear algebra

This section provides a brief summary of necessary concepts and results from linear algebra. For further discussion and proofs, see any linear algebra text, such as [6, 12, 19].

A.1.1 Vector spaces and linear transformations

Definition A.1. *A* vector space *over a field* $\mathbb{F}$ *is a set* V *together with two binary operations, addition (denoted* $+$*) and scalar multiplication (denoted by juxtaposition), satisfying the following axioms for all* $a, b \in \mathbb{F}$ *and* $\mathbf{u}, \mathbf{v}, \mathbf{w} \in V$*:*

1. *there exists* $\mathbf{0} \in V$ *such that* $\mathbf{0} + \mathbf{v} = \mathbf{v} + \mathbf{0} = \mathbf{v}$*;*

2. *for each* $\mathbf{v} \in V$*, there exists* $-\mathbf{v} \in V$ *such that* $\mathbf{v} + (-\mathbf{v}) = \mathbf{0}$*;*

3. $\mathbf{u} + (\mathbf{v} + \mathbf{w}) = (\mathbf{u} + \mathbf{v}) + \mathbf{w}$*;*

4. $\mathbf{v} + \mathbf{w} = \mathbf{w} + \mathbf{v}$*;*

5. $a(b\mathbf{v}) = (ab)\mathbf{v}$*;*

6. $1\mathbf{v} = \mathbf{v}$*;*

7. $(a + b)\mathbf{v} = a\mathbf{v} + b\mathbf{v}$*;*

8. $a(\mathbf{v} + \mathbf{w}) = a\mathbf{v} + a\mathbf{w}$*.*

A nonempty subset $W \subset V$ *is a* vector subspace *if it is closed under addition and scalar multiplication in* V*, in which case* W *is also a vector space.*

Definition A.2. *Let* $S \subset V$ *be a subset of the vector space* V*. Then* S *is* linearly dependent *if there exists a nonempty finite set of vectors* $\mathbf{v}_1, \mathbf{v}_2, \ldots, \mathbf{v}_n \in S$ *and nonzero scalars* $a_1, a_2, \ldots, a_n \in \mathbb{F}$*, such that*

$$a_1\mathbf{v}_1 + a_2\mathbf{v}_2 + \cdots + a_n\mathbf{v}_n = 0.$$

If no such vectors and scalars exist, the set S *is* linearly independent. *The set*

231

232   *Symmetry and Quantum Mechanics*

S spans *the vector space V if for every $\mathbf{v} \in V$ there exist finitely many vectors* $\mathbf{v}_i \in S$ *and scalars $a_i \in \mathbb{F}$ such that*

$$\mathbf{v} = a_1\mathbf{v}_1 + a_2\mathbf{v}_2 + \cdots + a_n\mathbf{v}_n.$$

In this case, $\mathbf{v}$ is said to be a linear combination *of the vectors $\mathbf{v}_i$.*

Definition A.3. *A subset $S \subset V$ is a* basis *of V if S is linearly independent and spans V. This means that every vector $\mathbf{v} \in V$ may be written uniquely as a linear combination of elements from S.*

Theorem A.4. *Every vector space has a basis. Moreover, if S and S' are two bases for the vector space V, then S and S' have the same cardinality. The cardinality of any basis for V is the* dimension *of V, denoted $\dim(V)$.*

Definition A.5. *A vector space V is* finite-dimensional *if it has a finite basis.*

Definition A.6. *Suppose that V and W are vector spaces over $\mathbb{F}$. The* direct sum *of V and W, denoted $V \oplus W$, is the vector space over $\mathbb{F}$ consisting of ordered pairs $(\mathbf{v}, \mathbf{w}) \in V \times W$ with component-wise addition and scalar multiplication. If V and W are finite dimensional, then so is $V \oplus W$, and $\dim(V \oplus W) = \dim(V) + \dim(W)$.*

Definition A.7. *Suppose that V and W are vector spaces over $\mathbb{F}$. A* linear transformation *from V to W is a function $L\colon V \to W$ such that for all $a \in \mathbb{F}$ and $\mathbf{u}, \mathbf{v} \in V$:*

$$L(a\mathbf{u} + \mathbf{v}) = aL\mathbf{u} + L\mathbf{v}.$$

If $V = W$, the transformation L is called a linear operator *on V.*

    The following two subsets associated to L are vector subspaces of V and W respectively:

$$\ker(L) := \{\mathbf{v} \in V \mid L\mathbf{v} = 0\} \subset V \qquad (\textit{kernel of } L);$$

$$\operatorname{im}(L) := \{L\mathbf{v} \in W \mid \mathbf{v} \in V\} \subset W \qquad (\textit{image of } L).$$

The rank *of L is the dimension of the image of L, and the* nullity *of L is the dimension of the kernel of L. Note that L is injective if and only if $\ker(L) = \{\mathbf{0}\}$, and L is surjective if and only if $\operatorname{im}(L) = W$. A bijective linear transformation is an* isomorphism *of vector spaces.*

Theorem A.8 (Rank-Nullity). *Suppose that $L\colon V \to W$ is a linear transformation with V finite dimensional. Then $\ker(L)$ and $\operatorname{im}(L)$ are also finite-dimensional, and*

$$\dim(\ker(L)) + \dim(\operatorname{im}(L)) = \dim(V).$$

Definition A.9. *Suppose that V and W are finite-dimensional vector spaces over $\mathbb{F}$, and let $\beta := \{\mathbf{v}_1, \ldots, \mathbf{v}_n\}$ and $\gamma := \{\mathbf{w}_1, \ldots, \mathbf{w}_m\}$ be bases for V and W respectively. Suppose that $L\colon V \to W$ is a linear transformation. For each*

Appendices
233

$j = 1, \ldots, n$, apply L to the jth basis vector $\mathbf{v}_j$ and express the result as a linear combination of the basis vectors $\mathbf{w}_i$:

$$L\mathbf{v}_j = a_{1j}\mathbf{w}_1 + a_{2j}\mathbf{w}_2 + \cdots + a_{mj}\mathbf{w}_m.$$

Note that the scalars $a_{ij} \in \mathbb{F}$ are uniquely determined. The matrix of L with respect to the bases β and γ is the $m \times n$ matrix

$$[L]_\beta^\gamma := \begin{bmatrix} a_{11} & a_{12} & \cdots & a_{1n} \\ a_{21} & a_{22} & \cdots & a_{2n} \\ \vdots & \vdots & \ddots & \vdots \\ a_{m1} & a_{m2} & \cdots & a_{mn} \end{bmatrix}.$$

If $\mathbf{v} = \sum_{j=1}^n b_j \mathbf{v}_j$ is an arbitrary vector in V, then the column vector of $\mathbf{v}$ in the basis β is the $n \times 1$ matrix

$$[\mathbf{v}]_\beta := \begin{bmatrix} b_1 \\ b_2 \\ \vdots \\ b_n \end{bmatrix}.$$

The correspondence $\mathbf{v} \mapsto [\mathbf{v}]_\beta$ defines an isomorphism $\varphi_\beta \colon V \xrightarrow{\sim} \mathbb{F}^n$.

Definition A.10. Suppose that $A = [a_{ij}]$ is an $m \times n$ matrix and $B = [b_{ij}]$ is an $k \times m$ matrix, both with entries from $\mathbb{F}$. Then the matrix product BA is the $k \times n$ matrix defined by taking the dot product of the rows of B with the columns of A:

$$BA := [c_{ij}] \qquad where \qquad c_{ij} := \sum_{l=1}^m b_{il}a_{lj}.$$

An $n \times n$ matrix A is invertible if there exists an $n \times n$ matrix A^{-1} such that $A^{-1}A = AA^{-1} = I_n$, where I_n is the $n \times n$ identity matrix:

$$I_n := \begin{bmatrix} 1 & 0 & \cdots & 0 \\ 0 & 1 & \cdots & 0 \\ \vdots & \vdots & \ddots & \vdots \\ 0 & 0 & \cdots & 1 \end{bmatrix}.$$

The identity matrix satisfies $I_n A = A I_n = A$ for all $n \times n$ matrices A.

Proposition A.11. Retaining the notation from definition A.9, we have

$$[L\mathbf{v}]_\gamma = [L]_\beta^\gamma [\mathbf{v}]_\beta,$$

the product of the matrix $[L]_\beta^\gamma$ with the column vector $[\mathbf{v}]_\beta$. This result is

234   *Symmetry and Quantum Mechanics*

recorded in the following commutative diagram, where the vertical arrows are the isomorphisms φ_β and φ_γ from definition A.9:

$$
\begin{array}{ccc}
V & \xrightarrow{\;L\;} & W \\
\varphi_\beta \downarrow & & \downarrow \varphi_\gamma \\
\mathbb{F}^n & \xrightarrow{[L]_\beta^\gamma} & \mathbb{F}^m.
\end{array}
$$

Proposition A.12. *Suppose that $L\colon V \to W$ and $M\colon W \to X$ are linear transformations between finite-dimensional vector spaces over $\mathbb{F}$. Choose bases β, γ, δ for V, W, X respectively. Then the matrix of the composition $M \circ L\colon V \to X$ is the product of the matrices of M and L:*

$$[M \circ L]_\beta^\delta = [M]_\gamma^\delta [L]_\beta^\gamma.$$

Definition A.13. *Let V be a finite-dimensional vector space, and suppose that $\beta = \{\mathbf{v}_j\}$ and $\gamma = \{\mathbf{w}_j\}$ are two different bases for V. The change of basis matrix from β to γ is the matrix of the identity operator on V with respect to the bases β and γ:*

$$Q := [\mathrm{id}]_\beta^\gamma.$$

Explicitly, the jth column of Q is the column vector $[\mathbf{v}_j]_\gamma$ expressing the jth basis vector $\mathbf{v}_j \in \beta$ as a linear combination of the vectors $\mathbf{w}_i \in \gamma$. The inverse matrix $Q^{-1} = [\mathrm{id}]_\gamma^\beta$ is the change of basis matrix from γ to β.

Proposition A.14. *Let $L\colon V \to V$ be a linear operator on a finite-dimensional vector space. Suppose that β and γ are two different bases for V, and denote by $[L]_\beta := [L]_\beta^\beta$ the matrix of L with respect to β, and similarly for $[L]_\gamma := [L]_\gamma^\gamma$. These two matrices representing L are related by conjugation with the change of basis matrix $Q = [\mathrm{id}]_\beta^\gamma$:*

$$[L]_\gamma = Q[L]_\beta Q^{-1}.$$

Definition A.15. *Suppose that $L\colon V \to V$ is a linear operator on a finite-dimensional vector space. The scalar $\lambda \in \mathbb{F}$ is an eigenvalue for L if there exists a nonzero vector $\mathbf{v} \in V$ such that $L\mathbf{v} = \lambda\mathbf{v}$; in this case $\mathbf{v}$ is an eigenvector for L with eigenvalue λ. The operator L is diagonalizable if there exists a basis for V consisting of eigenvectors for L. Equivalently, L is diagonalizable if and only if there exists a basis $\gamma = \{\mathbf{v}_1, \ldots, \mathbf{v}_n\}$ for which the matrix $[L]_\gamma$ is diagonal:*

$$[L]_\gamma := \begin{bmatrix} \lambda_1 & 0 & \cdots & 0 \\ 0 & \lambda_2 & \cdots & 0 \\ \vdots & \vdots & \ddots & \vdots \\ 0 & 0 & \cdots & \lambda_n \end{bmatrix}.$$

Appendices

Proposition A.16. *Let $M(n, \mathbb{F})$ denote the vector space of $n \times n$ matrices with entries in the field $\mathbb{F}$. Define the* trace *function* $\operatorname{tr}\colon M(n, \mathbb{F}) \to \mathbb{F}$ *as the sum of the diagonal entries:*

$$\operatorname{tr}[a_{ij}] := a_{11} + a_{22} + \cdots + a_{nn}.$$

The trace is a linear transformation satisfying the additional cyclic *property that* $\operatorname{tr}(AB) = \operatorname{tr}(BA)$ *for all* $A, B \in M(n, \mathbb{F})$. *In particular, if* $L\colon V \to V$ *is a linear operator on a finite-dimensional vector space, then we may choose any basis β for V and define the trace of L to be the trace of the matrix $[L]_\beta$; this is well defined by applying the cyclic property of the trace to proposition A.14. In particular, if L is diagonalizable, then the trace of L is the sum of its eigenvalues.*

Proposition A.17. *There is a unique function* $\det\colon M(n, \mathbb{F}) \to \mathbb{F}$ *satisfying the following three conditions for all row vectors* $\mathbf{a}_j, \mathbf{b} \in \mathbb{F}^n$ *and scalars* $c \in \mathbb{F}$:

$$i)\ \det \begin{bmatrix} \mathbf{a}_1 \\ \vdots \\ \mathbf{a}_{i-1} \\ c\mathbf{a}_i + \mathbf{b} \\ \mathbf{a}_{i+1} \\ \vdots \\ \mathbf{a}_n \end{bmatrix} = c\det \begin{bmatrix} \mathbf{a}_1 \\ \vdots \\ \mathbf{a}_{i-1} \\ \mathbf{a}_i \\ \mathbf{a}_{i+1} \\ \vdots \\ \mathbf{a}_n \end{bmatrix} + \det \begin{bmatrix} \mathbf{a}_1 \\ \vdots \\ \mathbf{a}_{i-1} \\ \mathbf{b} \\ \mathbf{a}_{i+1} \\ \vdots \\ \mathbf{a}_n \end{bmatrix}, \qquad 1 \le i \le n;$$

$$ii)\ \det \begin{bmatrix} \mathbf{a}_1 \\ \vdots \\ \mathbf{a}_{i-1} \\ \mathbf{b} \\ \mathbf{b} \\ \vdots \\ \mathbf{a}_n \end{bmatrix} = 0, \qquad 1 \le i \le n-1;$$

iii) $\det(I_n) = 1$, *where* $I_n = \operatorname{diag}(1, 1, \dots, 1)$ *is the identity matrix.*

That is, $\det\colon M(n, \mathbb{F}) \to \mathbb{F}$ *is the* unique *multi-linear, alternating, normalized function. The determinant is multiplicative:*

$$\det(AB) = \det(A)\det(B).$$

If $L\colon V \to V$ *is a linear operator on a finite-dimensional vector space, then we may choose any basis β for V and define the determinant of L to be the determinant of the matrix $[L]_\beta$; this is well defined by applying the multiplicative property of the determinant to proposition A.14. In particular, if L is diagonalizable, then the determinant of L is the product of its eigenvalues.*

236   *Symmetry and Quantum Mechanics*

Proposition A.18. *An $n \times n$ matrix A is invertible if and only if $\det(A) \neq 0$. It follows that the scalar $\lambda \in \mathbb{F}$ is an eigenvalue of A if and only if $\det(A - \lambda I_n) = 0$. Hence, the eigenvalues of a matrix A are the roots of its* characteristic polynomial

$$p_A(\lambda) := \det(A - \lambda I_n),$$

which is a polynomial of degree n in the variable λ with coefficients in the field $\mathbb{F}$.

Proposition A.19. *Define the* transpose *of a matrix to be the matrix obtained by reflecting across the diagonal: $[a_{ij}]^T := [a_{ji}]$. If A is a square matrix, then $\det(A^T) = \det(A)$.*

Proposition A.20. *The determinant has the following explicit formula in terms of the permutation group $\mathfrak{S}_n$ consisting of bijections $\sigma \colon \{1, 2, \ldots, n\} \to \{1, 2, \ldots, n\}$:*

$$\det[a_{ij}] = \sum_{\sigma \in \mathfrak{S}_n} \operatorname{sgn}(\sigma) a_{1\sigma(1)} a_{2\sigma(2)} \cdots a_{n\sigma(n)}.$$

Here, the sign function $\operatorname{sgn} \colon \mathfrak{S}_n \to \{\pm 1\}$ *is defined as*

$$\operatorname{sgn}(\sigma) = \begin{cases} +1 & \text{if } \sigma \text{ is the product of an even number of transpositions} \\ -1 & \text{if } \sigma \text{ is the product of an odd number of transpositions.} \end{cases}$$

In the 2×2 case

$$\det \begin{bmatrix} a & b \\ c & d \end{bmatrix} = ad - bc,$$

and in the 3×3 case:

$$\det \begin{bmatrix} a & b & c \\ d & e & f \\ g & h & i \end{bmatrix} = aei + bfg + dhc - gec - hfa - dbi.$$

Proposition A.21 (Cramer's Rule). *Suppose that $A = [a_{ij}]$ is an $n \times n$ matrix with entries from $\mathbb{F}$. For $1 \leq i, j \leq n$, let $A^{(ij)}$ denote the matrix obtained from A by removing the ith row and jth column, and set $c_{ij} = (-1)^{i+j} \det(A^{(ij)})$, called the ij-cofactor of A. Define $C := [c_{ij}]$ to be the* matrix of cofactors *of A. If A is invertible, then the inverse of A is given by the transpose of C divided by the determinant of A:*

$$A^{-1} = \frac{1}{\det(A)} C^T.$$

A.1.2   Inner product spaces and adjoints

In this section, the field of scalars $\mathbb{F}$ is either $\mathbb{R}$ or $\mathbb{C}$, and a^* denotes the complex conjugate of $a \in \mathbb{F}$.

Appendices 237

Definition A.22. *An* inner product *on a vector space V is a function $\langle,\rangle \colon V \times V \to \mathbb{F}$ such that if $a \in \mathbb{F}$ and $\mathbf{v}, \mathbf{v}', \mathbf{w} \in V$, then:*

i) $\langle a\mathbf{v} + \mathbf{v}', \mathbf{w} \rangle = a^* \langle \mathbf{v}, \mathbf{w} \rangle + \langle \mathbf{v}', \mathbf{w} \rangle;$

ii) $\langle \mathbf{v}, \mathbf{w} \rangle = \langle \mathbf{w}, \mathbf{v} \rangle^*;$

iii) $\langle \mathbf{v}, \mathbf{v} \rangle \geq 0$ *with equality if and only if $\mathbf{v} = \mathbf{0}$.*

A vector space endowed with an inner product is called an inner product space. *The* norm *of a vector is defined as* $\|\mathbf{v}\| = \sqrt{\langle \mathbf{v}, \mathbf{v} \rangle}$.

Theorem A.23 (Cauchy-Schwarz). *Let $(V, \langle,\rangle)$ be an inner product space. Then the following inequality holds for all vectors $\mathbf{v}, \mathbf{w} \in V$:*

$$|\langle \mathbf{v}, \mathbf{w} \rangle| \leq \|\mathbf{v}\| \|\mathbf{w}\|.$$

Definition A.24. *A subset S of an inner product space V is* orthonormal *if (1) every vector in S has norm 1, and (2) every pair of distinct vectors $\mathbf{v}, \mathbf{w} \in S$ satisfies $\langle \mathbf{v}, \mathbf{w} \rangle = 0$.*

Theorem A.25 (Gram-Schmidt). *Every finite-dimensional inner product space possesses an orthonormal basis. In fact, there is an algorithm which produces an orthonormal basis for $(V, \langle,\rangle)$ starting with an arbitrary finite spanning set for V.*

Proposition A.26. *Suppose that $(V, \langle,\rangle)$ is an inner product space and $W \subset V$ is a finite-dimensional subspace. Then $V = W \oplus W^\perp$, where $W^\perp$ is the orthogonal complement to W in V:*

$$W^\perp := \{\mathbf{v} \in V \mid \langle \mathbf{v}, \mathbf{w} \rangle = 0 \text{ for all } \mathbf{w} \in W\}.$$

Proposition A.27. *Suppose that $(V, \langle,\rangle)$ is a finite-dimensional inner product space and $L \colon V \to V$ is a linear operator on V. Let $\beta = \{\mathbf{u}_1, \ldots, \mathbf{u}_n\}$ be an orthonormal basis for V, and consider $[L]_\beta = [a_{ij}]$, the matrix of L in the basis β. Then the matrix entries $a_{ij} \in \mathbb{F}$ may be computed via the inner product:*

$$a_{ij} = \langle \mathbf{u}_i, L\mathbf{u}_j \rangle \qquad \text{for all } 1 \leq i, j \leq n.$$

Similarly, the coefficients of the expansion of a vector $\mathbf{v}$ in the basis β may be computed as inner products:

$$\mathbf{v} = \sum_{i=1}^{n} \langle \mathbf{u}_i, \mathbf{v} \rangle \mathbf{u}_i.$$

Proposition A.28. *Suppose that $(V, \langle,\rangle)$ is a finite-dimensional inner product space and $L \colon V \to V$ is a linear operator on V. There exists a unique linear operator $L^\dagger \colon V \to V$, called the* adjoint *of L, with the property that*

$$\langle L\mathbf{u}, \mathbf{v} \rangle = \langle \mathbf{u}, L^\dagger \mathbf{v} \rangle \qquad \text{for all } \mathbf{u}, \mathbf{v} \in V.$$

238 *Symmetry and Quantum Mechanics*

If β is an orthonormal basis for V, then the matrix of $L^\dagger$ with respect to β is the conjugate transpose of the matrix $[L]_\beta = [a_{ij}]$:

$$[L^\dagger]_\beta = [a_{ij}]^\dagger := [a^*_{ji}].$$

Note that, if $\mathbb{F} = \mathbb{R}$, then the conjugate transpose is just the ordinary transpose: $[L^\dagger]_\beta = [a_{ij}]^T = [a_{ji}]$.

Definition A.29. *A linear operator $L\colon V \to V$ on a finite-dimensional inner product space is called* self-adjoint *if $L^\dagger = L$. If β is an orthonormal basis for V, then L is self-adjoint if and only if the matrix of L with respect to β is* Hermitian, *i.e., equal to its conjugate transpose:*

$$[L]^\dagger_\beta = [L]_\beta.$$

When $\mathbb{F} = \mathbb{R}$, this is the statement that $[L]^T_\beta = [L]_\beta$, and such matrices are called symmetric.

Similarly, L is skew-adjoint *if $L^\dagger = -L$, which is equivalent to its matrix with respect to an orthonormal basis β being* skew-Hermitian:

$$[L]^\dagger_\beta = -[L]_\beta.$$

Again, when $\mathbb{F} = \mathbb{R}$ this means that $[L]^T_\beta = -[L]_\beta$, and such matrices are called skew-symmetric.

Definition A.30. *Suppose that $(V, \langle, \rangle)$ and $(W, \langle, \rangle')$ are inner product spaces. A linear transformation $L\colon V \to W$ is* orthogonal *(if $\mathbb{F} = \mathbb{R}$) or* unitary *(if $\mathbb{F} = \mathbb{C}$) if it is surjective and preserves the inner product:*

$$\langle L\mathbf{u}, L\mathbf{v} \rangle = \langle \mathbf{u}, \mathbf{v} \rangle' \qquad \text{for all } \mathbf{u}, \mathbf{v} \in V.$$

Note that the preservation of the inner product implies injectivity, so that orthogonal/unitary transformations are isomorphisms. Moreover, in the finite-dimensional case, injectivity implies surjectivity by rank-nullity (theorem A.8), so we many define finite-dimensional orthogonal/unitary transformations simply as those which preserve the inner product.

Proposition A.31. *Suppose that $(V, \langle, \rangle)$ is a finite dimensional inner product space and $L\colon V \to V$ is a linear operator on V. Then L is orthogonal/unitary if and only if $L^\dagger = L^{-1}$. If β is an orthonormal basis for V, then L is orthogonal/unitary if and only if the conjugate transpose of L is equal to its inverse:*

$$[L]^\dagger_\beta = [L]^{-1}_\beta.$$

Such matrices are called orthogonal/unitary *according to whether $\mathbb{F} = \mathbb{R}$ or $\mathbb{F} = \mathbb{C}$.*

Appendices 239

Theorem A.32 (Spectral Theorem). *Suppose that* $L\colon V \to V$ *is a self-adjoint linear operator on a finite-dimensional inner product space. Then the eigenvalues* λ_i *of* L *are real, and there exists an orthonormal basis of* V *consisting of eigenvectors for* L.

In terms of matrices, we may formulate this result as follows: let β *be an arbitrary orthonormal basis for* V, *and let* γ *be an orthonormal basis of eigenvectors for* L. *Set* $Q := [\mathrm{id}]_\beta^\gamma$, *the change of basis matrix from* β *to* γ. *Then* Q *is an orthogonal/unitary matrix and*

$$Q[L]_\beta Q^{-1} = [L]_\gamma = \mathrm{diag}(\lambda_1, \lambda_2, \ldots, \lambda_n).$$

That is: Hermitian matrices have real eigenvalues and may be diagonalized by conjugation with orthogonal/unitary matrices.

Theorem A.33 (Spectral Theorem for Normal Operators). *Suppose that* $\mathbb{F} = \mathbb{C}$ *and* $L\colon V \to V$ *is a* normal *operator on a finite-dimensional complex inner product space, meaning that* L *commutes with its adjoint:* $L^\dagger L = L L^\dagger$. *Then there exists an orthonormal basis of* V *consisting of eigenvectors for* L.

In terms of matrices, we may formulate this result as follows: let β *be an arbitrary orthonormal basis for* V, *and let* γ *be an orthonormal basis of eigenvectors for* L. *Set* $Q := [\mathrm{id}]_\beta^\gamma$, *the change of basis matrix from* β *to* γ. *Then* Q *is a unitary matrix and*

$$Q[L]_\beta Q^{-1} = [L]_\gamma = \mathrm{diag}(\lambda_1, \lambda_2, \ldots, \lambda_n).$$

That is: normal matrices (i.e., matrices that commute with their conjugate transpose) may be diagonalized by conjugation with unitary matrices.

A.2 Multivariable calculus

This section is a brief review of some topics in the calculus of functions on Euclidean n-space $(\mathbb{R}^n, \cdot)$. This is the space of column vectors with inner product given by the dot product:

$$\mathbf{v} \cdot \mathbf{w} := v_1 w_1 + v_2 w_2 + \cdots + v_n w_n.$$

We denote the Euclidean norm by $|\mathbf{v}| := \sqrt{\mathbf{v} \cdot \mathbf{v}}$. The *standard basis* of $\mathbb{R}^n$ is $\varepsilon := \{\mathbf{e}_1, \ldots, \mathbf{e}_n\}$, where $\mathbf{e}_j$ has a 1 in the jth slot and zeros elsewhere; it is an orthonormal basis for Euclidean space.

Our main goal is to describe and compute the derivative as a linear operator. The classic source for this material is [16], which includes much more, including the extension of these ideas to submanifolds of Euclidean space.

240　　　　　　　*Symmetry and Quantum Mechanics*

Definition A.34. *Let $F\colon \mathbb{R}^n \to \mathbb{R}^m$ be a function and $\mathbf{a} \in \mathbb{R}^n$. Then F is differentiable at $\mathbf{a}$ if and only if there exists a linear transformation $L\colon \mathbb{R}^n \to \mathbb{R}^m$ such that*

$$\lim_{\mathbf{h} \to 0} \frac{|F(\mathbf{a} + \mathbf{h}) - F(\mathbf{a}) - L(\mathbf{h})|}{|\mathbf{h}|} = 0.$$

If such an L exists, then it is unique, and we denote it by $(DF)_\mathbf{a}$ and call it the derivative of F at $\mathbf{a}$. The derivative $(DF)_\mathbf{a}$ should be thought of as the best linear approximation to the function $F(\mathbf{x}) - F(\mathbf{a})$ near $\mathbf{x} = \mathbf{a}$. In particular, if $F = L$ is a linear transformation, then $(DL)_\mathbf{a} = L$ for all $\mathbf{a}$.

Theorem A.35. *Suppose that $F\colon \mathbb{R}^n \to \mathbb{R}^m$ is differentiable at $\mathbf{a} \in \mathbb{R}^n$. Then the matrix of $(DF)_\mathbf{a}$ with respect to the standard bases is the Jacobian matrix of partial derivatives of F:*

$$[(DF)_\mathbf{a}]_\varepsilon^{\varepsilon'} = \left[\frac{\partial F^i}{\partial x_j}(\mathbf{a})\right].$$

Here, ε and ε' denote the standard bases on $\mathbb{R}^n$ and $\mathbb{R}^m$ respectively; F^i denotes the ith component function of F for $i = 1, \ldots, m$; and x_j denotes the jth coordinate function on $\mathbb{R}^n$ for $j = 1, \ldots, n$.

Proof. Before getting started, note the following relationship between various pieces of the notation: if $\varepsilon = \{\mathbf{e}_1, \ldots, \mathbf{e}_n\}$ is the standard basis on $\mathbb{R}^n$, then every $\mathbf{x} \in \mathbb{R}^n$ can be written uniquely as $\mathbf{x} = \sum_{j=1}^n c_j \mathbf{e}_j$ with $c_j \in \mathbb{R}$. The coordinate function $x_j\colon \mathbb{R}^n \to \mathbb{R}$ is just the function $\mathbf{x} \mapsto c_j$. Then $F\colon \mathbb{R}^n \to \mathbb{R}^m$ can be written more explicitly as

$$F(x_1, \ldots, x_n) = (F^1(x_1, \ldots, x_n), \ldots, F^m(x_1, \ldots, x_n))$$

for component functions $F^i\colon \mathbb{R}^n \to \mathbb{R}$.

To simplify the notation, set $L = (DF)_\mathbf{a}$ and $B := [b_{ij}] = [(DF)_\mathbf{a}]_\varepsilon^{\varepsilon'}$. From definition A.9, the jth column of B is the column vector $L(\mathbf{e}_j) \in \mathbb{R}^m$. Hence, the ith component of $L(\mathbf{e}_j)$ is the matrix entry b_{ij}. Now apply the limit property characterizing L to vectors of the form $\mathbf{h} = h\mathbf{e}_j$ for $h \in \mathbb{R}$ to obtain:

$$\begin{aligned}
0 &= \lim_{h \to 0} \frac{|F(\mathbf{a} + h\mathbf{e}_j) - F(\mathbf{a}) - L(h\mathbf{e}_j)|}{|h|} \\
&= \lim_{h \to 0} \left| \frac{F(\mathbf{a} + h\mathbf{e}_j) - F(\mathbf{a})}{h} - L(\mathbf{e}_j) \right|,
\end{aligned}$$

where we have used the fact that L is linear to write $L(h\mathbf{e}_j) = hL(\mathbf{e}_j)$. It follows that

$$\lim_{h \to 0} \frac{F(\mathbf{a} + h\mathbf{e}_j) - F(\mathbf{a})}{h} = L(\mathbf{e}_j) \in \mathbb{R}^m.$$

Appendices

Consider the ith component of this vector equation:

$$\lim_{h \to 0} \frac{F^i(\mathbf{a} + h\mathbf{e}_j) - F^i(\mathbf{a})}{h} = b_{ij} \in \mathbb{R}.$$

But the left-hand side of this equation is by definition the partial derivative of the function $F^i \colon \mathbb{R}^n \to \mathbb{R}$ with respect to its jth variable x_j:

$$\frac{\partial F^i}{\partial x_j}(\mathbf{a}) := \lim_{h \to 0} \frac{F^i(\mathbf{a} + h\mathbf{e}_j) - F^i(\mathbf{a})}{h}.$$

$\square$

Example A.36. *Suppose that $c \colon \mathbb{R} \to \mathbb{R}^m$ is a parametrized curve. Then c is differentiable at $a \in \mathbb{R}$ if and only if $\dot{c}(a) := \frac{d}{dt}c(t)|_{t=a}$ exists. In this case, the derivative $(Dc)_a \colon \mathbb{R} \to \mathbb{R}^m$ may be identified with the vector $\dot{c}(a) \in \mathbb{R}^m$ in the sense that*

$$(Dc)_a h = h\dot{c}(a) \qquad \text{for all } h \in \mathbb{R}.$$

Theorem A.37 (Chain Rule). *Suppose that $F \colon \mathbb{R}^n \to \mathbb{R}^m$ is differentiable at $\mathbf{a} \in \mathbb{R}^n$, and $G \colon \mathbb{R}^m \to \mathbb{R}^k$ is differentiable at $F(\mathbf{a}) \in \mathbb{R}^m$. Then the composition $G \circ F \colon \mathbb{R}^n \to \mathbb{R}^k$ is differentiable at $\mathbf{a}$ with derivative*

$$(D(G \circ F))_{\mathbf{a}} = (DG)_{F(\mathbf{a})}(DF)_{\mathbf{a}}.$$

Proposition A.38. *Suppose that $F \colon \mathbb{R}^n \to \mathbb{R}^m$ is differentiable at $\mathbf{a} \in \mathbb{R}^n$. Let $\mathbf{v} \in \mathbb{R}^n$ be a vector, thought of as a tangent vector at the point $\mathbf{a}$. The image of $\mathbf{v}$ under the linear transformation $(DF)_{\mathbf{a}}$ may be computed as follows: choose any differentiable curve $c \colon \mathbb{R} \to \mathbb{R}^n$ satisfying $c(0) = \mathbf{a}$ and $\dot{c}(0) = \mathbf{v}$. Then*

$$(DF)_{\mathbf{a}}\mathbf{v} = \frac{d}{dt}F(c(t))|_{t=0} \in \mathbb{R}^m.$$

Proof. Apply the Chain Rule to the composition $F \circ c \colon \mathbb{R} \to \mathbb{R}^m$, using example A.36 to identify the derivative of c with the vector $\dot{c}(0) = \mathbf{v}$:

$$(D(F \circ c))_0 = (DF)_{c(0)}(Dc)_0 = (DF)_{\mathbf{a}}\dot{c}(0) = (DF)_{\mathbf{a}}\mathbf{v}.$$

But $F(c(t))$ is also a parametrized curve in $\mathbb{R}^m$, so again by example A.36 we have $(D(F \circ c))_0 = \frac{d}{dt}F(c(t))|_{t=0}$. $\square$

Proposition A.39. *Let $F \colon \mathbb{R}^n \to \mathbb{R}^m$ with component functions $F^i(x_1, \ldots, x_n)$ for $i = 1, \ldots, m$. Suppose that all partial derivatives $\frac{\partial F^i}{\partial x_j}$ exist and are continuous in an open neighborhood of a point $\mathbf{a} \in \mathbb{R}^n$. Then F is differentiable at $\mathbf{a}$, and we say that F is* continuously differentiable *near $\mathbf{a}$.*

Theorem A.40 (Implicit Function Theorem). *Consider a function $F \colon \mathbb{R}^n \times \mathbb{R}^m \to \mathbb{R}^m$, so that*

$$F(x_1, \ldots, x_{n+m}) = (F^1(x_1, \ldots, x_{n+m}), \ldots, F^m(x_1, \ldots, x_{n+m})).$$

242   *Symmetry and Quantum Mechanics*

Suppose that F is continuously differentiable near $(\mathbf{a}, \mathbf{b}) \in \mathbb{R}^n \times \mathbb{R}^m$ with $F(\mathbf{a}, \mathbf{b}) = 0$. Let M denote the $m \times m$ matrix formed from the final m columns of the Jacobian matrix of partial derivatives:

$$M := \left[\frac{\partial F^i}{\partial x_j}(\mathbf{a}, \mathbf{b}) \right]_{i=1,\dots,m; j=n+1,\dots,n+m}.$$

If $\det(M) \neq 0$, then there exist open sets $\mathbf{a} \in A \subset \mathbb{R}^n$ and $\mathbf{b} \in B \subset \mathbb{R}^m$, and a differentiable function $g \colon A \to B$ with the property that $F(\mathbf{x}, g(\mathbf{x})) = 0$ for all $\mathbf{x} \in A$. Moreover, we have the equality

$$\{(\mathbf{x}, g(\mathbf{x})) \mid \mathbf{x} \in A\} = (A \times B) \cap \{(\mathbf{x}, \mathbf{y}) \in \mathbb{R}^n \times \mathbb{R}^m \mid F(\mathbf{x}, \mathbf{y}) = 0\}.$$

Theorem A.41 (Change of Variable). *Suppose that $A \subset \mathbb{R}^n$ is an open set, and $g \colon A \to \mathbb{R}^n$ is a one-to-one, continuously differentiable function with the property that $\det((DF)_{\mathbf{a}}) \neq 0$ for all $\mathbf{a} \in A$. If $f \colon g(A) \to \mathbb{R}$ is an integrable function, then*

$$\int_{g(A)} f = \int_A (f \circ g) |\det(Dg)|.$$

The Change of Variable Theorem is proved for the Riemann integral as theorem 3-13 of [16], but a formally identical statement holds for the Lebesgue integral.

A.3   Analysis

In this section we introduce some basic definitions concerning Hilbert spaces and state some major theorems that are mentioned in the main text. Throughout, we take the field of scalars to be $\mathbb{F} = \mathbb{C}$.

A.3.1   Hilbert spaces and adjoints

Definition A.42. *A Hilbert space is a complex inner product space $(\mathbf{H}, \langle, \rangle)$ that is complete in the norm topology: every Cauchy sequence in $\mathbf{H}$ converges.*

Definition A.43. *A subset $S \subset \mathbf{H}$ is a Hilbert space basis if it is linearly independent and the set of finite linear combinations of elements from S is dense in $\mathbf{H}$.*

Definition A.44. *A Hilbert space is separable if it contains a countable dense subset.*

Proposition A.45. *A Hilbert space $\mathbf{H}$ has a countable orthonormal Hilbert space basis if and only if it is separable. Moreover, the Gram-Schmidt procedure produces an orthonormal Hilbert space basis starting with an arbitrary countable dense subset of $\mathbf{H}$.*

Appendices 243

Proposition A.46. *Suppose that $\{\mathbf{u}_i\}$ is a countable orthonormal Hilbert space basis for $\mathbf{H}$, and let $\mathbf{v} \in \mathbf{H}$ be an arbitrary vector. Then $\mathbf{v}$ may be written uniquely as a convergent linear combination of the basis vectors $\mathbf{u}_i$, with coefficients determined by the inner product:*

$$\mathbf{v} = \sum_{i=1}^{\infty} \langle \mathbf{u}_i, \mathbf{v} \rangle \mathbf{u}_i.$$

Definition A.47. *A linear operator L on a Hilbert space $\mathbf{H}$ is a linear transformation $L \colon \mathcal{D}(L) \to \mathbf{H}$, where $\mathcal{D}(L)$ is a dense vector subspace of $\mathbf{H}$, called the* domain *of L.*

Definition A.48. *Suppose that $L \colon \mathcal{D}(L) \to \mathbf{H}$ is a linear operator on the Hilbert space $\mathbf{H}$. Define a subspace $\mathcal{D}(L^\dagger) \subset \mathbf{H}$ consisting of the vectors $\mathbf{v} \in \mathbf{H}$ for which there exists $\mathbf{w} \in \mathbf{H}$ with the property that*

$$\langle L\mathbf{u}, \mathbf{v} \rangle = \langle \mathbf{u}, \mathbf{w} \rangle \qquad \text{for all } \mathbf{u} \in \mathcal{D}(L).$$

For each $\mathbf{v} \in \mathcal{D}(L^\dagger)$, the vector $\mathbf{w}$ is unique by the density of $\mathcal{D}(L)$. Hence, we may define a linear transformation $L^\dagger \colon \mathcal{D}(L^\dagger) \to \mathbf{H}$ by setting $L^\dagger \mathbf{v} = \mathbf{w}$. When $\mathcal{D}(L^\dagger)$ is dense in $\mathbf{H}$, we obtain a linear operator $L^\dagger$ on $\mathbf{H}$, called the adjoint *of L. Note that we have*

$$\langle L\mathbf{u}, \mathbf{v} \rangle = \langle \mathbf{u}, L^\dagger \mathbf{v} \rangle \qquad \text{for all } \mathbf{u} \in \mathcal{D}(L) \text{ and } \mathbf{v} \in \mathcal{D}(L^\dagger).$$

Definition A.49. *A linear operator L on the Hilbert space $\mathbf{H}$ is* self-adjoint *if $\mathcal{D}(L) = \mathcal{D}(L^\dagger)$ and $L = L^\dagger$ on their common domain.*

A.3.2 Some big theorems

As in the finite-dimensional case, there is a Spectral Theorem for self-adjoint operators on separable Hilbert spaces; for the statement, proof, and substantial discussion in the context of quantum mechanics see [10]. In particular, the Spectral Theorem provides a *functional calculus* which allows for the definition of a unitary exponential operator $e^{iL} \colon \mathbf{H} \to \mathbf{H}$ associated to any self-adjoint operator L on $\mathbf{H}$. Even if L is defined only on a proper dense subspace $\mathcal{D}(L)$, the unitary operator e^{iL} is defined on the entire Hilbert space $\mathbf{H}$. Moreover, since L is a self-adjoint linear operator on $\mathbf{H}$, so is tL for all $t \in \mathbf{R}$. Thus, we obtain a family of unitary operators e^{itL} on the Hilbert space $\mathbf{H}$.

Definition A.50. *Let $U(\mathbf{H})$ denote the group of unitary operators on the Hilbert space $\mathbf{H}$. A* one-parameter unitary group *on $\mathbf{H}$ is a group homomorphism $\mathcal{U} \colon (\mathbb{R}, +) \to U(\mathbf{H})$. The homomorphism $\mathcal{U}$ is* strongly continuous *if*

$$\lim_{t \to t_0} \|\mathcal{U}(t)\mathbf{v} - \mathcal{U}(t_0)\mathbf{v}\| = 0$$

for all $\mathbf{v} \in \mathbf{H}$ and $t_0 \in \mathbf{R}$.

244   *Symmetry and Quantum Mechanics*

Theorem A.51 (Stone's Theorem). *Suppose that L is a self-adjoint operator on the separable Hilbert space $\boldsymbol{H}$. Then e^{itL} is a strongly continuous one-parameter unitary group on $\boldsymbol{H}$. Moreover, for any $\mathbf{v} \in \boldsymbol{H}$, the following limit exists in the norm-topology if and only if $\mathbf{v} \in \mathcal{D}(L)$, in which case it is equal to $L\mathbf{v}$:*

$$\lim_{t \to 0} \frac{1}{i} \frac{e^{itL}\mathbf{v} - \mathbf{v}}{t} = L\mathbf{v}.$$

Hence, we may recover the self-adjoint operator L from the one-parameter unitary group e^{itL} via a process of differentiation.

Conversely, suppose that $\mathcal{U}(t)$ is an arbitrary strongly continuous one-parameter unitary group on $\boldsymbol{H}$, and consider the linear transformation $A \colon \mathcal{D}(A) \to \boldsymbol{H}$ defined by the limit

$$A\mathbf{v} := \frac{1}{i} \frac{\mathcal{U}(t)\mathbf{v} - \mathbf{v}}{t}.$$

Here, the domain $\mathcal{D}(A)$ is given by the vectors $\mathbf{v} \in \boldsymbol{H}$ for which the limit exists in the norm-topology. Then $\mathcal{D}(A)$ is dense in $\boldsymbol{H}$, so that A is a linear operator on $\boldsymbol{H}$. Moreover, A is self-adjoint and $\mathcal{U}(t) = e^{itA}$ for all $t \in \mathbb{R}$. Hence, every strongly continuous one-parameter unitary group arises from a self-adjoint operator via exponentiation.

Definition A.52. *The* Heisenberg group H_n *is the Lie group consisting of the set $\mathbb{R}^n \times \mathbb{R}^n \times \mathbb{R}$ with operation*

$$(\mathbf{v}, \mathbf{w}, \alpha) \bullet (\mathbf{v}', \mathbf{w}', \alpha') = (\mathbf{v} + \mathbf{v}', \mathbf{w} + \mathbf{w}', \alpha + \alpha' + \mathbf{v} \cdot \mathbf{w}).$$

Fix a real number $b \in \mathbb{R}$. The Schrödinger representation *of H_n with central parameter b is the H_n-action on the Hilbert space $L^2(\mathbb{R}^n)$ defined by*

$$(\mathbf{v}, \mathbf{w}, \alpha) \star \psi(\mathbf{r}) := e^{-ib(\alpha + \mathbf{v} \cdot (\mathbf{r} - \mathbf{w}))} \psi(\mathbf{r} - \mathbf{w}).$$

Proposition A.53. *For each $b \in \mathbb{R}$, the Schrödinger representation with central parameter b is an irreducible strongly continuous unitary representation of the Heisenberg group H_n. Here, strong continuity means that*

$$\lim_{h \to h_0} \|h \star \psi - h_0 \star \psi\| = 0$$

for all $\psi \in L^2(\mathbb{R}^n)$ and $h_0 \in H_n$. Irreducibility means that the only closed H_n-invariant subspaces are $\{0\}$ and $L^2(\mathbb{R}^n)$.

Theorem A.54 (Stone-von Neumann Theorem). *Let $\mathcal{U} \colon H_n \to U(\boldsymbol{H})$ be an irreducible strongly continuous unitary representation of the Heisenberg group H_n. Then there exists a unique value of $b \in \mathbb{R}$ and a unitary isomorphism of H_n-representations, $\varphi \colon \boldsymbol{H} \to L^2(\mathbf{R}^n)$, where H_n acts on $L^2(\mathbb{R}^n)$ via the Schrödinger representation with central parameter b:*

$$
\begin{array}{ccc}
H_n \times \boldsymbol{H} & \xrightarrow{\ \mathcal{U}\ } & \boldsymbol{H} \\[4pt]
{\scriptstyle \mathrm{id} \times \varphi} \big\downarrow & & \big\downarrow {\scriptstyle \varphi} \\[4pt]
H_n \times L^2(\mathbb{R}^n) & \xrightarrow{\ \star\ } & L^2(\mathbb{R}^n).
\end{array}
$$

Appendices 245

Explicitly, this commutative diagram means that for all $h \in H_n$ and $\mathbf{v} \in \mathbf{H}$, we have

$$\mathcal{U}(h)\mathbf{v} = \varphi^{-1}(h \star \varphi(\mathbf{v})).$$

The unitary isomorphism φ is unique up to multiplication by a phase $e^{i\theta} \in U(1)$.

In particular, choosing $b = \frac{1}{\hbar}$ for the parameter in the Schrödinger representation, the Stone-von Neumann theorem implies that there is a unique irreducible strongly continuous unitary representation of H_n in which the central element $(\mathbf{0}, \mathbf{0}, 1)$ acts as the scalar $e^{-\frac{i}{\hbar}}$.

A.4   Solutions to selected exercises

2.5 For any $\mathbf{a} \in W$, we have

$$
\begin{aligned}
\langle \psi | \mathbf{a} \rangle &= \langle \mathbf{a} | \psi \rangle^* \\
&= \langle \mathbf{a} | c_1 | \phi_1 \rangle^* + \langle \mathbf{a} | c_2 | \phi_2 \rangle^* \\
&= c_1^* \langle \mathbf{a} | \phi_1 \rangle^* + c_2^* \langle \mathbf{a} | \phi_2 \rangle^* \\
&= c_1^* \langle \phi_1 | \mathbf{a} \rangle + c_2^* \langle \phi_2 | \mathbf{a} \rangle \\
&= (c_1^* \langle \phi_1 | + c_2^* \langle \phi_2 |)(\mathbf{a}).
\end{aligned}
$$

2.36 Suppose that $A \in O(3)$ and $c \colon [0,1] \to O(3)$ is a continuous path with $c(0) = I$ and $c(1) = A$. Then the composition $\det \circ c \colon [0,1] \to \{\pm 1\}$ is continuous, hence constant. But $\det(c(0)) = \det(I) = 1$, so $\det(A) = \det(c(1)) = 1$ and $A \in SO(3)$.

Conversely, suppose that $A \in SO(3)$. By proposition 1.19, A is the rotation through some angle θ around a fixed line $\ell \subset \mathbb{R}^3$. For each $0 \leq t \leq 1$, let $A_t \in SO(3)$ denote the rotation through the angle $t\theta$ around the line ℓ. Then $t \mapsto A_t$ is a continuous path in $SO(3)$ connecting $A_0 = I$ to $A_1 = A$.

2.38 Using the fact that $\mathbf{u}$ is a unit vector, a direct check verifies that $B^\dagger = B^{-1}$ and $\det(B) = 1$, so $B \in SU(2)$. Note that $B = \cos(\theta)I - i\sin(\theta)(\mathbf{u} \cdot \boldsymbol{\sigma})$, where $\boldsymbol{\sigma} = (\sigma_1, \sigma_2, \sigma_3)$ is the vector of Pauli matrices. It follows that B commutes with $\mathbf{u} \cdot \boldsymbol{\sigma}$, so that

$$
B \begin{bmatrix} u_3 & u_1 - iu_2 \\ u_1 + iu_2 & -u_3 \end{bmatrix} B^{-1} = \begin{bmatrix} u_3 & u_1 - iu_2 \\ u_1 + iu_2 & -u_3 \end{bmatrix},
$$

which means that $\mathbf{u}$ is fixed by $f(B)$. Now choose any unit vector $\mathbf{v}$ that is orthogonal to $\mathbf{u}$, and set $\mathbf{w} = \mathbf{u} \times \mathbf{v}$. Then $\{\mathbf{u}, \mathbf{v}, \mathbf{w}\}$ forms a right-handed orthonormal basis for $(\mathbb{R}^3, \cdot)$. The following computation will be useful, in

246   *Symmetry and Quantum Mechanics*

which we use the fact that the Pauli matrices anti-commute with each other and square to I:

$$
\begin{aligned}
(\mathbf{u}\cdot\boldsymbol{\sigma})(\mathbf{v}\cdot\boldsymbol{\sigma}) &= \sum_{j,k=1}^{3} u_j v_k \sigma_j \sigma_k \\
&= (\mathbf{u}\cdot\mathbf{v})I + \sum_{j<k}(u_j v_k - v_j u_k)\sigma_j \sigma_k \\
&= i(\mathbf{u}\times\mathbf{v})\cdot\boldsymbol{\sigma} \\
&= i\mathbf{w}\cdot\boldsymbol{\sigma}.
\end{aligned}
$$

In the third line we have used the orthogonality of $\mathbf{u}$ and $\mathbf{v}$, the fact that $\sigma_1 \sigma_2 = i\sigma_3$, and the corresponding equalities obtained by cyclic permutation of the indices.

Now compute

$$
\begin{aligned}
B(\mathbf{v}\cdot\boldsymbol{\sigma})B^{-1} &= (\cos(\theta)I - i\sin(\theta)(\mathbf{u}\cdot\boldsymbol{\sigma}))(\mathbf{v}\cdot\boldsymbol{\sigma})(\cos(\theta)I + i\sin(\theta)(\mathbf{u}\cdot\boldsymbol{\sigma})) \\
&= \cos^2(\theta)(\mathbf{v}\cdot\boldsymbol{\sigma}) - \frac{i}{2}\sin(2\theta)((\mathbf{u}\cdot\boldsymbol{\sigma})(\mathbf{v}\cdot\boldsymbol{\sigma}) - (\mathbf{v}\cdot\boldsymbol{\sigma})(\mathbf{u}\cdot\boldsymbol{\sigma})) \\
&\quad + i\sin^2(\theta)(\mathbf{w}\cdot\boldsymbol{\sigma})(\mathbf{u}\cdot\boldsymbol{\sigma}) \\
&= \cos^2(\theta)(\mathbf{v}\cdot\boldsymbol{\sigma}) + \sin(2\theta)(\mathbf{w}\cdot\boldsymbol{\sigma}) - \sin^2(\theta)(\mathbf{w}\times\mathbf{u})\cdot\boldsymbol{\sigma} \\
&= (\cos^2(\theta) - \sin^2(\theta))(\mathbf{v}\cdot\boldsymbol{\sigma}) + \sin(2\theta)(\mathbf{w}\cdot\boldsymbol{\sigma}) \\
&= \cos(2\theta)(\mathbf{v}\cdot\boldsymbol{\sigma}) + \sin(2\theta)(\mathbf{w}\cdot\boldsymbol{\sigma}).
\end{aligned}
$$

This means that $f(B)$ rotates the vector $\mathbf{v}$ through an angle of 2θ in the plane spanned by $\mathbf{v}$ and $\mathbf{w}$.

3.12 For existence, check that $H := \frac{1}{2}(M+M^\dagger)$ and $K := \frac{1}{2i}(M-M^\dagger)$ are Hermitian and satisfy $H + iK = M$. For uniqueness, suppose that $H' + iK' = M$ is another decomposition with H' and K' Hermitian. Bringing the H's to one side and the K's to the other, we have $H - H' = i(K' - K)$. The left-hand side is Hermitian while the right-hand side is skew-Hermitian; it follows that both sides are zero, so $H = H'$ and $K = K'$. Finally, $\mathrm{tr}(M) = 0$ if and only if $\mathrm{tr}(H) = -i\mathrm{tr}(K)$ if and only if $\mathrm{tr}(H) = \mathrm{tr}(K) = 0$, since traces of Hermitian matrices are real.

3.17 Let ϕ_1 and ϕ_2 be the eigenstates for H, with real eigenvalues λ_1, λ_2. Let $|\phi\rangle = c_1|\phi_1\rangle + c_2|\phi_2\rangle$ be an arbitrary spin-state and compute

$$
\begin{aligned}
(\Delta_\phi H)^2 &= \langle\phi|H^2|\phi\rangle - \langle\phi|H|\phi\rangle^2 \\
&= |c_1|^2\lambda_1^2 + |c_2|^2\lambda_2^2 - (|c_1|^2\lambda_1 + |c_2|^2\lambda_2)^2 \\
&= (1 - |c_1|^2)|c_1|^2\lambda_1^2 + (1 - |c_2|^2)|c_2|^2\lambda_2^2 - 2|c_1 c_2|^2\lambda_1\lambda_2 \\
&= |c_1 c_2|^2(\lambda_1 - \lambda_2)^2,
\end{aligned}
$$

where we have used the fact that $|c_1|^2 + |c_2|^2 = 1$. If $\lambda_1 = \lambda_2$, then H is a

Appendices 247

scalar operator, and so ϕ is an eigenvector for H. If $\lambda_1 \neq \lambda_2$, then $\Delta_\phi H = 0$ if and only if $c_1 c_2 = 0$ if and only if $c_1 = 0, |c_2| = 1$ or $|c_1| = 1, c_2 = 0$. In either case, ϕ is an eigenstate for H.

3.22 The matrix $M^\dagger M$ is clearly Hermitian, so by the Spectral Theorem it is diagonalized by a unitary matrix Q:

$$\text{diag}(\lambda_1, \ldots, \lambda_n) = QM^\dagger MQ^{-1}.$$

The eigenvalues λ_i are real. To show that they are in fact positive, let $\mathbf{v} \in \mathbb{C}^n$ be an eigenvector with eigenvalue $\lambda = \lambda_i$. Thinking of $\mathbf{v}$ as a column vector, we have

$$\lambda \|\mathbf{v}\|^2 = \lambda \mathbf{v}^\dagger \mathbf{v} = \mathbf{v}^\dagger \lambda \mathbf{v} = \mathbf{v}^\dagger M^\dagger M \mathbf{v} = (M\mathbf{v})^\dagger M\mathbf{v} = \|M\mathbf{v}\|^2 > 0.$$

The last inequality is strict because M is invertible and $\mathbf{v} \neq 0$. Dividing by $\|\mathbf{v}\|^2$ yields $\lambda > 0$.

Now define $P := Q^{-1} \text{diag}(\sqrt{\lambda_1}, \ldots, \sqrt{\lambda_n})Q$. Then P is positive definite and

$$P^2 = Q^{-1} \text{diag}(\lambda_1, \ldots, \lambda_n)Q = M^\dagger M.$$

5.19 The map $\varphi_{\mathbb{C}}$ is clearly additive, and it is also $\mathbb{C}$-linear:

$$
\begin{aligned}
\varphi_{\mathbb{C}}((a + bi)(X + iY)) &= \varphi_{\mathbb{C}}((aX - bY) + i(bX + aY)) \\
&= \varphi(aX - bY) + i\varphi(bX + aY) \\
&= a\varphi(X) - b\varphi(Y) + i(b\varphi(X) + a\varphi(Y)) \\
&= (a + bi)(\varphi(X) + i\varphi(Y)) \\
&= (a + bi)\varphi_{\mathbb{C}}(X + iY).
\end{aligned}
$$

To show that $\varphi_{\mathbb{C}}$ preserves the Lie bracket, we compute

$$
\begin{aligned}
\varphi_{\mathbb{C}}([X + iY, X' + iY']) &= \varphi_{\mathbb{C}}([X, X'] - [Y, Y'] + i([X, Y'] + [Y, X'])) \\
&= \varphi([X, X'] - [Y, Y']) + i\varphi([X, Y'] + [Y, X']) \\
&= [\varphi(X), \varphi(X')] - [\varphi(Y), \varphi(Y')] \\
&\quad + i([\varphi(X), \varphi(Y')] + [\varphi(Y), \varphi(X')]) \\
&= [\varphi(X) + i\varphi(Y), \varphi(X') + i\varphi(Y')] \\
&= [\varphi_{\mathbb{C}}(X + iY), \varphi_{\mathbb{C}}(X' + iY')].
\end{aligned}
$$

Now suppose that $\rho \colon \mathfrak{g}_{\mathbb{C}} \to \mathfrak{gl}_n(\mathbb{C})$ is a $\mathbb{C}$-linear representation. Define φ to be the restriction of ρ to the real subalgebra $\mathfrak{g} \subset \mathfrak{g}_{\mathbb{C}}$:

$$\varphi(X) := \rho(X) \qquad \text{for all } X \in \mathfrak{g}.$$

Then φ is an $\mathbb{R}$-linear representation of $\mathfrak{g}$, with $\varphi_{\mathbb{C}} = \rho$:

$$\varphi_{\mathbb{C}}(X + iY) = \varphi(X) + i\varphi(Y) = \rho(X) + i\rho(Y) = \rho(X + iY).$$

248   *Symmetry and Quantum Mechanics*

Finally, note that $W \subset \mathbb{C}^n$ is $\mathfrak{g}$-invariant if and only if $\varphi(X)W \subset W$ for all $X \in \mathfrak{g}$ if and only if $(\varphi(X) + i\varphi(Y))W \subset W$ for all $X, Y \in \mathfrak{g}$ if and only if $\varphi_{\mathbb{C}}(X + iY)W \subset W$ for all $X + iY \in \mathfrak{g}_{\mathbb{C}}$ if and only if W is $\mathfrak{g}_{\mathbb{C}}$-invariant.

5.31 For the base case $k = 0$ we have $\|w_1^m\|^2 = 1$ by our choice of normalization. Suppose that the claim holds for k, so that

$$\|w_1^{m-k} w_2^k\|^2 = \binom{m}{k}^{-1}.$$

Now compute

$$\langle X w_1^{m-k} w_2^k | X w_1^{m-k} w_2^k \rangle = 4(m - k)^2 \|w_1^{m-k-1} w_2^{k+1}\|^2.$$

But we also have

$$\begin{aligned}
\langle X w_1^{m-k} w_2^k | X w_1^{m-k} w_2^k \rangle &= \langle w_1^{m-k} w_2^k | Y X w_1^{m-k} w_2^k \rangle \\
&= -2(m - k)\langle w_1^{m-k} w_2^k | Y w_1^{m-k-1} w_2^{k+1} \rangle \\
&= 4(m - k)(k + 1)\langle w_1^{m-k} w_2^k | w_1^{m-k} w_2^k \rangle \\
&= 4(m - k)(k + 1)\binom{m}{k}^{-1}.
\end{aligned}$$

Combining these two computations, we obtain the result for $k + 1$:

$$\begin{aligned}
\|w_1^{m-k-1} w_2^{k+1}\|^2 &= \frac{k + 1}{m - k}\binom{m}{k}^{-1} \\
&= \frac{k + 1}{m - k} \frac{k!(m - k)!}{m!} \\
&= \frac{(k + 1)!(m - k - 1)!}{m!} \\
&= \binom{m}{k + 1}^{-1}.
\end{aligned}$$

5.32 Each monomial $w_1^{m-k} w_2^k$ is an eigenvector for σ_3 with eigenvalue $2k - m$, so it follows that each $|j\rangle$ is a σ_3-eigenket with eigenvalue j. We will check the formula for the action of $\sigma_1 + i\sigma_2$, leaving the similar computation for $\sigma_1 - i\sigma_2$ to the reader. Given a value of $j = m, m - 2, \ldots, -m$, set $k = \frac{1}{2}(j + m)$, so that $2k - m = j$. Then the ket $(-1)^k |j\rangle$ corresponds to the scaled monomial $\binom{m}{k}^{\frac{1}{2}} w_1^{m-k} w_2^k$. So we compute:

$$(\sigma_1 + i\sigma_2) \star \binom{m}{k}^{\frac{1}{2}} w_1^{m-k} w_2^k = -2\binom{m}{k}^{\frac{1}{2}} (m - k) w_1^{m-k-1} w_2^{k+1}.$$

But

$$\binom{m}{k}^{\frac{1}{2}}(m-k) = \binom{m}{k+1}^{\frac{1}{2}}\sqrt{(m-k)(k+1)}$$

$$= \frac{1}{2}\binom{m}{k+1}^{\frac{1}{2}}\sqrt{m(m+2)-j(j+2)}.$$

Thus, we see that $\sigma_1 + i\sigma_2$ sends $(-1)^k\binom{m}{k}^{\frac{1}{2}}w_1^{m-k}w_2^k$ to

$$(-1)^{k+1}\binom{m}{k+1}^{\frac{1}{2}}\sqrt{m(m+2)-j(j+2)}\,w_1^{m-k-1}w_2^{k+1},$$

thus establishing the formula for the action of $\sigma_1 + i\sigma_2$ on the ket $|j\rangle$.

5.41 The number of monomials of exact degree k in two variables is $k+1$. Hence the total number of monomials of degree at most d in two variables is the $(d+1)$st triangular number

$$1 + 2 + \cdots + (d+1) = \frac{(d+2)(d+1)}{2}.$$

5.47 This derivation may be accomplished via a long and tedious computation involving repeated applications of the chain rule. For an alternative approach, see [1, Chapter 10, Section 9].

6.8 We compute

$$\frac{d}{dt}(\mathbf{a}(t) \otimes \mathbf{b}(t)) = \lim_{h \to 0} \frac{\mathbf{a}(t+h) \otimes \mathbf{b}(t+h) - \mathbf{a}(t) \otimes \mathbf{b}(t)}{h}$$

$$= \lim_{h \to 0} \frac{1}{h}\left(\mathbf{a}(t+h) \otimes \mathbf{b}(t+h) - \mathbf{a}(t) \otimes \mathbf{b}(t+h)\right.$$

$$\left. + \mathbf{a}(t) \otimes \mathbf{b}(t+h) - \mathbf{a}(t) \otimes \mathbf{b}(t)\right)$$

$$= \lim_{h \to 0} \left(\frac{\mathbf{a}(t+h) - \mathbf{a}(t)}{h} \otimes \mathbf{b}(t+h)\right.$$

$$\left. + \mathbf{a}(t) \otimes \frac{\mathbf{b}(t+h) - \mathbf{b}(t)}{h}\right)$$

$$= \dot{\mathbf{a}}(t) \otimes \mathbf{b}(t) + \mathbf{a}(t) \otimes \dot{\mathbf{b}}(t).$$

8.15 Suppose that $(m\mathbf{v}, \mathbf{w}, \alpha, A) \in G$ and $\psi(\mathbf{r})$ is P's description of a position state. Then acting via the inverse element yields a new wavefunction:

$$\psi'(\mathbf{r}) = (m\mathbf{v}, \mathbf{w}, \alpha, A)^{-1} \star \psi(\mathbf{r})$$

$$= (-mA^{-1}\mathbf{v}, -A^{-1}\mathbf{w}, -\alpha + m\mathbf{v} \cdot \mathbf{w}, A^{-1}) \star \psi(\mathbf{r})$$

$$= e^{-\frac{i}{\hbar}(-\alpha + m\mathbf{v} \cdot \mathbf{w} - mA^{-1}\mathbf{v} \cdot (\mathbf{r} + A^{-1}\mathbf{w}))}\psi(A(\mathbf{r} + A^{-1}\mathbf{w}))$$

$$= e^{\frac{i}{\hbar}(\alpha + m\mathbf{v} \cdot A\mathbf{r})}\psi(A\mathbf{r} + \mathbf{w}),$$

250   *Symmetry and Quantum Mechanics*

where in the last line we use the orthogonality of A:

$$A^{-1}\mathbf{v} \cdot A^{-1}\mathbf{w} = \mathbf{v} \cdot \mathbf{w} \quad \text{and} \quad A^{-1}\mathbf{v} \cdot \mathbf{r} = \mathbf{v} \cdot A\mathbf{r}.$$

If observer M is rotated according to A and translated by $\mathbf{w}$, then her position-values are related to P's as $\mathbf{r}' = A^{-1}(\mathbf{r} - \mathbf{w})$. It follows that

$$\psi'(\mathbf{r}') = e^{\frac{i}{\hbar}(\alpha + m\mathbf{v} \cdot (\mathbf{r} - \mathbf{w}))}\psi(\mathbf{r}),$$

so the position-probability densities agree: $|\psi'(\mathbf{r}')|^2 = |\psi(\mathbf{r})|^2$.

The position-dependent phase ensures that M and P will also agree on momentum-probability densities when M is rotated according to A and moving toward P with velocity $-\mathbf{v}$. To see this, consider the Fourier transform of P's wavefunction:

$$\psi(\mathbf{r}) = \frac{1}{(2\pi\hbar)^{\frac{3}{2}}} \int_{\mathbb{R}^3} \widetilde{\psi}(\mathbf{p}) e^{\frac{i}{\hbar}\mathbf{p} \cdot \mathbf{r}} d\mathbf{p}.$$

It follows that

$$
\begin{aligned}
\psi'(\mathbf{r}) &= \frac{1}{(2\pi\hbar)^{\frac{3}{2}}} e^{\frac{i}{\hbar}(\alpha + m\mathbf{v} \cdot A\mathbf{r})} \int \widetilde{\psi}(\mathbf{p}) e^{\frac{i}{\hbar}\mathbf{p} \cdot (A\mathbf{r} + \mathbf{w})} d\mathbf{p} \\
&= \frac{1}{(2\pi\hbar)^{\frac{3}{2}}} e^{\frac{i}{\hbar}\alpha} \int \widetilde{\psi}(\mathbf{p}) e^{\frac{i}{\hbar}(\mathbf{p} + m\mathbf{v}) \cdot A\mathbf{r}} e^{\frac{i}{\hbar}\mathbf{p} \cdot \mathbf{w}} d\mathbf{p} \\
&= \frac{1}{(2\pi\hbar)^{\frac{3}{2}}} e^{\frac{i}{\hbar}\alpha} \int \widetilde{\psi}(A\mathbf{p} - m\mathbf{v}) e^{\frac{i}{\hbar}\mathbf{p} \cdot \mathbf{r}} e^{\frac{i}{\hbar}(A\mathbf{p} - m\mathbf{v}) \cdot \mathbf{w}} d\mathbf{p} \\
&= e^{\frac{i}{\hbar}\alpha} \int e^{\frac{i}{\hbar}(A\mathbf{p} - m\mathbf{v}) \cdot \mathbf{w}} \widetilde{\psi}(A\mathbf{p} - m\mathbf{v}) \psi_{\mathbf{p}}(\mathbf{r}) d\mathbf{p}.
\end{aligned}
$$

Thus, the Fourier transform of ψ' is related to the Fourier transform of ψ as follows:

$$\widetilde{\psi}'(\mathbf{p}) = e^{\frac{i}{\hbar}(\alpha + (A\mathbf{p} - m\mathbf{v}) \cdot \mathbf{w})} \widetilde{\psi}(A\mathbf{p} - m\mathbf{v}).$$

Now note that, if observer M is rotated according to A and moving toward P with constant velocity $-\mathbf{v}$, then M's momentum-values are related to P's via $\mathbf{p}' = A^{-1}(\mathbf{p} + m\mathbf{v})$. Hence, we have

$$\widetilde{\psi}'(\mathbf{p}') = e^{\frac{i}{\hbar}(\alpha + m\mathbf{v}) \cdot \mathbf{w}} \widetilde{\psi}(\mathbf{p}).$$

Taking the squared modulus we find that

$$|\widetilde{\psi}'(\mathbf{p}')|^2 = |\widetilde{\psi}(\mathbf{p})|^2,$$

which means that M and P agree on the momentum-probability densities as claimed.

Bibliography

[1] M.L. Boas. *Mathematical Methods in the Physical Sciences*. John Wiley & Sons, New Jersey, 3rd edition, 2006.

[2] P.A.M. Dirac. *The Principles of Quantum Mechanics*. Oxford UP, London, 4th edition, 1958.

[3] L.D. Faddeev and O.A. Yakubovskiĭ. *Lectures on Quantum Mechanics for Mathematics Students*, volume 47 of *Student Mathematical Library*. American Mathematical Society, Providence, 2009.

[4] G.B. Folland. *Real Analysis: Modern Techniques and Their Applications*. Pure and Applied Mathematics. Wiley-Interscience, New York, 2nd edition, 1999.

[5] G.B Folland. *Quantum Field Theory: A Tourist Guide for Mathematicians*, volume 149 of *Mathematical Surveys and Monographs*. American Mathematical Society, Providence, 2008.

[6] S. Friedberg, A. Insel, and L. Spence. *Linear Algebra*. Prentice Hall, New Jersey, 4th edition, 2003.

[7] W. Fulton and J. Harris. *Representation Theory: A First Course*. Springer, New York, 1991.

[8] J. Gallier and J. Quaintance. Notes on differential geometry and lie groups. Available at http://www.cis.upenn.edu/~jean/gbooks/manif.html. Book in progress.

[9] D.J. Griffiths. *Introduction to Quantum Mechanics*. Prentice Hall, New Jersey, 1995.

[10] B.C. Hall. *Quantum Theory for Mathematicians*, volume 267 of *Graduate Texts in Mathematics*. Springer, New York, 2013.

[11] B.C. Hall. *Lie Groups, Lie Algebras, and Representations: An Elementary Introduction*, volume 222 of *Graduate Texts in Mathematics*. Springer, New York, 2nd edition, 2015.

[12] P.R. Halmos. *Finite-Dimensional Vector Spaces*. Undergraduate Texts in Mathematics. Springer, New York, 1987.

[13] J.H. Hubbard and B.H. West. *Differential Equations: A Dynamical Systems Approach (Higher-Dimensional Systems)*, volume 18 of *Texts in Applied Mathematics*. Springer, New York, 1995.

[14] T. Lancaster and S.J. Blundell. *Quantum Field Theory for the Gifted Amateur*. Oxford UP, Oxford, 2014.

[15] S.F. Singer. *Linearity, Symmetry, and Prediction in the Hydrogen Atom*. Undergraduate Texts in Mathematics. Springer, New York, 2005.

[16] M. Spivak. *Calculus on Manifolds*. Addison-Wesley, New York, 1965.

[17] M. Spivak. *A Comprehensive Introduction to Differential Geometry*, volume 1. Publish or Perish, Houston, 3rd edition, 1999.

[18] J. Stillwell. *Naive Lie Theory*. Undergraduate Texts in Mathematics. Springer, New York, 2008.

[19] G. Strang. *Introduction to Linear Algebra*. Wellesley-Cambridge Press, Wellesley, 5th edition, 2016.

[20] L. Takhtajan. *Quantum Mechanics for Mathematicians*, volume 95 of *Graduate Studies in Mathematics*. American Mathematical Society, Providence, 2008.

[21] B. Tekin. Stern-Gerlach experiment with higher spins. *arXiv preprint physics.atom-ph*, 2015. http://arxiv.org/abs/1506.04632.

[22] J.S. Townsend. *A Modern Approach to Quantum Mechanics*. McGraw-Hill, New York, 1992.

[23] S. Weinberg. *The Quantum Theory of Fields*, volume I. Cambridge University Press, Cambridge, 1995.

[24] E. Zeidler. *Quantum Field Theory I: Basics in Mathematics and Physics*. Springer, Berlin, 2006.

[25] E. Zeidler. *Quantum Field Theory II: Quantum Electrodynamics*. Springer, Berlin, 2009.

[26] E. Zeidler. *Quantum Field Theory III: Gauge Theory*. Springer, Berlin, 2011.

Index

absolutely convergent, 37
absolutely integrable, 141
adjoint, 237, 243
adjoint action, 51
affine space, 135, 161
angle, 4
angular velocity, 18
annihilation operator, 156
automorphism, 7
    inner product space, 7, 29
azimuthal quantum number, 176

basis, 232
    Hilbert space, 242
    orthonormal, 5, 237, 242
    positively oriented, 5
    right-handed, 5
    standard, 4, 239
Bessel equation, spherical, 178
Bessel functions, spherical, 178
Bohr radius, 184
boson, 131, 191
bound state, 183
bra, 25

Cartesian decomposition, 56, 58
Cauchy-Schwarz inequality, 4, 24, 58, 237
central potential, 176
chain rule, 241
change of basis, 234
Change of Variable Theorem, 242
characteristic polynomial, 236
classical region, 153, 159
Clebsch-Gordan coefficients, 127
Clebsch-Gordan problem, 128
commutator, 52, 142, 165

complete reducibility, 82
complexification, 64
conjugate transpose, 27, 238
connected
    path, 31, 42
    simply, 31, 43, 66
conserved quantity, 77
Coulomb potential, 182–183, 189
Cramer's Rule, 236
creation operator, 156
cross product, 53

degeneracy, 176, 177, 190
derivative, 51, 239
determinant, 235
diagonalizable, 234
differentiable, 239
    continuously, 241
dimension, 232
    finite, 232
Dirac delta function, 138, 163
Dirac equation, 222, 223, 225, 227
Dirac Hamiltonian, 222
Dirac spinor, 225
direct sum, 232
distribution, 138
dot product, 4, 24

effective potential, 183
Ehrenfest's Theorem, 149
eigenvalue, 234
eigenvector, 234
energy, 71, 147
energy-momentum 4-vector, 221
entangled state, 123
Euclidean n-space
    complex, 24, 27

253

254 *Index*

real, 4, 7, 239
Euclidean boost, 199
Euclidean group, 198
Euclidean transformation, 198
expectation value, 57, 142, 165
exponential, matrix, 34, 54, 90

fermion, 131, 191
fine structure, 189
fine structure constant, 184, 190
Fourier Inversion Theorem, 141, 165
Fourier transform, 141, 146, 164, 174
free particle, 149, 174
functional calculus, 118, 243

Galilean group, 199
    central extension, 203
gamma matrices, 223
Gram-Schmidt algorithm, 237, 242
group, 8
    action, 14
        simply transitive, 135, 161
    general linear
        complex, 27
        real, 8
    orthogonal, 9
    semi-direct product, 168
    special linear, 62–65
    special orthogonal, 9, 12
    special unitary, 28, 31
    subgroup, 9
    unitary, 28, 40, 67
gyromagnetic ratio, 189

Hamiltonian, 71, 74, 104, 147
harmonic oscillator, 152–158, 176
harmonic polynomial, 107
Heisenberg group, 81, 143, 145, 166, 244
Heisenberg Lie algebra, 143, 166
Hermite polynomial, 157
Hermitian, 33, 40, 48, 238
Hilbert space, 118, 242
    separable, 242
hydrogen atom, 189–190
hyperfine splitting, 127

hyperfine structure, 189

identical particles, 130, 191
image, 232
Implicit Function Theorem, 60, 63, 241
infinite spherical well, 178
infinite square well, 150
inner product space, 237
    complex, 23
    indefinite, 208
    real, 3
        oriented, 6
isomorphism
    inner product space, 6, 29, 41
    vector space, 232

Jacobian matrix, 60, 240

kernel, 232
ket, 24
Klein-Gordan equation, 221

Laguerre polynomial, 184
Laplacian operator, 107
Legendre equation, general, 115
Legendre function, associated, 116
Legendre polynomial, 116
length, 4
Lie algebra, 53, 60
    action, 88
    bracket, 53
    commutation relations, 53
    complexification, 55, 89
    homomorphism, 87
    representation, matrix, 87
Lie group, 50, 62
    matrix, 87
    representation, matrix, 81
linear operator, 232, 243
linear transformation, 232
linearly dependent, 231
linearly independent, 231
Lorentz boost, 212
Lorentz contraction, 213
Lorentz group, 209

Index 255

orthochronous, 209
   restricted, 210
   special, 209
lowering operator, 93, 154, 156

magnetic dipole moment, 19, 72, 127, 189
magnetic field, 19, 72, 75, 127, 189
magnetic quantum number, 173, 176
manifold, smooth, 50
matrix, 232
   identity, 233
   invertible, 233
   multiplication, 233
measurement, 48
minimum uncertainty state, 143
Minkowski space, 209
moment of inertia, 18
momentum, angular
   classical, 17–20
   quantum, 169
     total, 187
momentum, linear
   classical, 17
   quantum, 138, 140, 164

Newton's Second Law, 149, 152
norm
   matrix, 35
   vector, 24, 237
number operator, 156

observable
   classical, 47
   quantum, 48, 104
one-parameter unitary group, 243
orbitals, 193
orientation, 6
orientation preserving, 9
orthogonal complement, 237
orthogonal transformation, 238
orthogonal vectors, 24
orthonormal, 237

Pauli exclusion principle, 191
Pauli matrices, 41, 49, 222

periodic table, 191–194
phase, 24, 100
   global, 144
physical space, 3, 100, 135, 161
Planck constant, reduced, 21
Poincaré group, 220
   restricted, 220
   universal cover, 226
polar decomposition, 64, 214
position observable, 162
position operator, 138
position space, 137, 161
positive definite
   inner product, 4
   matrix, 64
   norm, 35
probability amplitude, 27
Probability Interpretation, 26, 98

quantization, 71

raising operator, 93, 154, 156
rank, 232
rank-nullity theorem, 232
rapidity, 213
reduced mass, 182
representation
   direct sum, 81
   group, 14, 80
     isomorphism, 80
     morphism, 80, 108
     invariant subspace, 81, 88
     irreducible, 81, 88
     Lie algebra, 87
     subrepresentation, 81
rest-energy, 221
Rodrigues' formula, 116
rotation
   generators of, 54
   three dimensions, 12
   two dimensions, 10
Runge-Lenz vector, 185

scattering state, 185
Schrödinger equation, 71, 74, 104, 147, 173–174, 220

256 *Index*

time-independent, 71
Schrödinger representation, 144, 168, 172, 244
self-adjoint operator, 118, 238, 243
skew-Hermitian, 33, 40, 67, 166, 238
skew-symmetric, 96, 238
solid angle measure, 117, 170
span, 231
spectral decomposition, 48
Spectral Theorem, 40, 48, 57, 118, 175, 239, 243
    for normal operators, 239
spherical coordinates, 111, 170
spherical harmonics, 117
spin, 23, 99, 100
    precession, 73
spin observable, 49, 99, 104
spin-orbit coupling, 189
spin-spin coupling, 190
Spin-Statistics Theorem, 131, 191
spinor space, 23, 185
square-integrable, 117, 137, 161, 183
squared total spin operator, 104, 125
stationary states, 71
Stern-Gerlach device, 20, 22, 26, 100
Stone's Theorem, 118, 138, 139, 244
Stone-von Neumann Theorem, 144, 168, 244
strongly continuous, 118, 137, 244
superposition, 27, 72, 100, 127
symmetric, 238

tangent space, 31
tensor
    antisymmetric, 130
    decomposable, 122
    symmetric, 130
tensor product, 122
time dilation, 213
time-evolution, 69, 72, 147, 173
trace, 235
translation action, 137, 145, 162
transpose, 236

uncertainty, 57, 142, 165

Uncertainty Principle, 58–59, 104
    energy-time, 77
    Heisenberg, 142, 165
unitary transformation, 238
universal double cover, 43, 216–218

vector space, 231
    subspace, 231

wavefunction, 137
Weyl spinor, 225